Strict Finitism and the Logic of Mathematical Applications

SYNTHESE LIBRARY

STUDIES IN EPISTEMOLOGY,
LOGIC, METHODOLOGY, AND PHILOSOPHY OF SCIENCE

VOLUME 355

For further volumes:
http://www.springer.com/series/6607

Strict Finitism and the Logic of Mathematical Applications

by

Feng Ye
Peking University, Beijing, P. R. China

Springer

Prof. Feng Ye
Department of Philosophy
Peking University
100871 Beijing
P. R. China
fengye63@gmail.com

ISBN 978-94-007-1346-8 e-ISBN 978-94-007-1347-5
DOI 10.1007/978-94-007-1347-5
Springer Dordrecht Heidelberg London New York

Library of Congress Control Number: 2011930748

Printed on acid-free paper

Springer is part of Springer Science+Business Media (www.springer.com)

Preface

In almost all mathematical applications, the physical entities we deal with are finite and discrete. Macroscopically, the universe is believed to be finite; microscopically, current well-established physics theories describe only things above the Planck scale (about 10^{-35} m, 10^{-45} s etc.). Except for the theories about the microscopic structure of spacetime below the Planck scale, all scientific theories in a broad sense, from physics to cognitive psychology and population studies, describe only finite things within the finite range from the Planck scale to the cosmological scale. In these theories, infinity and continuity in mathematics are idealizations to gloss over microscopic details or generalize beyond an unknown finite limit, in order to get simplified mathematical models of finite and discrete natural phenomena. Scientists are guided by their intuitions and experiences in searching for appropriate infinite and continuous mathematical models to simulate finite and discrete phenomena, and they rely on observations and experiments to confirm that their models can represent those phenomena sufficiently accurately.

However, as logicians and philosophers, we have a few questions:

1. What are the *logically minimum* premises that imply a scientific conclusion about a finite and discrete phenomenon in the universe, and in particular, are the mathematical axioms that apparently refer to infinite mathematical entities *logically strictly* indispensable for expressing natural laws and deriving literal truths about finite and discrete physical entities?
2. Is it possible to demonstrate, in plain logic, that applying an infinite mathematical model does in the end derive literal truths about a finite and discrete phenomenon?
3. How exactly does infinity bring simplifications in the mathematical models of finite and discrete phenomena?

These are questions about the logic of mathematical applications or questions for a logical explanation of the applicability of infinite mathematics to finite physical things. They should deserve some attention from philosophers and logicians.

This monograph presents a research into some of these questions. I will focus on the questions (1) and (2) above. More specifically, I will show that some applied *classical* mathematical theories can be developed within *strict finitism*, a fragment of quantifier-free primitive recursive arithmetic (**PRA**) with the accepted functions limited to elementary recursive functions. Strict finitism is therefore elementary recursive mathematics. It can be interpreted as a theory about concrete and *finite* computational devices. This then implies that the applications of those classical mathematical theories in the sciences are in principle reducible to the applications of strict finitism, and therefore the apparent references to mathematical entities in applying those classical mathematical theories are not strictly indispensable. This answers the question (1) for the applications of those theories. It also implies that the applications of those classical theories to *finite* physical things can in principle be translated into valid logical deductions from literally true premises about finite physical entities alone to literally true conclusions about them, which demonstrates the applicability of those classical mathematical theories in plain logic and answers the question (2) for them.

The classical theories covered in this monograph are still limited, but they do include some advanced applied mathematics, for instance, the basic theory of unbounded linear operators on Hilbert spaces for the applications in classical quantum mechanics and the basics of semi-Riemannian geometry for the applications in general relativity. Moreover, the techniques used here suggest that more applied mathematics can be developed within strict finitism in similar ways. Therefore, this monograph can perhaps show that this is also a feasible strategy for answering the questions (1) and (2) for other applications of classical mathematics to finite things. Certainly, the logic of mathematical applications should be a big research topic, and the work in this monograph is only a small part of it and only a start.

Chapter 1 of this monograph first introduces the philosophical motivation for this work. My general philosophical position about mathematics is radical naturalism and nominalism. See Ye [43, 45, 46, 48–50] for more details. This chapter then characterizes the problem of applicability of mathematics under naturalism and shows that the problem of applicability becomes a logical problem after abstracting away some details. My strategy for explaining applicability and answering the questions (1) and (2) above is then introduced and explained informally. The chapter also explains how this explanation of applicability supports nominalism and radical naturalism. Chapter 1 is an expanded version of the article Ye [44]. Chapter 2 presents the logical framework for strict finitism, and Chap. 3 develops the basics of calculus within strict finitism. Then, the rest of the monograph develops some other more advanced applied mathematical theories within strict finitism. Chapters 3 and 8 each contains a case study of demonstrating applicability by reducing to strict finitism.

Since this is not a textbook, I will not give all the technical details. In particular, I will assume that readers are already familiar with calculus and other materials in classical mathematics of the same level. However, I have tried to make this book self-contained on more advanced topics, for instance, on Hilbert space and semi-Riemannian geometry, so that logicians and philosophers who are not very familiar with these advanced topics can also follow the arguments in this book. To make the technical work more accessible, I will not present mathematical theories in their most general and abstract format. For instance, the integration theory in Chap. 6 will be restricted to Lebesgue integration for real functions of one variable. An extension to multiple variable functions is straightforward though tedious.

Developing mathematics within strict finitism is very close to developing mathematics in Errett Bishop's constructive mathematics. This monograph will follow many ideas in the book *Constructive Analysis* [6] by E. Bishop and D. Bridges. In particular, Chaps. 4 and 5 on metric space and complex analysis and most of Chap. 3 on calculus follow Bishop and Bridges rather closely. My task here is to demonstrate that the main ideas of Bishop and Bridges can actually be realized with strict finitism, that is, elementary recursive mathematics. For this, sometimes we have to restate the definitions and unravel the recursive constructions in the original proofs. The integration theory in Chap. 6 simplifies the approach by Bishop and Bridges. This will allow us to see more clearly the finitistic content of Lebesgue integration and see its applicability to finite things. The theory of bounded linear operators on Hilbert spaces in Chap. 7 is also based on the ideas from Bishop and Bridges, while the proofs of the spectral theorem and Stone's theorem for *unbounded* linear operators are revisions of the early work in my Ph.D. dissertation (Ye [40, 41]), which uses a logical framework less restrictive than strict finitism. In general, Chaps. 3–7 are revisions and improvements of my dissertation to fit into the current more restrictive framework of strict finitism.[1] Chapter 8 on semi-Riemannian geometry is a recent addition to strict finitism. It is not based on any existent constructive theory. With it, strict finitism now covers the basics of applied mathematics required for developing both quantum mechanics and general relativity.

I would like to thank Princeton University and her philosophy department for the graduate fellowship they offered me during my graduate studies in Princeton from Year 1994 to 1999, and I am deeply grateful to my advisors John P. Burgess and Paul Benacerraf for their great help and enduring encouragement during these years. Without them, this research would not have started. Readers can easily see that my work is deeply influenced by Benacerraf and Burgess. In a sense, I take their criticisms of contemporary philosophies of mathematics more seriously than perhaps many other philosophers do, and I try to respond to their criticisms in a more thorough, honest and down-to-earth manner. More specifically, after many years of thinking about the issues and debates surrounding the epistemological difficulty of abstract entities raised by Benacerraf, I realize that an honest, down-to-earth naturalism or physicalism about human cognitive subjects and cognitive processes is the true philosophical foundation for nominalism and the key for resolving puzzles and conflicting intuitions about alleged abstract entities. (See my article [45] for the details.) On the other side, studies on the criticisms of contemporary nominalization programs by Burgess make me realize that the real problem of contemporary nominalistic philosophies of mathematics is that they haven't offered any literally truthful and completely scientific and naturalistic account of human mathematical practices

[1] I have adopted a slightly different philosophical position since my dissertation work, although the basic inclination, namely, nominalism and naturalism, has not changed. On the technical side, my dissertation work is based on a framework that is essentially (quantifier-free) primitive recursive mathematics, and strict finitism adopted in this book is essentially (quantifier-free) elementary recursive mathematics.

including mathematical applications. (See my article [43] for the details.) This then leads to the idea that the real philosophical problem of applicability of mathematics, from the naturalistic point of view, is the *logical* problem of how proofs in classical, infinite mathematics (conducted in human brains) can apply to derive literal truths about strictly *finite* physical things in the universe. This logical problem of applicability is the subject matter of this book. Resolving the problem will also resolve the puzzles and conflicting intuitions about the indispensability argument, and it will turn out to support nominalism.

I would also like to thank Professors Solomon Feferman and Edward Nelson for their comments on my dissertation many years ago. Many thanks are also due to Professors Geoffrey Hellman, Michael Liston, Robert Thomas, and another anonymous referee for the comments on my paper [44]. This paper becomes a part of Chap. 1 of this book. Moreover, Professor James Tappenden's review of the manuscript of this book encourages me greatly, and I am very grateful for that. I would also like to mention my indebtedness to Professor David Papineau, whose book *Philosophical Naturalism* was the first inspiration for me in pursuing this radically naturalistic philosophy of mathematics. I am also indebted to the managing editor Ms. Ingrid van Laarhoven and the language editor of this book. Without their hard work and great help, this book would not have come out and would not be in such a good shape.

Finally, I want to thank my wife Jingjuan and mother-in-law Wu Yonglian for their patience and support during these years, and thank my baby daughter Tiantian for the joy she brings.

Beijing
March 2011

Feng Ye

Acknowledgements

This research is financially supported by the **Chinese National Social Science Foundation** (grant no. 05BZX049), and it is also supported by **Peking University, China**.

Chapter 1 is an expanded version of my article 'The applicability of mathematics as a scientific and a logical problem', *Philosophia Mathematica (3)* **18**, 144–165 (2010). I would like to thank the publisher **Oxford University Press** for the permission to reuse the materials here.

Contents

Chapter 1
Introduction

This chapter will first introduce my general philosophical position, which is radical naturalism and nominalism. Then, I will explain how the problem of applicability of mathematics can be *naturalized*, that is, formulated as a problem about some natural regularity in a class of natural phenomena (i.e., the phenomena involving human brains and their physical interactions with other physical entities in human environments in mathematical practices). Applicability becomes a logical problem after we abstract away psychological, physiological, physical, and many other details. I will argue that there are some genuine logical puzzles regarding applicability, due to the gap between infinity in mathematics and the finitude of the physical things to which we apply mathematics. No current philosophy of mathematics, neither Platonism nor nominalism, has resolved these logical puzzles. Finally, a strategy for resolving some of the puzzles and explaining applicability is introduced. I will also discuss how this solution is a naturalistic solution and how it supports nominalism.

This chapter is an expanded version of Ye [44]. This monograph focuses on logical and technical issues. The philosophical introduction in Sect. 1.1 will be brief. It is not intended to be a defense of my position or a refutation of its opponents or alternatives. Readers interested in the details of my philosophical position can consult other related articles of mine. A brief introduction to these articles is given at the end of Sect. 1.1.

1.1 A Naturalistic Philosophy of Mathematics

I will start with an examination of the status of infinity in mathematics and in the sciences, and I will conclude first that a naturalistic and nominalistic philosophy of mathematics should not assume the reality of infinity in any place, neither in the physical world, nor in a mathematical world independent of the physical world, nor in a world that only a non-physical mind can grasp. That is, a naturalistic and nominalistic philosophy of mathematics should offer a strictly finitistic account of mathematical practices. Then, I will point out a problem in some recent nominal-

F. Ye, *Strict Finitism and the Logic of Mathematical Applications*, Synthese Library 355,
DOI 10.1007/978-94-007-1347-5_1,

istic accounts of mathematical practices, which motivates radical naturalism as the philosophical foundation for a coherent naturalistic and nominalistic philosophy of mathematics. Radical naturalism naturally endorses a strictly finitistic account of human mathematical practices. I will also caution that this is not meant to suggest abandoning the practices of classical mathematics. On the contrary, this is meant to be a truly naturalistic description of human mathematical practices, including the applications of classical mathematics in the sciences.

1.1.1 Infinity and Nominalism

The part of this universe about which scientists have confident knowledge today is strictly finite. Beyond some finite range, for instance, above the cosmological scale recognized today (about 10^{45}m) or below the Planck scale (about 10^{-35} m, 10^{-45} s etc.), things are still unknown to scientists (if there *are* things beyond that range). One thing we *are* confident about today is that physical entities at one scale (e.g., sub-atomic particles with quantum effects) can look wildly different from physical entities at another scale (e.g., the medium size physical objects). In particular, things far away from us can be quite beyond our imagination. For instance, some physicists suggest that the microscopic spacetime may be more than 4-dimensional or may be discrete. Modern physics teaches us that we should not quickly generalize our observations about things within a finite range to things beyond that range straightforwardly. For instance, people used to think that we can 'cut a rod into halves *forever*'. However, in fact, before cutting for a hundred and twenty times, we will reach the Planck scale (for 2^{-120} m $< 10^{-36}$ m) and spacetime might be discrete there and 'cutting into halves' might become meaningless. In the real world, we are never very confident about the result of repeating any operation 'following the same pattern *forever*'.

On the other side, recall how confident we are about infinity in classical mathematics. For instance, one way to get infinity in mathematics is just to repeat an operation 'forever' following a pattern or a rule. This is how we conceive of simple infinite sequences such as the sequence of natural numbers. We never wonder if the sequence of natural numbers has to stop somewhere, and we never wonder if unexpected and unimaginable things may happen when we repeatedly add 1 to a natural number to get the next number.

This contrast naturally suggests that perhaps infinity in mathematics is only our imagination, or a manner of speech. The question whether this physical universe is infinite and the question what will happen if we cut a rod into halves repeatedly are questions about objective things, things independent of our minds. That is why we are not sure about the answers. On the other side, we can always *imagine* that some operation is repeated infinitely many times following exactly the same pattern. (We do not really imagine each of the operations; we only imagine *that* the operation is repeated infinitely many times.) This depends only on our decision about how to use the imagination. For instance, even after conceding the possibility of discrete spacetime, we can still *imagine* that a rod is cut into halves repeatedly forever and ignore what will happen if we actually do it in the real world. Similarly, we never suspect that unexpected things can happen when we repeatedly add 1 to a natural number to get the next number, because we ourselves decide how to imagine the sequence of natural numbers. This explains why we are confident about infinity in mathematics. That is, infinity in mathematics is perhaps not real. It is perhaps merely our imagination.

In contemporary philosophy of mathematics, *Platonism* (or realism[1]) claims that mathematical entities and structures literally exist and that there are infinitely many mathematical entities and there are infinite mathematical structures. In that sense, Platonism (and realism) is committed to the reality of infinity in mathematics. Mathematical entities are allegedly *abstract entities*, which means that they do not exist in spacetime, and therefore the existence of infinity in mathematics is independent of the existence of infinity in this physical universe. Philosophically, Platonism faces the difficult task of explaining how human beings can have knowledge about abstract entities, for the description of abstract entities as mind independent entities not existing in spacetime makes it difficult to see how human beings living in spacetime could have knowledge about abstract entities (Benacerraf [4]). This is called 'the epistemological difficulty for Platonism'. This epistemological difficulty drives some philosophers to deny the existence of abstract entities. Their view is *nominalism*. However, many contemporary nominalists are ambiguous about the status of infinity. Some of them seem to concede the reality of infinity. For example, Field [14] takes a metaphysical hypothesis about the infinity of spacetime in order to save the objectivity of arithmetic truths involving infinity; and Yablo [38, 39] takes arithmetic truths involving infinity as logical truths. Some other nominalists are silent about the status of infinity under nominalism.

I believe that it is incoherent for a nominalistic philosophy of mathematics to assume the reality of infinity in either mathematics or the physical world. First, physicists are indeed still undecided about whether the universe is finite or infinite. However, we do not want our philosophical account of human mathematical practices to depend on such assumptions about this physical universe. Practices in pure mathematics are obviously independent of whether the physical universe is infinite. We apply infinite and continuous mathematical models in areas such as economics and population studies, where the subject matter is obviously finite and discrete. The applicability of infinite mathematics to finite things in the universe does not depend on whether the universe is ultimately infinite. An account of human mathematical practices that relies on assuming the infinity of this physical universe must have missed something essential about mathematics. Second, if a nominalist claims that there is real infinity independent of this physical universe, then she is already committed to something that does not exist in spacetime and is therefore abstract. This will contradict nominalism.

Some nominalists (e.g., Chihara [11] and Hellman [17]) suggest that we can accept the *possibility* of infinity in mathematics. This is similar to accepting the reality

[1] I will ignore the subtle differences between 'Platonism' and 'realism' (or between 'nominalism' and 'anti-realism') in the literature, since my position is straight nominalism *and* anti-realism.

of *potential* infinity by intuitionists. I will come back to these views later, but it is certainly preferable if a nominalist can offer a philosophical account of mathematics without assuming the reality of infinity in any format, not even a potential infinity or possibility of infinity, unless it is somehow reducible to finite physical facts about finite physical things. This is a challenge for nominalists, because it is relatively easy to find concrete things in spacetime as substitutes for *finite* abstract entities. For instance, instead of talking about a word type as an abstract entity, we can talk about structurally similar concrete word tokens (namely, tokens belonging to the same type), and instead of talking about the number 3 as an abstract entity, we can talk about the numerical properties of concrete things, such as '3-pieces-of', '3-inches', and so on, or we can talk about numerals as ink marks on paper or bits-and-bytes in computers. That is, we can paraphrase assertions that are on apparent about finite abstract entities into assertions about finite concrete objects as surrogates. Then, we can claim that we never really refer to finite abstract entities. However, infinite mathematical entities or structures may have absolutely no instances in this physical universe. There is no obvious linguistic maneuver for replacing statements about infinite mathematical entities or structures with statements about finite concrete things in spacetime. A well-known *indispensability argument* for realism in philosophy of mathematics then claims that, since references to abstract mathematical entities are indispensable in the sciences, scientific practices confirm that abstract mathematical entities literally exist.

On the other side, infinity is not a problem only for nominalism. Otherwise, it will be a strong reason against nominalism. There are some logical and technical puzzles about the role of infinity in the mathematical applications to finite physical things above the Planck scale in this universe, due to the fact that infinite mathematical models do not represent finite physical things strictly truthfully. For instance, are infinite models really logically indispensable for describing finite and discrete phenomena in the sciences, and why using infinite models, which do not represent finite and discrete phenomena truthfully, can derive truths about finite and discrete phenomena? I will give more details about these logical puzzles of mathematical application in the next section. These are logical and technical issues in themselves. On apparent, they have nothing to do with philosophy. However, with a sound logical common sense, we naturally suspect that perhaps the apparent references to infinite mathematical entities and structures are actually not strictly indispensable for the applications to strictly finite physical things. We also suspect that perhaps mathematical proofs in those applications can in principle be translated into valid logical deductions from premises about finite physical things alone to conclusions about them. That is perhaps why the applications preserve truths about strictly finite things. If these turn out to be true, they will support the nominalistic idea that infinity in mathematics is merely our imagination or a manner of speech, because the applicability of infinity is explained exactly by eliminating infinity.

Moreover, if these turn out to be true, we will have a logically plain demonstration of the applicability of infinity, because our scientific conclusions about finite physical things then logically follow from premises about finite physical things alone. This is exactly what I will attempt to demonstrate in this monograph. In other words, there are some technical puzzles about the role of infinity in mathematical applications and resolving the puzzles may turn out to support nominalism. Besides, this will also favor a strictly finitistic account of mathematics over those allegedly nominalistic accounts that assume the possibility of infinity or potential infinity, because this will show that the so-called possibility of infinity or potential infinity is also merely our imagination or a manner of speech. An explanation of their applicability to strictly finite things in the universe is also achieved by eliminating them.

Of course, if spacetime *is* continuous, then there *is* infinity in the physical world, but this does not contradict nominalism by itself, for what is infinite is then in the physical world. On the other side, the idea of defending the reality of infinity *in mathematics* based on mathematical applications should be that current mathematical applications in well-established scientific theories about strictly finite physical things above the Planck scale *already* confirm the existence of infinite mathematical entities to some degree. This idea will be wrong if the applicability of those applications is explained exactly by eliminating infinity and turning those applications into valid logical deductions from literally true premises about finite physical things alone.

There may be widespread doubts among philosophers about the possibility of a strictly finitistic account of the practices and applications of classical mathematics. The work in this monograph can perhaps dispel or at least ease those doubts. However, I must clarify that I am not suggesting that scientists should abandon classical mathematics and adopt a finitistic version of mathematics. On the contrary, classical mathematics is an ingenious human invention for representing finite physical things in human environments in approximate but simple and human-tractable ways. The task of a logician and philosopher is to resolve logical and philosophical puzzles about how exactly that ingenious invention works, not to suggest replacing it with another clumsy tool. This leads to my general philosophical view on the nature of mathematics.

1.1.2 Naturalism

The nominalistic view on the nature of mathematics is sometimes identified with mathematical instrumentalism, which usually implies that nominalists treat mathematics as an instrument and accept the truths about physical things that mathematics helps to derive but reject mathematical truths themselves. This position is sometimes accused of being ‘intellectually dishonest’ and is alleged to be based on a metaphysical (and hence anti-naturalistic) prejudice against abstract entities. For instance, Burgess and Rosen [7, 8, 32] have some forceful criticisms of nominalism so construed. I agree that their criticisms are strong as far as they go, but they do not affect the kind of radically naturalistic nominalism that I hold.

To explain why, let me first point out a problem in the common characterization of mathematical instrumentalism. A thermometer is really an instrument, but a thermometer is a real physical object, and there are natural laws governing how a thermometer works as an instrument, and there is a literally true scientific theory about how a thermometer works as an instrument. In other words, a true instrument should be a real thing and a description of how an instrument works should be a literally true description. The instruments in mathematical applications cannot be mathematical entities since there is no such entity. Moreover, it is not accurate to say that a mathematical proposition or theory is an instrument. What is a mathematical proposition or theory? If a proposition or theory is an abstract entity, then this contradicts nominalism already. If a proposition or theory consists of some concrete word tokens printed on paper, then how can these ink marks work as an instrument and what is the literally true description of how these ink marks work as an instrument in mathematical applications? If a proposition or theory is a non-physical, mental entity, then does this mean that mathematical instrumentalism must be committed to dualism, that is, the view that there are non-physical, mental entities? Therefore, this characterization of mathematical instrumentalism is only a vague and figurative description, not a literally truthful description of the real situation of human mathematical practices. It does not say what exactly the instruments in mathematical applications are, and more importantly it has not given any literally true description of how exactly those instruments (as real things) work in human mathematical applications.

Similarly, some nominalists (e.g., Hoffman [19] and Leng [21]) identify their position as fictionalism, which says that mathematical entities are fictional entities and mathematical theories are fictions. Leng's basic idea is that we can use fictional entities to build models for real things in the applications. This has the same problem. There are really no such things as 'fictional entities'. There are only mental processes in human imaginative activities or words printed on paper seen and understood by humans as fictions. In particular, it is *literally false* to say that scientists *use* fictional mathematical entities to build models to simulate real physical things in mathematical applications, because literally there are no such entities for scientists to use and there are no such models. Offering such an explanation for human mathematical applications only shows that one has not really explained how human mathematical applications work. More generally, in explaining how a fictional discourse works in human cognitive activities, one should not again refer to 'fictional entities mentioned in the discourse' as if they were real. Instead, one should give a realistic and literally truthful description of what *really* exist in practicing that fictional discourse by humans.

These problems in the current nominalistic philosophies of mathematics motivate my radically naturalistic approach to philosophy of mathematics. More specifically, I start from methodological naturalism (e.g., that adopted by Maddy [22]), which suggests that we accept our scientific knowledge as our starting point in philosophy. I emphasize that we should first of all accept what mainstream scientists confidently assert, and I emphasize that this includes the thesis that we humans are physical things ourselves and results of evolution. This means that human mental processes are ultimately neural processes in human brains and human mathematical practices are the cognitive activities of human brains in their physical interactions with their physical environments. Then, what really exist in human mathematical practices and applications are neural activities inside brains and their physical interactions with other physical things in human environments. The instruments in human mathematical practices are actually human brains (as well as paper-and-pencils or computers used by human brains), which are indeed the most sophisticated instruments in the universe. A study of human mathematical practices is then essentially a scientific and very realistic study of how such instruments work. The study can abstract away psychological, biological or physical details and focus on the aspects that interest logicians and philosophers, but it is still a scientific study of a kind of natural phenomena and is not essentially different from any other branches of science.

This philosophical position is close to philosophical naturalism or physicalism in contemporary philosophy of mind. See, for instance, Papineau [28], which originally inspired my research. My research project is meant to be an improvement upon Papineau's naturalistic account of mathematics, for Papineau's account is still based on the locution of fictionalism and is not strictly finitistic and physicalistic. There are subtle differences between 'naturalism' and 'physicalism' in philosophy of mind, and there are different species of physicalism. I will ignore them here. Personally, I am more sympathetic with the kind of strongly reductive physicalism that Papineau [28] holds, and I am inclined to think that this is the only coherent naturalism. However, a naturalistic philosophical account of human mathematical practices perhaps does not have to rely on such a strong understanding of naturalism. For instance, David Chalmers, one of the representatives of anti-physicalistic naturalists, also agrees that everything about the cognitive *functions* of a brain can in principle be explained in physicalistic terms, and that only phenomenal consciousness may be beyond physicalism (Chalmers [9]). Now, in our naturalistic philosophical account of human mathematical practices, we are interested only in the functions, not the phenomenal experiences, of doing mathematics. Therefore, it is harmless to use physicalism as a working assumption.

My position is also close to Maddy's [22] radical methodological naturalism, but I try to emphasize that holding to methodological naturalism implies accepting that we human cognitive subjects are physical things ourselves, which will have very significant philosophical consequences. For instance, I believe that naturalism in this sense implies straight nominalism. The basic reason is that a physicalistic description of human mathematical practices as the cognitive activities of human brains will not need to say which abstract entities human brains 'refer to', or 'posit', or 'are committed to'. Actually we should not use such terms as 'refer to', 'posit', or 'be committed to' in a *physicalistic* description, since they are not physicalistic terms and they may have to presuppose a non-physical mind with irreducible intentionality, unless they can be naturalized (see the next section). I will not argue for this implication from naturalism to nominalism here. See Ye [45] for an argument. Note that Quine holds both naturalism and realism in mathematics. My view is that when Quine talks about neurons and stimuli, and when he admits that humans are physical denizens in the physical world (Quine [30], p. 16) and that epistemology is a branch of psychology (Quine [29]), he does abide by naturalism. However, when he starts to talk about 'positing abstract entities', he actually slips away from naturalism. That is, there is an internal inconsistency in his philosophy. See Ye [45] for my argument on this. That is also why I call my position 'radical naturalism' to distinguish it from Quine's position.

Therefore, my approach is to explore a radically naturalistic and down-to-earth realistic (i.e., literally truthful) account of human mathematical practices and applications, by referring to what *really* exist in human mathematical practices, that is, human brains and their activities and interactions with physical things in human environments. It turns out to be nominalistic, but the accusation of 'intellectual dishonesty' or 'holding metaphysical prejudice' does not apply to it, because it is meant to be literally truthful and truly naturalistic. It is not 'a subject with no object' (cf. Burgess and Rosen [8]). Rather, all are physical objects, including we humans ourselves, and there is no non-physical 'subject', and therefore there are no abstract objects *seen from the point of view of a 'subject' facing an 'external world'*. It is 'objects with no subject'. That should be what a true naturalistic worldview is. Naturalism is not the view that scientific methods are the best methods for a non-physical 'subject' to know objects in a world 'external to the subject'. A naturalist philosopher may start her philosophical thinking with a tacit and vague assumption that she is a 'subject' herself and scientific methods are the best methods for her to know things in a world 'external' to her. However, after accepting the mainstream scientific theories, she should admit that she is a physical system herself and there is no 'subject' dwelling inside her brain and trying to know a world 'external to the subject' by utilizing her brain. She should admit that human cognitive subjects themselves are natural objects, namely, physical or biological systems, and that cognitive processes are natural processes. Then, human mathematical practices are *literally* neural activities inside human brains and physical interactions between brains and the environments.

Quine used to say, 'Physicalism, on the other hand, is materialism, bluntly monistic except for the abstract objects of mathematics' (Quine [30], p. 15). My view is that there is no exception and assuming the exception is actually inconsistent with naturalism, mostly because we ourselves are merely physical systems and it is redundant or even meaningless to say that a physical system is 'committed to a so and so abstract entity' (Ye [45]). Quine thinks that there has to be an exception, possibly because his early effort to nominalize mathematics did not succeed (Goodman and Quine [15]). Unable to get rid of infinity seemed to be the main reason why the nominalization program by Goodman and Quine did not succeed. Therefore, a goal of this monograph is to show that a strictly finitistic account of the applicability of classical mathematics is possible.

However, I must also clarify that I didn't mean that radical naturalism is the only consistent worldview. An honest dualistic worldview like Gödel's may also be consistent. What may be inconsistent is a view that claims to be naturalism, but that still conceives of a 'subject' as something that is not just a physical system in physical interactions with its physical environments, and that still talks about how a 'subject' is 'committed to' a so and so object in an 'external world' (Ye [45]). Moreover, radical naturalism is not a dogma or faith, and it is not meant to be extremism. To me, it is a modest, cautious and down-to-earth attitude. We are not sure if there are immaterial minds that can somehow 'grasp' mathematical concepts by some sort of intuition, as Gödel believes, and we do not know how to start studying such minds if we hypothesize them. However, we are quite sure that there *are* brains and that the neural networks in human brains *can* do pattern recognition, language parsing, memory association, concept formation, logical inference, and so on and so forth. Therefore, why don't we start from our mainstream scientific description of human cognitive activities and investigate, from the logical and philosophical point of view, whether or not this is sufficient to give an account of human mathematical practices? This will require some difficult and perhaps tedious technical work, as this book and the related researches will show, but I take that to be an advantage of a philosophical working framework, not a disadvantage. As long as we are sure that we are analyzing real things in the real world, not making up things, we are entitled to be confident that the work will be rewarding, no matter how difficult or tedious it is.

This monograph therefore concerns explaining the applicability of mathematics under radical naturalism and nominalism. Several other articles of mine address other issues in this philosophy of mathematics and other aspects of mathematical practices. More specifically, Ye [43] motivates this approach by evaluating its alternatives and arguing that an approach like this is unavoidable for anyone who wants to hold a coherent naturalistic and nominalistic philosophy of mathematics. Ye [45] argues that naturalism implies nominalism and it discusses the problem in Quine's philosophy. Then, Ye [48–50] examine several aspects of mathematical practices from this naturalistic point of view. Ye [48] tries to explain what an understanding of mathematical statements consists in, what knowledge, intuition, and experience in mathematical practices consist in, and what the relationships between the mathematical and the physical consist in. Ye [49] discusses various aspects of objectivity in mathematical practices and explains why admitting objectivity does not imply admitting the objective existence of abstract mathematical entities (or the objective existence of concepts as 'brain-independent' entities). Ye [50] discusses the apparent universality, apriority and necessity of logic and elementary arithmetic from the naturalistic point of view.

1.2 The Applicability of Mathematics Under Naturalism

This section will characterize the problem of applicability of mathematics under radical naturalism. It means formulating the problem as a problem about human brains and their interactions with their physical environments. It will be called '*naturalizing the applicability of mathematics*'. For that, we must first explain how the notions of semantic reference, truth, and logical validity can become naturalistic notions, that is, how they can be *naturalized*. This is a big research topic in itself and I can only briefly introduce relevant notions here. My focus will be on explaining how the problem of applicability of mathematics becomes a logical problem after we abstract away psychological and many other details. I will also argue that there are some genuine logical puzzles regarding applicability.

1.2.1 Naturalizing Reference, Truth and Validity

To naturalize reference, truth, and logical validity, we need some general assumptions about human cognitive architecture. We know very little about human cognitive architecture. Fortunately, since our interests are only in the philosophical and logical aspect of human mathematical practices, we can ignore psychological details and rely on a greatly simplified model of human cognitive architecture, as long as we have reasons to believe that these simplifications will not invalidate our answers to the philosophical and logical questions regarding human mathematical practices. In particular, I will assume the *Representational Theory of Mind.* It means that brains create *inner representations* realized as neural structures in brains. Brains associate linguistic expressions with inner representations in order to communicate them. We say that linguistic expressions *express* inner representations.

Some inner representations are *concepts*, which are typically expressed by nominal phrases. Note that concepts here are concrete neural structures in individual brains, not the Fregean concepts or senses as public and abstract entities. Some concepts *semantically represent* or *refer to* physical objects or their properties. For instance, a concept RABBIT in someone's brain expressed by the word 'rabbit' may represent rabbits. I will call these *realistic concepts.* This semantic representation (or reference) relation is a sort of physical connection between a neural structure in a brain and other physical entities or properties.

Characterizing this representation relation in naturalistic terms (i.e., without using intentional or semantic terms such as 'represent', 'mean', 'refer to', etc.) is called *naturalizing content* in philosophy of mind. See Adams [1] and Neander [26] for some surveys. There *are* difficulties in the current theories for naturalizing content and I proposed a new theory that can perhaps resolve those difficulties (Ye [47]). I will not go into the details here, and certainly naturalizing content is still an unfinished undertaking in philosophy of mind. However, I will assume that this representation relation *can* be naturalized. That is, there may be many serious technical difficulties to overcome, but there is no essential philosophical obstacle to it. This also means naturalizing the semantic norm in the representation relation, or giving an account of what semantic misrepresentations (or semantic errors) are under naturalism.

Some other concepts do not represent anything directly. They have more flexible and abstract cognitive functions inside a brain, and they can connect with physical things outside the brain indirectly. I will call these *abstract concepts.* Mathematical concepts are abstract concepts. For instance, a mathematical concept expressed by the word '2' in a brain can combine with a realistic concept RABBIT to form a composite concept 2-RABBIT, which does represent physical things outside the brain directly. Similarly, in applying a geometrical theory to the physical space, a mathematical concept POINT in the geometrical theory in a brain is translated into a realistic concept representing small space regions directly. Such translations of abstract concepts into realistic concepts are neural processes in brains. Abstract concepts are abstract representational tools (as neural structures) inside brains, which allow the brains to represent things in more flexible and abstract ways.

Thoughts are another type of inner representation and are typically expressed by declarative sentences. For instance, a simple thought expressed by 'rabbits are animals' in a brain is composed of two concepts RABBIT and ANIMAL in that brain. This thought is *true in the naturalized* sense if the entities represented by RABBIT are among the entities represented by ANIMAL. Since the representation relation for realistic concepts is a naturalized relation, this is *naturalized truth* or *naturalized semantic correspondence* between thoughts and the environments, and it is ultimately a physical connection between neural structures and other physical things as well.

Note that this naturalized 'true' applies to thoughts composed of realistic concepts only, which are *realistic thoughts*. Thoughts composed of abstract concepts do not represent any state of affair directly. They are *abstract thoughts*. They have more flexible and abstract cognitive functions inside a brain, and they can similarly connect with physical things outside the brain indirectly. Mathematical thoughts are abstract thoughts. For instance, an abstract thought in a geometrical theory in a brain is translated into a realistic thought about the physical space when the geometrical theory is applied to the physical space, and then that realistic thought can connect with the physical space by the naturalized representation relation. Mathematical concepts and thoughts constitute a rich pool of representational tools, allowing a brain to represent, organize, and process its inner representations of physical things efficiently, in some flexible and abstract manner. They are similar to the flexible and abstract software tool packages that a software system (or a robot) uses in processing data about real physical things (e.g., the personnel data of a university). See Ye [48] for more details on the cognitive functions of mathematical concepts and thoughts.

There are also logically composite thoughts composed of other thoughts and logical concepts. I will assume that the naturalized truth for thoughts composed of realistic simple thoughts will respect common logical rules. For instance, a thought 'p AND q' will be true just in case both the thoughts p and q are true. Note that in naturalism, we do not intend to offer any foundational justification of logical truths or any non-circular definition of logical constants. We simply describe how human brains work, based on all our scientific knowledge, including our logical knowledge. This is the stance of methodological naturalism (see, for instance, Maddy [22]).

A *logical-inference rule* is an inference process pattern in brains, which produces a thought in some format as the conclusion from some other thoughts in some formats as the premises. As a logical-inference pattern, we fix logical concepts in the pattern and consider other realistic or abstract concepts variable. Therefore, a realistic thought may share the same format as an abstract thought, and an inference process instance involving abstract thoughts can share the same inference pattern with an inference process instance consisting of realistic thoughts exclusively. A logical-inference rule is *valid in the naturalized sense*, if for any inference process instance with that pattern and with realistic thoughts as the premises and conclusion, whenever the premises are true in the naturalized sense, the conclusion is also true in the naturalized sense. Therefore, this is a naturalistic notion as well. An assertion about the naturalized validity of a logical-inference rule is an assertion about some regularity in a class of natural processes.

Note that while a logical-inference pattern may apply to abstract thoughts as well as realistic thoughts, naturalized validity is characterized by the effects on realistic thoughts only, because naturalized truth is meaningful only for realistic thoughts. Moreover, note that a brain may frequently conduct logical inferences that are not valid according to this characterization. Therefore, there is normativeness in naturalized logical validity. It is naturalized normativeness coming from naturalized semantic normativeness in the representation relation for realistic concepts. In other words, for a logical rule, being valid is not simply equivalent to being used by most (or many, or some) people as a natural fact, and the common criticism of psychologism in logic does not apply here, although validity *is* determined by total natural facts (which has to be the case under naturalism).

All logical rules in classical first-order logic are valid in this naturalized sense. Again, remember that we are not offering any foundational justification of the validity of logical rules. We reach this conclusion based on our scientific knowledge. Some of these logical rules are a priori and necessary in a naturalized sense. That is, as a result of evolution, human brains have an innate cognitive architecture adapted to human environments so that some patterns of inferences are universally valid *and* a human brain has the innate tendency to accept these rules after a normal maturation and learning process. See Ye [50] for a discussion on the apriority and necessity of logic under naturalism.[2]

On the other side, if there are only finitely many physical objects in the universe, then, for some numerical expression N, it may happen that all realistic thought instances of the format 'there are only N Ps' (for a predicate variable P) are in fact universally true. Then, this becomes a logically valid thought pattern in this naturalized sense.[3] However, this thought pattern is not a priori and necessary in the above naturalized sense, because brains do not have any innate tendency (as a result of evolution) to accept such a thought pattern (after a normal maturation process). It seems that traditional logical truths are those naturalistically valid thought patterns that are also a priori and necessary in the naturalized sense (and are therefore knowable to brains), together with some idealizations of these, for instance, idealizations by ignoring any limitation on the complexity of thoughts that could be produced by brains.

Finally, some philosophers (e.g., those who hold eliminativism on intentional or semantic notions) may deny that there is anything like a concept or thought inside a brain. However, it is a fact that humans can do symbolic inferences, at least on paper with pencils if not inside their skulls. Therefore, for our purpose here, we can ignore psychological details and consider a bigger physical system consisting of a brain together with the words produced by the brain, printed on paper and used by the brain to assist its work. We can view mathematical applications as interactions

[2] Maddy [22] has another kind of naturalistic description of logic. It is not based on naturalized reference and truth and is not radical naturalism in my sense.

[3] I would like to thank a referee for raising this question and the question discussed in the next paragraph.

between this bigger system and its environments. Then, we can take those linguistic expressions (printed on paper) inside such an 'extended brain' as concepts and thoughts. In other words, we can take all books in our libraries as a part of an extended brain. Certainly, it is still an interesting *psychological* question how neurons inside a brain interact with those word tokens printed on paper and whether anything inside a brain can be seen as a representation of those word tokens (or other physical things outside the brain). However, this is a question about the details of human psychological mechanisms and it does not concern us logicians and philosophers.

Note that it is physicalism that allows us to say this. Since a cognitive subject is just a physical system, when we consider how to delineate the boundary of that system for the purpose of studying its cognitive interactions with its environments, we naturally take those pieces of paper with words printed on by that brain as a part of the system, instead of a part of its environments. That is, since books are produced by the brain and used by the brain to assist its work, they should belong to the cognitive system. This is similar to the fact that a knowledge base stored on a mobile hard disk should be treated as a part of a robot's internal representational system when describing the cognitive interactions between the robot and its environments, and a human is essentially a robot under physicalism. If we were speculating about how a non-physical 'subject' knows things in an 'external world', we would probably think that this idea of an 'extended brain' is ridiculous, because those word tokens printed on paper clearly belong to the 'external world', not the 'subject'. Therefore, it is naturalism that makes the debate between eliminativism and representationalism irrelevant for a *logical and philosophical* study of the applicability of mathematics. We can simply assume that there *are* concepts and thoughts inside brains.

1.2.2 Naturalizing the Applicability of Mathematics

In a typical mathematical application, we first have some realistic thoughts representing observed data. These are our *realistic premises*. For instance,

that object is 98°C now

can be such a realistic premise. Here, '98°C' expresses a composite concept representing a physical property. Then, we choose a collection of mathematical concepts and thoughts for modeling the phenomenon and we translate realistic premises into mathematical thoughts, which are our *mathematical premises*. For instance, we use an abstract concept 'representing a mathematical function' to summarize the temperatures of the object at different moments and we translate the premise above into the thought

$$T(0) = 98,$$

which is one of our mathematical premises. We may directly adopt a mathematical premise as a representation of some natural regularity. For instance, another mathematical premise here may be a differential equation on T, which is intended to represent the regularity in heat conduction. Then, by a mathematical proof, we draw a *mathematical conclusion* from these mathematical premises, together with other mathematical axioms as *extra mathematical premises*. For instance, we may derive this mathematical thought as the mathematical conclusion

$$T(3) = 53.$$

Finally, we translate that mathematical conclusion back into a realistic thought as our *realistic conclusion* about the phenomenon. For instance, translating the mathematical conclusion above, we will get

that object will be 53°C in 3-seconds.

Here, '3-seconds' again expresses a composite concept representing a real physical quantity. The entire application process can then be illustrated below:

$$\begin{array}{rcl} \text{realistic premises} & \longrightarrow & \text{mathematical premises} \\ & & \Downarrow \text{(mathematical proof)} \\ \text{realistic conclusion} & \longleftarrow & \text{mathematical conclusion} \end{array}$$

The translations between realistic thoughts and abstract mathematical thoughts are frequently expressed as *bridging postulations*, which are thoughts connecting realistic concepts and abstract concepts. For instance, the thought 'the function T represents the temperatures of the object' is such a bridging thought. More accurately, a bridging postulation is usually expressed as a biconditional:

for any t, y, that object will be y°C in t-seconds,
iff $T(t) = y$.

Since physical quantities are only meaningful up to some finite precision, we can understand the quantification for t, y here as a substitutional quantification ranging over decimal numerals with a fixed finite number of decimal points. Then, 'y°C' and 't-seconds' again express realistic concepts representing physical quantities. Note that such bridging postulations are abstract thoughts, since they include abstract concepts as essential components.

With translations between realistic thoughts and abstract mathematical thoughts expressed as bridging postulations, an application process becomes an inference process from some realistic premises, mathematical premises, and bridging postulations to a realistic conclusion:

$$\frac{\begin{array}{c} \text{realistic premises} \\ \text{mathematical premises + bridging postulations} \\ \cdots\cdots\cdots \end{array}}{\text{realistic conclusion}} \qquad \text{(appl)}$$

A more detailed analysis of an example of application will be given in the next section.

Note that in this inference process on thoughts (as neural structures in a brain), only the realistic premises and the realistic conclusion at the beginning and end can be true in the naturalized sense. Mathematical thoughts at the beginning or the intermediate stages do not represent anything in the naturalized sense. An explanation of the applicability of mathematics means explaining why the realistic conclusions are true (in the naturalized sense) in ordinary valid mathematical applications in the sciences. This is similar to explaining why a physical property, for instance, a property about mass or energy, is present at the end of some physical processes (while it is neither present at the beginning nor preserved at the intermediate stages). It is a scientific question. It asks for an explanation of some regularity among a class of natural processes.

Since it is a scientific question, an answer to it has to be a scientific answer. In particular, it has to consist of literally true scientific assertions about what *really* exist, for instance, about those neural structures, their functions in brains, and their naturalized representation relation with physical entities in the environments. The correctness or value of a putative explanation should be judged by ordinary scientific standards, like any other scientific explanation of natural regularity. The applicability of mathematics is thus naturalized. Moreover, remember that under naturalism an explanation of applicability is not meant to be a foundational justification of applicability.

1.2.3 Applicability as a Logical Problem

We want to ignore details as much as it is reasonable in studying the problem of applicability. In particular, we can distinguish the logical aspect from the psychological aspect. The psychological aspect is about the psychological mechanisms involved in the mathematical activities of a brain. For instance, how does a brain invent and operate on mathematical concepts, and what human cognitive architecture enables a brain to do this? Lakoff and Núñez [20] is an example of such research. In contrast, the logical aspect is about the logical structures of mathematical concepts, thoughts, and inference patterns, and about how these structures allow finally producing a literally true realistic conclusion about physical things in an application scenario.

In studying this logical aspect, we can abstract away psychological and other details. For instance, as an approximation, we can assume that the structures of inner representations are just the syntactical structures of linguistic expressions expressing them. Then, instead of talking about concepts and thoughts as inner representations realized as neural structures in a brain, we can talk about linguistic expressions, as if those words and sentences were themselves in the brain. As a further simplification, I will assume that inner representations are syntactical entities in a language with a clearly defined basic vocabulary and syntax, such as a first-order language. Then, concepts are terms in the language, and thoughts are sentences there, and inference processes are syntactical inferences on sentences. These assumptions appear reasonable for our specific purpose here. We may assume that these syntactical entities are inside an 'extended brain' as mentioned above. Then, these assumptions amount to assuming that only the structures that are encoded in these syntactical entities are really relevant for explaining, from the logical and philosophical point of view, how an application finally produces a literally true realistic conclusion about physical entities. Other details beyond those structures, psychological or otherwise, are not really relevant.

More specifically, we can distinguish between two vocabularies in the language. The realistic vocabulary is for expressing realistic concepts and thoughts. It makes up a sub-language $\mathscr{L}_r$. The abstract vocabulary is for expressing abstract mathematical concepts and thoughts and makes up another sub-language $\mathscr{L}_m$. Terms and sentences in $\mathscr{L}_r$ are *realistic terms and sentences*, and those in $\mathscr{L}_m$ are *abstract terms and sentences*. Bridging sentences will use both vocabularies. Realistic terms and predicates in $\mathscr{L}_r$ have fixed semantic references consisting of physical entities or their properties in the real world, based on the naturalized representation relation. We will ignore the details in that naturalized representation relation and treat that relation as a satisfaction relation between a formal language and a semantic model consisting of physical entities. Therefore, $\mathscr{L}_r$ has a fixed semantic model $\mathfrak{A}_r$ consisting of real physical entities.

Let Γ_r be the collection of realistic premises in a specific application instance; let Γ_m be the collection of mathematical premises, including the premises expressing scientific laws and the mathematical axioms of classical mathematics; and let Γ_b be the collection of bridging postulations in that application. The application is then a purely logical inference

$$\Gamma_r \cup \Gamma_m \cup \Gamma_b \vdash \varphi$$

from these premises to a realistic sentence φ in $\mathscr{L}_r$ as the realistic conclusion. We may assume that $\mathfrak{A}_r \models \Gamma_r$ when this application is scientifically valid. However, $\mathfrak{A}_r \models \Gamma_m \cup \Gamma_b$ does not hold, since the semantic model $\mathfrak{A}_r$ consists of only physical entities. Then, the applicability problem becomes this logical problem:

> **The Logical Problem of Applicability**: *In a scientifically valid application, assuming that $\mathfrak{A}_r \models \Gamma_r$, why does $\Gamma_r \cup \Gamma_m \cup \Gamma_b \vdash \varphi$ imply $\mathfrak{A}_r \models \varphi$, for φ that is in the language $\mathscr{L}_r$ and is scientifically meaningful?*

1.2.4 The Logical Puzzles of Applicability

It is sometimes claimed that realism in philosophy of mathematics has a ready explanation of the applicability of mathematics and that this is the advantage of realism over anti-realism. A realistic explanation claims that the conclusion in an inference process (appl) above is true, because all the premises there are true (although some of them are 'true of abstract mathematical entities'), and because the inferential steps there preserve truth. Under naturalism, this means that (i) there is a property 'true' that is applicable to both abstract thoughts and realistic thoughts in brains, and (ii) this 'true' property is consistent with the naturalized 'true' property for realistic thoughts, and (iii) this 'true' property is possessed by all thoughts in the inference process (appl) above.

From the naturalistic point of view, there are two problems in this alleged explanation; one is philosophical and the other is logical and technical. The philosophical problem is that realists have not offered any naturalistic characterization of the alleged 'true' property for abstract thoughts in brains satisfying the conditions (i) to (iii) above. In particular, Quine's philosophy did not offer any, nothing like the naturalized truth as a relation between *realistic* thoughts in brains and the physical environments. Indeed, Quine accepts disquotational reference and truth, but this cannot dodge the problem here if one takes naturalism seriously. I will not go into the details here. See Ye [45] for an argument that disquotational reference and truth cannot save abstract entities under naturalism.

The logical and technical problem is that even if we agree that abstract mathematical objects exist and mathematical theorems are true of them, in many cases, there is still no clear logical explanation of why the conclusion drawn in an application is true of physical things. This is because of a clear but constantly neglected fact about applying infinite mathematical models to the physical world: The physical things we deal with in current well-established scientific theories are strictly finite, from the Planck scale (about 10^{-35} m, 10^{-45} s etc.) to the cosmological scale; infinite mathematical models are only 'approximations' to finite physical things in these applications and the logic of these 'approximations' is sometimes unclear; the premises of an application are sometimes not literally true (of finite physical entities), and the application is not simply a series of valid logical deductions from literally true premises to a literally true conclusion. Ideally, a logically clear explanation of applicability should identify what literally true premises an application *really assumes*, and then it should demonstrate how the conclusion drawn in the application logically follows from these literally true premises. So far realists have not given this (even if we agree that mathematical axioms are literally true).

I am not saying that this is impossible. On the contrary, I believe that it is possible, but in doing this, we may in the end find out that our conclusions about finite physical things logically indispensably depend on literally true premises about finite physical things alone. That is, mathematical theorems about infinite mathematical entities are perhaps not really among the logically minimum premises required to imply our conclusions about strictly finite physical things above the Planck scale in current scientific theories.

For instance, consider the case of using a continuous model to simulate the motion of a fluid consisting of discrete particles. A mathematical premise here may claim that the mathematical model satisfies some differential equation. This comes from applying physics laws to the continuous model, pretending that mass in the fluid distributes continuously. Then, our conclusions about those discrete particles in the fluid appear to depend on mathematical theorems about that continuous model and depend on the hypothesis that the model 'approximately simulates' the fluid. However, physicists certainly believe that physics laws about the collisions between those discrete particles in the fluid (and physics laws about electromagnetic force, which finally account for the collision force) are the true fundamental physics premises that really imply our realistic conclusions about the fluid. Physicists do rely on experiments to confirm that a continuous mathematical model works fine for modeling the fluid, that is, to confirm that the model does 'approximately simulate' the fluid. However, they do not consider this to be discovering a new fundamental physics law of nature. They believe that this is only using experiments to confirm a simplified computation method. Our physics conclusions about the fluid should in principle follow from the fundamental physics laws (and observation data) about those discrete particles alone.

For instance, if we have a gigantic computer that can simulate the motion of each particle directly, by computing the forces exerted on each particle from all other particles (and from gravitation), then we will have a literally more accurate description of the motion of the fluid. This description will refer only to those discrete particles and their physical properties, and it will not assume infinity, continuity, or any abstract mathematical objects. A physics conclusion about those discrete particles will then follow from fundamental physics laws and other observation data about those particles alone. That is, a derivation of a conclusion will be a series of valid logical deductions from literally true premises about finite physical things alone to a literally true conclusion about them. This suggests that the same should be true for any physically valid conclusion about the fluid drawn by applying that continuous mathematical model: the conclusion, as long as it is physically valid, should not really depend logically on any mathematical theorem specific about the *continuous* model; it should depend only on premises about those finite and discrete particles, although applying the continuous model greatly simplifies our proofs and calculations.

This example is not peculiar. Other applications of mathematics in current scientific theories about natural phenomena above the Planck scale may be similar, since they similarly describe only finite and discrete things. For instance, general relativity is similar in that differentiable spacetime manifolds in general relativity are approximations to real physical spacetime only at the macroscopic scale. Smoothness of our mathematical model is used to gloss over microscopic details, not to mean exact accuracy. Intuitively we feel that infinity and differentiability conditions are not logically strictly indispensable for implying physically valid conclusions about real spacetime. Similarly, the standard mathematical formalism of classical quantum mechanics appears to refer to infinite mathematical entities such as wave functions. However, considering the fact that it is also accurate only above the Planck scale, our physics intuition seems to be that it is similar to using continuous models to simulate fluids. For instance, it is perhaps possible to discretize wave functions and Schrödinger's equation, and then, with a hypothetical gigantic computer, we can perhaps similarly simulate a system of quantum particles.

I admit that at this point it is still unclear whether infinite mathematical models are indeed in principle dispensable for the applications to finite physical things. More technical work is required to clarify it. However, the observations above do suggest that there are some genuine logical questions regarding the applicability of classical mathematics in current scientific theories about natural phenomena above the Planck scale.

1. What are the logically minimum premises implying our scientific conclusions about finite physical things in current scientific theories? In particular, are mathematical theorems about infinite mathematical entities really among the logically minimum premises?
2. How do we demonstrate in plain logic that using infinite mathematical models to simulate finite and discrete phenomena does preserve literal truths about them?
3. How do infinite models simplify our theories about finite things?

These are the logical puzzles of applicability. They are questions about the logic of mathematical applications or questions for a logical explanation of applicability. Realism has not answered these questions yet. The observations above suggest that we can perhaps eliminate infinity and transform the applications of infinite mathematics into logically valid deductions from literally true premises about finite physical things alone, to literally true conclusions about them. If this can succeed, then we will have answers to the first two questions. That is, mathematical theorems about infinite mathematical entities are *not* really among the logically minimum premises implying our scientific conclusions about finite physical things, and our scientific conclusions are literally true of finite physical things because they logically follow from literally true premises about finite physical things.

Note that current anti-realistic philosophies of mathematics have not resolved these logical puzzles of applicability either. Some of them accept the entire classical mathematics but do not address the issue of applicability. They merely label the facts of applicability by a novel name, e.g., 'nominalistic adequacy' or 'empirical adequacy' (e.g., Melia [23], Hoffman [19]). Some of them try to develop subsystems of classical mathematics as philosophically more justifiable mathematics (e.g., Field [13], Chihara [11], Hellman [17]). However, these subsystems all are committed to potential infinity, and applying them to strictly finite physical things in the universe above the Planck scale is still using infinite models to simulate strictly finite things. For instance, when we use a continuous function in intuitionistic mathematics to simulate the mass distribution of a fluid, the premises are again not literally true of those discrete particles in the fluid, and the logical puzzles of applicability remain the same. (See Ye [43] for more criticisms of other current anti-realistic philosophies of mathematics.)

Also note that I never assume that there is no real infinity in the physical world. Most physicists today agree that current well-established physics theories accurately describe only physical phenomena above the Planck scale. As for what phenomena are below the Planck scale, physicists are still considering several possible theories, including discrete or non-4-dimensional spacetime structures. This means that physicists believe that the successes of current theories above the Planck scale do not strictly imply the structure of spacetime below that scale. That is, current theories are only approximations above the Planck scale. Even if physical spacetime is in fact continuous, the validity of current theories above the Planck scale does not depend on this fact, and the logical puzzles of applicability for current theories are still the same. Moreover, intuitively it is still reasonable to think that infinity is not strictly indispensable in the current theories. If someday physicists do confidently assert that spacetime *is* continuous, then the logical puzzles of applicability in that future theory about spacetime might change or even disappear (in case there is no gap between the mathematical model and the physical spacetime in that future theory). This does not affect the logical puzzles of applicability in the current theories (or in future theories about other finite and discrete phenomena, such as the phenomenon of population growth).

1.3 A Logical Explanation of Applicability

Here I will give an informal introduction to the technical strategy for explaining applicability in this book. I will discuss the technical feasibility of the strategy and the naturalistic nature of it. I will also compare it with some related works. Finally, a concrete example is used to illustrate in more details the logical puzzles of applicability and the strategy for resolving the puzzles.

1.3.1 The Strategy

Here is a more detailed presentation of the strategy for answering the questions (1) and (2) in the last section. Using the notations in the last section, we suspect that in an application, there are realistic premises (about concrete physical things) that are not explicitly included in Γ_r, but are implicitly implied by $\Gamma_m \cup \Gamma_b$. Suppose that Γ_r' is the collection of realistic premises that we can excavate from $\Gamma_m \cup \Gamma_b$. Then, we suspect that for any scientifically meaningful realistic conclusion φ drawn by scientists, we actually have $\Gamma_r \cup \Gamma_r' \vdash \varphi$, where the deductions are valid in the naturalized sense. The truth of the realistic premises in Γ_r' is implicitly accepted by scientists when they use $\Gamma_m \cup \Gamma_b$ to model real physical entities in that application. This, together with $\Gamma_r \cup \Gamma_r' \vdash \varphi$, then implies that the realistic conclusion φ must also be literally true (of concrete physical things). That is, our logical explanation of applicability will go like this:

$$\mathfrak{A}_r \models \varphi, \text{ because } \mathfrak{A}_r \models \Gamma_r \cup \Gamma_r' \text{ and } \Gamma_r \cup \Gamma_r' \vdash \varphi.$$

To explain applicability, we must then excavate and identify such implicit realistic premises Γ_r' and show that the original mathematical proof from $\Gamma_r \cup \Gamma_m \cup \Gamma_b$ to a scientifically meaningful conclusion φ can be transformed into a series of valid logical deductions from $\Gamma_r \cup \Gamma_r'$ to φ.

When infinite and continuous mathematical models are used to simulate finite and discrete phenomena, apparently we cannot translate mathematical premises into literally true realistic sentences about finite and discrete physical entities with the logical structures of those premises preserved. The mathematical premise stating the differentiability of the mass distribution function in a continuous model of fluid, for instance, cannot be so translated, since a real fluid consists of discrete particles. Therefore, we cannot obtain Γ_r' by translating mathematical premises and bridging postulations in $\Gamma_m \cup \Gamma_b$ into realistic assertions about physical entities straightforwardly. The case is even more complex when we use mathematical entities to simulate physical entities indirectly. For instance, we use vectors in Hilbert spaces to simulate the states of quantum particles. Here, we appear to be using something alien to a physical system to encode information about the system. Then, it is a genuine challenge to extract true realistic premises about physical things implicit in those infinite mathematical models and to transform proofs on mathematical models into logical deductions from realistic premises about discrete and finite real things alone.

To solve this problem, I will use the following technical strategy in this monograph. First, Chap. 2 introduces a logical framework called *strict finitism*, for developing a kind of mathematics without infinity. Strict finitism is essentially a fragment of quantifier-free primitive recursive arithmetic (i.e., **PRA**), with the accepted functions restricted to elementary recursive functions. Closed statements in strict finitism are reducible to the format $t = s$, where t and s are closed terms constructed from numerals and base elementary recursive functions by composition, bounded primitive recursion, finite sum, and finite product. We can interpret closed terms as programs (with fixed inputs) in computational devices (including brains). Then, $t = s$ says that two such programs will produce the same output. Some closed instances of an axiom in strict finitism can be interpreted as literally true statements about such programs in a finite computational device. Note that not all instances of an axiom schema can be so interpreted for a real computational device, because a real computational device has physical limitations and cannot handle very large numerals properly. However, as long as the numerals involved are not too large and a computational device is functioning properly, an instance of an axiom can become literally true when interpreted as an assertion about that computational device.

Applying mathematics in strict finitism is essentially using a computational device (including a brain) to simulate other physical entities and their properties. We also have realistic premises, mathematical premises and axioms, and bridging postulations here. However, mathematical premises and the axioms of strict finitism are interpreted as statements about a computational device, and bridging postulations are interpreted as statements about how the computational device simulates other physical entities. These are all realistic statements. Therefore, an application is a series of logical deductions from realistic premises to a realistic conclusion.

Then, to explain the applicability of classical mathematics in an application instance, we can try to show that the application is in principle reducible to an application of strict finitism. Instead of translating the applications of classical mathematics into the applications of strict finitism directly, my strategy is to develop applied classical mathematical theories within strict finitism. The syntactic structure of a theorem in strict finitism is very similar to the syntactic structure of the corresponding theorem in classical mathematics. Therefore, for instance, after a branch of applied mathematics is developed within strict finitism, we can rather straightforwardly translate a physics textbook written with that branch of applied classical mathematics into a textbook written with strict finitism. A physics theory will then have the same formal structure. Moreover, recall that we will need only finite precisions in representing physical quantities above the Planck scale. Therefore, we have reasons to believe that physical quantities and states in the actual applications can all be represented by the functions available to strict finitism. Since the formal structure of a physics theory is preserved and real physical quantities can be represented, a physics theory formulated with strict finitism actually states the same physical facts and regularities as the original one formulated with classical mathematics. They are actually the same physics theory with different mathematical formalisms. Therefore, the development of an applied classical mathematical theory within strict finitism implies that an application of that theory can be automatically translated into an application of strict finitism.

This means that we can in principle reformulate mathematical premises and bridging postulations in those applications as assertions about computational devices and their simulation relations with other physical entities. This should not be very surprising. After all, from the point of view of a naturalistic observer, humans are actually using their brains, assisted by paper-and-pencils or computers, to simulate other physical entities when they apply classical mathematics to those physical entities. The only puzzle for logicians is that when humans use classical mathematical concepts and thoughts that appear committed to infinity, the logic of how those concepts and thoughts simulate *finite* physical entities is not very clear. Then, the idea here is that the convoluted logic in those abstract mathematical thoughts in classical mathematics can in principle be straightened, to get logically simpler and more transparent (but much lengthier and more tedious) thoughts directly about finite computational devices and their simulation relations with physical entities.

Using the symbolic notations above, here is how the explanation of applicability goes. For an application instance $\Gamma_r \cup \Gamma_m \cup \Gamma_b \vdash \varphi$, the implicit realistic premises Γ_r' implied by $\Gamma_m \cup \Gamma_b$ include the axioms of strict finitism, which state how brains or computers work as computational devices, and they also include other statements about how these computational devices simulate physical entities in the application. The fact that the classical mathematical theory in $\Gamma_m \cup \Gamma_b$ can in principle be developed within strict finitism will imply that we then have $\Gamma_r \cup \Gamma_r' \vdash \varphi$. This is the explanation mentioned in the opening paragraph of this section. This will also show that classical mathematical theorems about mathematical entities are not among the logically minimum premises implying this conclusion φ, and it also demonstrates, in plain logic, how this conclusion φ about physical entities logically follows from true premises about finite physical entities alone.

1.3.2 The Conjecture of Finitism

Whether or not this strategy can work for all applications of mathematics in current scientific theories depends on the status of the following technical conjecture:

The Conjecture of Finitism: *Strict finitism is in principle sufficient for formulating current scientific theories about natural phenomena above the Planck scale and for conducting proofs and calculations in those theories.*

There are reasons supporting this conjecture. First, an impressive part of applied mathematics has been developed within strict finitism. Starting from Chap. 3, this monograph will develop the basics of calculus, metric space theory, complex analysis, Lebesgue integration theory, the theory of bounded and unbounded linear operators on Hilbert spaces, and semi-Riemannian geometry. These cover a significant part of *Constructive Analysis* by Bishop and Bridges [6] , as well as the mathematical theories needed for the applications in classical quantum mechanics and general relativity. Moreover, the general techniques used here seem to show that applied mathematics within strict finitism can advance much further.

Second, there are also other intuitive reasons supporting the conjecture. For instance, the ratio between the linear cosmological scale to the Planck scale is less than 10^{100}. It seems that number theoretic functions not essentially bounded by a few iterations of the power function do not have any chance of representing real physical quantities. This suggests that elementary recursive functions available to strict finitism may already be sufficient for encoding real numbers, functions of real numbers and so on for realistic applications.

Moreover, since infinity and continuity are only approximations to finite and discrete things above the Planck scale in the applications, intuitively we expect that infinity ought not to be strictly indispensable for deriving a physically meaningful conclusion. Otherwise, the conclusion may be physically unreliable, for it may need the infinite model to be exactly isomorphic with real things, not merely an approximation at the macro-scale. This suggests that physically valid applications are perhaps in principle reducible to the applications of strict finitism. On the other side, being essentially finitistic can explain why an application respects the fact that our model is merely an approximation at the macro-scale. This is merely a vague and intuitive idea at this point, but we will see this more clearly later in developing mathematics within strict finitism. That is, we will see that there is a close connection between a mathematical proof's being finitistic and its being sound logical deductions on statements about strictly finite things.

For instance, in general relativity, we model spacetime by differentiable manifolds and some classical proofs on the existence of singularities in the spacetime manifolds are non-constructive. However, in general relativity, differentiable manifolds are meant to be approximations to real spacetime only at the macroscopic scale. We expect our physical conclusions in general relativity to remain valid even if real spacetime turns out to be discrete at the microscopic scale. Now, if an existence proof about our spacetime model is strictly irreducibly non-constructive, then we have intuitive reason to think that the proof takes infinity and continuity of the model too literally and does not respect the fact that the model is merely an approximation at the macroscopic scale. On the other side, in Chap. 3, we will see that a continuity condition in strict finitism can be translated into literally true claims about the smoothness of a physical quantity at a macro-scale, when that quantity is actually discrete at the micro-scale. That is, strict finitism can respect the fact that our continuous models are only approximations to real things at the macro-scale. Therefore, we have good reason to think that physically meaningful proofs about our models should be essentially finitistic. In Chap. 8, I will actually present a case study of a version of Hawking's singularity theorem, whose common classical proofs in general relativity are non-constructive. I will show that the essential steps of a classical proof of the theorem can be conducted within strict finitism. This then shows that the classical proof can be transformed into valid logical deductions on statements about real spacetime from literally true premises about real spacetime, even if real spacetime turns out to be discrete at the microscopic scale.

Similarly, the Jordan Curve Theorem in its original format may not be provable in strict finitism. However, considering the fact that spacetime structure below the Planck scale is still unknown (and may be discrete or non-4-dimensional), we can expect that if the theorem is applicable in a real situation, what is really relevant for the application must be some approximate version of the theorem that does not take continuity of space too literally (e.g., a discretized version that allows space to be a discrete lattice of points). Such a version is likely to be essentially finitistic.

Third, consider what the possible counter-examples to the conjecture are. Design a computer simulating the proofs in **ZFC**. We believe that the machine will never output $0 = 1$ as a theorem, which follows from our belief that **ZFC** is consistent. This belief about a concrete machine is perhaps not obtainable without entertaining the concepts and axioms in set theory. This appears to be a counter-example to the conjecture. However, from a strictly naturalistic point of view, our belief in the consistency of **ZFC** is inductive in nature. In other words, after human brains practice entertaining concepts in set theory for a long time, and after obvious paradoxes are eliminated, brains come to believe that no paradoxes will be derived in the future. This is essentially an inductive belief achieved by a brain based on reflections upon (i.e., observing) its own activities. It should not be surprising that such a belief is not obtainable without entertaining those abstract mathematical concepts in brains, because it is just about what will happen in entertaining those concepts. It is much like recognizing a complex pattern among a class of natural phenomena. As a case of scientific application, I take it that our inductive belief in the consistency of **ZFC** is among the premises for deriving our belief about that machine. Now, the derivation from the belief of consistency to that conclusion about a concrete machine is finitistic. Therefore, this is not a counter-example to the conjecture. On the contrary, it is another way to justify our beliefs about finite natural processes in the universe (e.g., the computational process of that **ZFC** machine) under naturalism and strict finitism. I will use this strategy in the case study on Hawking's singularity theorem in Chap. 8.

Similarly, recall that for a Π_1^0 arithmetic sentence φ in **PRA**, we have

$$\mathbf{PRA} \vdash Con_{\mathbf{ZFC}} \wedge Pr_{\mathbf{ZFC}}(\ulcorner \varphi \urcorner) \rightarrow \varphi,$$

where $Con_{\mathbf{ZFC}}$ states the consistency of **ZFC**, and $Pr_{\mathbf{ZFC}}$ is the proof predicate of **ZFC**, and $\ulcorner \varphi \urcorner$ is the Gödel number of the formula φ. In strict finitism, the consistency of **ZFC** similarly implies a (quantifier-free) arithmetic formula (of strict finitism) derivable from **ZFC**. As for an arbitrary first-order arithmetic formula, if it is to be meaningful for real things in this universe, from the Planck scale to the cosmological scale whose (linear) ratio is $< 10^{100}$, we can perhaps expect that all its quantifiers are actually bounded by elementary recursive functions. Then, it is reduced to an essentially quantifier-free arithmetic formula of strict finitism. This means that beliefs about strictly finite, concrete things in this universe obtained by applying first-order arithmetic formulas provable from **ZFC** are accountable as finitistic consequences of the inductive belief in the consistency of **ZFC**. Therefore, we will not get any counter-example to the conjecture by applying **ZFC** in this manner.

These reasons are still far from conclusive. More work has to be done in developing applied mathematics within strict finitism, as well as in analyzing what could be a counter-example to the conjecture, in order to support the conjecture better. However, based on the reasons we already have, a positive answer to the conjecture seems plausible. Then, this strategy for explaining applicability seems feasible.

Finally, remember that I take a completely naturalistic and scientific attitude here. I am not trying to look for an a priori argument that the conjecture of finitism must be true. A priori arguments in the traditional sense are meaningless under radical naturalism, since we ourselves are just physical systems in the physical world and the products of evolution. Traditional a priori arguments have to assume a transcendental cognitive subject, which is alien to naturalism. On the other side, even if we end up with a negative answer to the conjecture of finitism, it will still be a valuable thing to know where exactly infinity is strictly logically indispensable for an application to finite things in this physical world and how that can happen. Moreover, recall that I never assume that there is no infinity in the physical world. That question should be left for physicists to answer. The real job here is to explain how infinite mathematics is applicable in *current* scientific theories about a finite part of the physical world.

1.3.3 The Naturalistic Nature of the Strategy

To see how this explanation of applicability is naturalistic, consider the following type of explanation of physical phenomena. Suppose that we have a physical system, and suppose that in a class A of state transition processes for the system which we frequently observe, the end states always have a property T. Suppose that this regularity is not obvious from the known physics laws, because the property T is not present at the initial states of those processes in A, and it is not preserved at the intermediate states in those processes either. Therefore, we have a puzzle. To resolve the puzzle, we analyze those processes and find that a state transition process σ in A can be transformed into another state transition process σ' ending at the same end state. σ' has an initial state with the property T, and it follows from known physics laws that the state transitions in σ' preserve the property T. Therefore, it follows that the end state of σ' will have T. This then demonstrates that the property T will present itself at the end state of the original process σ.

Our strategy for explaining the applicability of classical mathematics is of the same kind, with the system being the brain of a scientist (or a brain plus a paper-and-pencil or a computer), the processes being the inference processes in the brain in valid applications of classical mathematics, and the property T being the naturalized 'true' property for relevant realistic thoughts in the brain. The property T regularly presents itself at the end states of those applications of classical mathematics. However, it does not present itself at (or is not applicable to) the initial states, because the initial states involve abstract mathematical premises and bridging postulations, to which the naturalized 'true' property does not apply. Moreover, the property T is not preserved in the intermediate inference steps, because those steps may involve abstract mathematical thoughts as well. Our explanation then says that such an inference process σ in a brain can *in principle* be transformed into another inference process σ' in the brain. σ' reaches the same realistic conclusion, but it starts from true realistic premises about physical entities and a computational device. Moreover, σ' uses only valid logical inference rules (in the naturalized sense). Then, according to the known laws about the naturalized property 'true' for realistic thoughts and valid inference rules, the end state of σ' will have the property 'true'. This then demonstrates that the property 'true' will present itself at the end state of the original inference process σ.

A few clarifications are in order.[4] First, note that the hypothetical inference process σ' and the computational device mentioned above do not really exist in the actual world. Is it legitimate for nominalism and radical naturalism to refer to them? To see that we never go beyond nominalism and naturalism here, note that if a logician really wants to explain applicability in a specific instance of applying classical mathematics, she can study our technical work in developing applied mathematics within strict finitism and then go ahead and translate the application into an application of strict finitism. This will involve some tedious logical and mathematical work, but accomplishing the work will actually realize the process σ' above by her brain and realize the virtual computational device (as a paper-and-pencil or a computer). In other words, we offer a schema, and then anyone with sufficient patience can instantiate it into a concrete logically plain demonstration of why the conclusion drawn in a specific application of classical mathematics is true (in the naturalized sense). This follows the spirit of strict finitism. That is, we use concrete schemas to achieve generality and we never really 'refer to' non-existent, hypothetical things.

We did not even say that those non-existent hypothetical inferences are 'possible'. The hypothetical inference process σ' and the hypothetical computational device *are* physically possible, but we do not really rely on the notion of possibility in any irreducible manner. Instead, we provide concrete instructions for creating the inference process σ' and the computational device. Therefore, this is also different from the philosophical approaches by Chihara [10, 11], and Hellman [16, 17], which seem to rely on the notion of modality in an essential and irreducible manner. Under radical naturalism, assuming an irreducible modality will cause difficulties. For instance, there is no naturalistic notion of truth for irreducible modal statements (vs.

[4] I want to thank several referees for raising some of the questions discussed here.

the naturalized truth for realistic thoughts), and it is difficult to explain how brains as physical systems in the *actual* world can know irreducible modal truths. Only a naturalized modality is meaningful under naturalism, much like the naturalized truth. See Ye [42] for an attempt to naturalize modality, and see Ye [43] for more criticisms of the modal approaches in philosophy of mathematics. Different from these modal approaches, the strategy here is within nominalism, radical naturalism, and strict finitism.

On the other side, in demonstrating that an instance of application of classical mathematics *can in principle* be translated into sound logical deductions on realistic statements about finite real things, we can resort to our inductive belief on the consistency of classical mathematics, as the example of **ZFC** machine in the last subsection shows. We may rely on our inductive beliefs on consistency to convince ourselves that a series of sound logical deductions on realistic statements about finite real things will end up with a specific conclusion. The important thing is that they are logically valid deductions from literally true premises about finite real things to a conclusion about finite real things, where 'truth' is naturalized truth. No truth about alleged abstract entities is assumed. The inductive belief on consistency functions merely to predict that a natural process will have some specific outcome, which is similar to our inductive beliefs in other areas of science. The case study at the end of Chap. 8 will provide such an example.

Second, in explaining applicability, is it legitimate for nominalism and radical naturalism to resort to scientific laws (about brains and other things) that appear committed to abstract mathematical entities? To clarify this, note that if we really develop all scientific theories with strict finitism, we actually refer only to concrete computational devices and physical entities in stating scientific laws. Then, our explanations of applicability do not really refer to any abstract mathematical entities. That is, nominalism and radical naturalism can in principle be held thoroughly and consistently.

Third, in what sense does this explain applicability? Does it shift explanandum from the original process σ to a different process σ'? Admittedly, this is unlike an ordinary physics explanation of natural phenomena, where we explain an observed phenomenon by referring to an initial or boundary condition and a general law. However, if successful, this does help to resolve some of the logical puzzles of applicability, which is our real concern here. That is, from a given application of classical mathematics, this can demonstrate in plain logic how the conclusion logically follows from literally true premises. It also shows that the assumptions about abstract mathematical entities are not among the logically minimum premises implying literal truths about real physical things.

Fourth, I have no intention to suggest adopting strict finitism in place of classical mathematics for scientific applications. This is meant to be a scientific study of the successes of applying classical mathematics as natural phenomena, aiming to resolve some logical puzzles there. The reductions to strict finitism obviously make the applications much more complex. Scientists and mathematicians discover a simple and effective mathematical language of infinity for describing very complex finite things in the real world sufficiently accurately. This is their great ingenuity and achievement. A logician's job is to resolve the logical puzzles in such ingenious inventions, and strict finitism is invented as an assistant logical analytical tool for that purpose. It is certainly not meant to replace these ingenious inventions by scientists.

Moreover, I do not mean that strict finitism is the only true mathematics or the foundation of the only meaningful mathematics. Firstly, the axioms of strict finitism are abstract thoughts in brains and the naturalized property 'true' does not apply to them either. Secondly, some of these axioms *can* be interpreted into true realistic thoughts about concrete computational devices, but many of them have no such chance, because there are no large enough concrete computational devices in the universe. Finally, for a true naturalist, the quest for an absolutely certain foundation of knowledge is pointless. All knowledge in a brain comes from the innate cognitive architecture of the brain determined by genes and from the physical interactions between the brain and its environments. Moreover, under naturalism, a brain having so and so knowledge can only mean that the brain is in so and so neural states with respect to so and so historical and environmental states. The reliability of knowledge should also be naturalized, presumably by referring to natural regularities in the interactions between brains and their environments. A brain can reorganize its knowledge base and distinguish between its more reliable and less reliable knowledge. However, the idea of having some absolutely certain knowledge as the foundation of all knowledge has to presuppose an absolute, non-physical and transcendental cognitive subject, which is alien to naturalism.

1.3.4 Some Comparisons and Evaluations

Explaining applicability in this sense is similar to demonstrating the conservativeness of mathematics over its finitistic fragment. The latter was the goal of Hilbert's program [18]. Hilbert thought that classical mathematics might be conservative over its finitistic fragment, where the finitistic fragment for him is largely *quantifier-free* primitive recursive arithmetic (**PRA**) (see Tait [33]). That is, classical mathematics only facilitates the derivations of finitistic theorems, which are in principle provable without classical mathematics. We know today that classical mathematics is not conservative over its finitistic fragment understood as **PRA**. In a similar manner, we believe that classical mathematics facilitates the derivations of realistic conclusions about finite physical things in the universe in a mathematical application, but if we gather all realistic premises in the application, they will logically imply the realistic conclusion, without the help of classical mathematics. We try to demonstrate this by showing that a finitistic mathematical system is already in principle sufficient for current scientific applications. This will imply that the part of classical mathematics that is actually applied in the sciences *is* conservative over finitism. Hilbert's program intended to look for a proof of conservativeness once and for all. Our approach is piecemeal and is limited to the part of classical mathematics that is applied in the sciences (and therefore it does not suffer from the failure because of Gödel's Incompleteness Theorem).

Our strategy is also similar to Hartry Field's strategy of demonstrating the conservativeness of (a part of) classical mathematics over a nominalistic physics-cum-mathematics theory (Field [13, 14]). However, Field's alleged nominalistic mathematics assumes infinity, in particular, the infinity and continuity of spacetime. We do not want our philosophical and logical explanation of the applicability of mathematics to areas such as economics to rely on a physics assumption about spacetime, not to mention an assumption that is still undecided today. Moreover, recall that the logical puzzles of applicability are about how infinite and continuous mathematical models are applicable for deriving truths about *strictly finite* things in the universe. Field's strategy assumes an infinite and continuous spacetime structure and then uses infinitely many spacetime points and continuous space regions to replace mathematical entities and structures. Applying this allegedly nominalized mathematics is still using infinite models to simulate finite things and cannot really resolve the logical puzzles of applicability.

This study of applicability addresses only some problems in the logic of mathematical applications. For instance, I did not touch upon the problem of how infinite models simplify the applications. Moreover, it is conceivable that there are other approaches to explaining applicability. For instance, one may try to analyze directly how mathematical thoughts in classical mathematics 'approximately represent' finite physical entities and get a demonstration of applicability from there.

On the other side, this research seems to have its own values. It might potentially affect our mathematical practices. If the Conjecture of Finitism is correct, it suggests that the kind of complexity entertained by logicians in exploring logically more and more powerful mathematical axiomatic systems (e.g., extensions of **ZFC**) may not be relevant to the genuine complexities in this real world met by the sciences. In other words, logicians' complex imagination develops in a direction deviating from the genuine complexities in this real world. This thought might affect logicians and mathematicians in deciding what research topics to pursue.

If the Conjecture of Finitism is correct, it also encourages exploring new ways of practicing mathematics. For instance, while imagining infinite mathematical structures is so far the most efficient way to model complex finite and discrete phenomena in the world, with more and more powerful computers available, it is possible that another kind of language that can make use of computers more directly will become even more efficient for humans. If strict finitism is in principle sufficient for formulating a scientific theory, it means that we can *in principle* refer to programs in computers and refer to their simulation relations with physical entities and quantities in stating our scientific theories. That is, we can in principle express our scientific theories in a computer language. In other words, if classical mathematics does not consist of objective truths about a mind independent reality, and if its value consists in its rich pool of mathematical concepts and thoughts that allow us to model finite things in the world *approximately but efficiently*, then we will naturally abandon dogmatism about classical mathematics, and we will naturally try to explore other ways of representing and modeling this world when some powerful new tools (i.e., computers) are available.

On the philosophical side, as I have explained in Sect. 1.1, this research supports nominalism and naturalism. Note that I do not mean that the validity of nominalism depends on the status of the Conjecture of Finitism, and it is not the goal of this monograph to defend nominalism. If spacetime is continuous, then certainly some infinite mathematics is needed for physics, but that mathematics may have a nominalistic interpretation as a theory about physical structures. Therefore, this possibility does not invalidate nominalism, although it does show that nominalism is independent of the conjecture (and independent of whether there *is* infinity). I believe that nominalism follows from a coherent understanding of naturalism (Ye [45]). This research on explaining applicability supports nominalism and naturalism by showing that nominalism and naturalism *can* offer a coherent account of human mathematical applications that we have so far.

Finally, irrespective of any philosophical position one holds regarding mathematics and irrespective of the possible effects on mathematical practices, as a scientific and realistic study of current human mathematical practices, this research should have its own value. In particular, a logically clear demonstration of how infinite mathematics is applicable for deriving truths about finite things in the universe should help us in understanding the true logic of mathematical applications and the true nature of mathematics, no matter if one holds to Platonism, nominalism, or naturalism.

1.3.5 An Example

In the following, I will analyze an example of application and use it to illustrate more technical details regarding the logical puzzles of applicability and the strategy for resolving the puzzles in this monograph. I will come back to this example at the end of Chap. 3.

Consider the logistic model of the population growth on the Earth:

$$\frac{1}{p}\frac{\mathrm{d}p}{\mathrm{d}t} = \alpha(Q-p).$$

Intuitively, it says that when the population p approaches the maximum value Q (in billions) that the Earth can support, the population growth rate $\frac{1}{p}\frac{dp}{dt}$ approaches zero linearly.

There are two realistic premises specific for this application. First, there is a premise stating the population at some initial moment, for instance,

there are Q_0-billion people on January 1st, 1935. (R-1)

Q_0 here is a concrete decimal numeral and 'Q_0-billion people' is a composite concept representing the population (in billions). Second, there is a premise stating the population growth rate at various moments. Note that real population values are discrete, and a real population growth rate must be measured by a discrete formula like

$$\frac{p(t+\delta_0)-p(t)}{\delta_0 p(t)},$$

where δ_0 is the minimum meaningful temporal duration (with 'year' as the unit, for instance) for measuring the population growth. (δ_0 can be a positive rational number less than 1.) This temporal distance δ_0-years must be of an appropriate scale so that $p(t+\delta_0)-p(t)$ is significant but not too large. The sentence stating the population growth rate at all moments then quantifies over concrete temporal moments, for instance,

for each moment and q, if there are q-billion (R-2)
people at that moment, then there are
$\alpha(Q-q)q\delta_0$-billion more δ_0-years after.

Here, δ_0, α, Q are constant decimal numerals, and q ranges over decimal numerals with some fixed finite precision, and $\alpha(Q-q)q\delta_0$ is the result of arithmetic operations on decimal numerals. This is then a realistic statement. It is another realistic premise about the population growth. There are other realistic premises tacitly assumed for the application, for instance, premises about the relations between temporal moments, such as their linear order. We will ignore them here.

Then, we represent the population by a differentiable function $p(t)$. We translate the realistic premise on the initial population into an initial condition for a differential equation:

$$p(0) = Q_0, \tag{M-1}$$

and we translate the realistic premise on the population growth rate into a differential equation

$$p'(t) = \alpha(Q-p(t))p(t) \text{ for } t \geq 0. \tag{M-2}$$

The first translation can be achieved by the following bridging postulation:

for each t, q, there are q-billion people (B-1)
t-years after January 1st, 1935, iff $p(t) \approx q$.

Here, t, q ranges over decimal numerals with some finite precision. (M-1) follows from (R-1) and (B-1). However, there are no obvious bridging postulations to allow deriving the mathematical premise (M-2) from realistic premises. On the contrary, the realistic premise (R-2) follows from the mathematical premise (M-2) and the bridging postulation (B-1). This means that the mathematical premise (M-2) contains some idealization beyond the real physical states of affairs.

A mathematical proof then derives a mathematical conclusion, for instance,

$$p(t_1) = Q_1, \tag{M-3}$$

from the mathematical premises (M-1) and (M-2), plus other mathematical theorems. This mathematical conclusion can be translated into a realistic conclusion about the population at a moment, for instance,

there are Q_1-billion people t_1-years after January 1st, 1935. (R-3)

This follows from (M-3) and the bridging postulation (B-1).

Now, the mathematical premise (M-2) cannot be directly translated into a literally true realistic sentence about the population, with the logical structure of (M-2) preserved, because the population values are discrete and the population grows in discrete jumps. Similarly, many other mathematical theorems used in the proof of (M-3) cannot be translated into true realistic sentences about discrete population values or other finite concrete physical entities straightforwardly. Therefore, the mathematical proof that derives (M-3) from (M-1) and (M-2) cannot be straightforwardly translated into sound logical deductions from (R-1), (R-2) and other tacitly assumed realistic premises to the realistic conclusion (R-3). There is no logically obvious guarantee that the realistic conclusion (R-3) must be literally true if (R-1), (R-2) and other tacitly assumed realistic premises are literally true. This is the logical problem of applicability from the point of view of nominalism and naturalism.

A realist might claim that as a pure mathematical truth, there literally exists a mathematical function p satisfying (M-1) and (M-2), *and*

empirical data confirm the bridging postulation (B-1) for the mathematical function p satisfying (M-1) and (M-2). (B-1*)

We can derive the realistic conclusion (R-3) from (B-1*), or actually, from (M-1), (M-2) and (B-1). Such an explanation of applicability will assume the literal truth of mathematical premises (M-1) and (M-2). However, our intuition appears to be that this puts too much weight on the empirical claim (B-1*). It is certainly an empirical fact that a continuous model can simulate a discrete phenomenon sufficiently accurately, and we usually do use empirical data to justify a continuous model of a discrete phenomenon. Moreover, discovering a simple continuous model of a discrete phenomenon is indeed a great scientific achievement. Nevertheless, our intuition is that the mathematical premises (M-1) and (M-2) and the empirical claim (B-1*) are not really among the logically minimum premises that imply the realistic conclusion (R-3). What a continuous model offers is rather an indirect and approximate but simple and sufficiently accurate way of simulating the real population growth. Instead, the realistic premises (R-1) and (R-2) are the real reason why (B-1*) is empirically acceptable, and the realistic conclusion (R-3) should logically follow from realistic premises like (R-1) and (R-2) alone.

Alternatively, a realist might just start with the claim that there literally exists a mathematical function p that makes the bridging postulation (B-1) true. However, we cannot derive the differential equation (M-2) from (B-1) and the realistic premise

(R-2). Then, this is not sufficient to explain applicability without adding further assumptions, for instance, empirical assumptions like (B-1*).

This monograph will show that a sufficiently rich amount of calculus can be developed in strict finitism. Recall that strict finitism can be interpreted as a theory about concrete and finite computational devices. It will represent a rational number as two integer numerals, and represent a real number as a program generating an elementary recursive Cauchy sequence of rational numbers, and represent real functions as (elementary recursive) programs transforming the previous programs. Then, we can restate this application with calculus in strict finitism. The bridging postulation will say that a program (representing a real function) represents the population at various temporal moments. It will take the same format as (B-1) but p will be a program. We have good reasons to believe that there is such a program, because programs available to strict finitism are rich enough. This bridging postulation plus the realistic premise (R-1) will similarly imply (M-1), understood as a statement about the program p. The bridging postulation plus the realistic premise (R-2) will imply a finitistic version of the differential equation (M-2), which is basically a discretization of (M-2) and is our mathematical representation of the population growth in strict finitism. The fact that calculus can be developed within strict finitism means that we can similarly derive (M-3) from (M-1) and the finitistic version of (M-2), together with other axioms of strict finitism. Finally, the realistic conclusion (R-3) similarly follows from (M-3) and (B-1). This translates the original application of classical mathematics into an application of strict finitism. It means that the final realistic conclusion (R-3) logically follows from some literally true realistic premises about the programs in a concrete computational device, the population, and the simulation relation between the programs and the population.

Moreover, a finitistic version of the differential equation (M-2) actually has a very similar structure as (R-2). That is why the finitistic version of (M-2) can be derived from (R-2) and (B-1). We will see that in general, conditions such as continuity and differentiability in strict finitism are not committed to any idealization to infinity. They can be interpreted into literally true conditions about discrete physical quantities, as long as those discrete quantities are 'smooth at the macro-scale'. Note that this is the intuitive reason why we can simulate a population growth curve by a differentiable curve. That is, the population growth is discrete, but it looks smooth at the macro-scale. These imply that we can translate the derivation from (M-1), (M-2) to (M-3) in strict finitism rather straightforwardly into a derivation from (R-1), (R-2) to (R-3).

In Chap. 3 I will explain how a continuity or differentiability condition in strict finitism can be satisfied by discrete quantities, and I will come back to this example in the last section of Chap. 3, to add more details on the finitistic version of (M-2) and the finitistic derivation of (M-3) from (M-1) and (M-2).

Chapter 2
Strict Finitism

This chapter presents the logical framework for strict finitism. I will first present a logical formal system for strict finitism. Then, I will introduce a system of semi-formal notations to allow the presentation of the constructions and inferences in strict finitism in a simplified and more readable format. This will allow us to state ordinary mathematics in strict finitism in an informal way and make it look very similar to classical mathematics. This includes allowing us to talk about sets and functions in strict finitism, although we are not really committed to such entities.

2.1 The Formal System SF for Strict Finitism

Strict finitism is essentially a fragment of quantifier-free primitive recursive arithmetic (**PRA**) with the accepted functions restricted to elementary recursive functions. Elementary recursive functions are the functions constructed from some base arithmetic functions by composition and *bounded* primitive recursion. In classical mathematics, we can take the successor function and the power function as all the base functions, but in strict finitism, we will also need the base functions to include addition, multiplication, and the characteristic function of the relation $<$. To allow encoding real numbers, functions of real numbers and so on, we will use variables of higher types in strict finitism. However, you will see that this is not an essential extension. All numerical functions constructed in strict finitism are still only elementary recursive functions.

Some philosophers argue that primitive recursive arithmetic represents the scope of finitism (Tait [33]). The reason for restricting to elementary recursive functions here is to recognize the fact that in scientific applications, perhaps elementary recursive functions are all the functions we actually need, since science describes only things above the Planck scale in the universe. Remember that our goal here is to show that an application of classical mathematics to concrete things in the universe can in principle be reduced to valid logical deductions from premises about finite

F. Ye, *Strict Finitism and the Logic of Mathematical Applications*, Synthese Library 355, DOI 10.1007/978-94-007-1347-5_2, © Springer Science+Business Media B.V. 2011

concrete things in the universe alone to a conclusion about them. A more restrictive reduction base will get us closer to this goal.

The formal system for strict finitism will be denoted as **SF**. It is closely related to the system $\widehat{\mathbf{T}_0}$ in Avigad and Feferman [2], or the system $\mathbf{T}_0$ defined in Troelstra [35], with the recursion operator restricted to numerical functions. **SF** is a proper subsystem of these systems because it admits only bounded primitive recursions and thus only elementary recursive functions, not all primitive recursive functions.

2.1.1 The Language, Axioms and Rules of SF

The language of **SF** is the language of typed λ-calculus, plus constants for 0 and base elementary recursive functions, and plus operators for bounded primitive recursion, finite sum, finite product, and definition by cases. They are summarized as follows:

Types: o is a type, and if $\sigma_1, ..., \sigma_n, \sigma$ are types, then $(\sigma_1, ..., \sigma_n \to \sigma)$ is a type.
Variables: For each type σ, there are variables x_1^σ, x_2^σ, ... of the type.
Constants: 0, S, $+$, $\cdot$, pow, $I_<$.
Terms:

(1) 0 is a term of the type o; S is a term of the type $(o \to o)$; $+$, $\cdot$, pow, $I_<$ are terms of the type $(o, o \to o)$; each x_i^σ is a term of the type σ.
(2) If $t_1, ..., t_n, t$ are terms of the types $\sigma_1, ..., \sigma_n, (\sigma_1, ..., \sigma_n \to \sigma)$ respectively, then $Ap(t, t_1, ..., t_n)$ is a term of the type σ.
(3) If t is a term of the type σ, then $\lambda x_{i_1}^{\sigma_1} ... x_{i_n}^{\sigma_n}.t$ is a term of the type $(\sigma_1, ..., \sigma_n \to \sigma)$.
(4) If t is a term of the type o, and t_1, t_2 are terms of the type σ, then $J(t, t_1, t_2)$ is a term of the type σ.
(5) If $t[i, j]$, r, s are terms of the type o, and b is a term of the type $(o \to o)$, and i, j are variables of the type o, and i, j are not free in b, r, s, then $\mathrm{Re}\, ij(s, r, b, t[i, j])$ is a term of the type o.
(6) If $t[i]$, r are terms of the type o and i is not free in r, then $\sum_{i \le r} t[i]$, $\prod_{i \le r} t[i]$ are terms of the type o.

Formulas:

(1) If t, s are terms of the type o, then $t = s$ is an atomic formula.
(2) If φ, ψ are formulas, then $(\varphi \vee \psi)$, $(\varphi \wedge \psi)$, $(\varphi \to \psi)$ are formulas.
(3) $\neg\varphi$ is defined as $\varphi \to S0 = 0$.

o is the *base* or *numerical type*, and terms of the type o are *numerical terms*. S, $+$, $\cdot$, pow, $I_<$ are meant to represent the successor function, addition, multiplication, exponentiation, and the characteristic function of the relation $<$. Terms 0, $S0$, $SS0$, ... are numerals and are denoted by 0, 1, 2, ... respectively. $\overline{n}$ denotes the term $SS...S0$ with n occurrences of S. $pow(s, t)$ and $s \cdot t$ will be written as s^t and st. We use the convention that $I_<(m, n) = 0$ represents $m < n$, and we define

$$s < t \equiv_{df} I_< (s,t) = 0,$$
$$s \leq t \equiv_{df} s < t \vee s = t.$$

Note that while all elementary recursive functions can be constructed from the functions S and x^y by composition and bounded primitive recursion in classical mathematics, in **SF**, we need $+$, $\cdot$, and $I_<$ as primitives in order to state the primitive recursive equations for x^y and express the 'bound' in a bounded primitive recursion.

$Ap(t, s_1, ..., s_n)$ is usually denoted as $t(s_1, ..., s_n)$ and is to mean functional application, and correspondingly $\lambda x_{i_1}^{\sigma_1} ... x_{i_n}^{\sigma_n}.t$ is called a *λ-term* and is to mean λ-abstraction. In the classical typed λ-calculus, terms of the type o are evaluated into natural numbers, and terms of the type $(\sigma_1, ..., \sigma_n \to \sigma)$ represent functionals operating on entities of the types σ_1, ..., σ_n respectively and producing values of the type σ. However, remember that we do not recognize such abstract entities. Instead, we will treat terms as programs, and only terms of the type o can be evaluated, which will output one of the numerals 0, 1, 2, 3, ... A term t of the type $(\sigma_1, ..., \sigma_n \to \sigma)$ can be viewed, in a trivial manner, as a program operating on terms $s_1, ..., s_n$ of the types σ_1, ..., σ_n respectively and generating the term $t(s_1, ..., s_n)$ of the type σ.

$J(t, t_1, t_2)$ is called a *J-term* and is to mean definition by cases. That is, it is just t_1 or t_2 according to whether $t = 0$ or $t > 0$. See Selection Axioms below. In the classical typed λ-calculus with recursions on higher type entities, this is redundant, but we need it here because we have only recursions on the numerical type. Note that t_1, t_2 in $J(t, t_1, t_2)$ must be of the same type, which can be any type, but t must be a numerical term.

$\mathrm{Re}\, i j (s, r, b, t[i, j])$ is to mean bounded primitive recursion *restricted* to numerical terms, where s is the number of recursive construction steps (so far), and r is the initial value, and b gives the bound, and $t[i, j]$ gives the recursion pattern. See Recursion Axioms below. This, together with constants for base elementary recursive functions, allows constructing numerical terms representing all elementary recursive functions.

$\sum_{i \leq r} t[i]$ and $\prod_{i \leq r} t[i]$ are to mean finite sum and finite product. They are again redundant in the classical theory, but we need them here to allow easy development of basic arithmetic in strict finitism.

Bold face letters $\mathbf{x}, \mathbf{t}, \boldsymbol{\sigma}$ will denote sequences of variables, terms, and types respectively. The letters l, m, n, i, j, k are reserved for variables of the numerical type o. They are called *numerical variables*. The notions of *free variable*, *bound variable*, *substitution for free variables*, *closed formula* and *sentence* are as usual. In particular, $x_{i_1}^{\sigma_1}, ..., x_{i_n}^{\sigma_n}$ in (3) and i, j in (5) and (6) of the definition of terms above are bound variables in the terms. The notation $t[x]$ is used to indicate a free variable x in the term t, and then $t[r]$ will denote the result of substituting r for all free occurrences of x in t. The notations $\varphi[x]$, $\varphi[t]$ for a formula φ are similar. The notions of subterm and subformula are as usual. Note that the variables right after a λ or Re are not considered subterms. We use $t\{s\}$ to indicate a *single* occurrence of a subterm s in t. We will use $\equiv$ as a meta-language symbol to denote the syntactical identity.

A few points about the language of **SF** must be carefully noted. First, $t = s$ is a formula only if t, s are numerical terms. There are no equalities between higher type terms in the language of **SF**. Intuitively, an equality $t = s$ between two terms of the type $(o \to o)$ is implicitly committed to either infinity or abstract entities. More specifically, if $t = s$ is understood extensionally as $\forall n\,(t\,(n) = s\,(n))$, then it is committed to infinity, and if it is understood intensionally, then it is committed to a function of natural numbers as an abstract intensional object, since it assumes an equality relation between functions. With atomic formulas restricted to equalities between numerical terms, we will see that assertions in **SF** expressed by closed formulas are only assertions about the outcomes of computing elementary recursive functions for *concretely given* argument values. In particular, no assertion is made about the equality between two arbitrary elementary recursive functions, neither as an equality between two intensional abstract entities, nor as an assertion about the outcomes of computing two elementary recursive functions for *all* argument values.

Second, there is no quantifier in the language of **SF**. Generality has to be achieved by using free variables. For instance, when one proves $x + y = y + x$ in **SF**, one also proves its instances such as $3 + 2 = 2 + 3$, $5 + 3 = 3 + 5$, and so on. Generality can also be achieved by making schematic assertions about arbitrary formulas of some format, for instance, by claiming schematically that a formula of the format $t + s = s + t$ is provable. In other words, strict finitism allows abstraction but it is not committed to any idealization. For instance, suppose that the universe is finite and discrete, and therefore there is a limit on how many numerals could really exist. We can still interpret *some* formulas in strict finitism as assertions about numerals that really exist. If a formula has no such chance of being interpreted as an assertion about concretely existent numerals, then let it be. Formulas with free variables can be interpreted as schematic assertions about those concretely existent numerals. This is abstraction, which allows us to talk about concretely existent things in a schematic and hence more abstract format. On the other side, since we never use quantifiers to refer to '*all* numerals', we are not committed to the literal existence of 'all numerals' when we make assertions about those concretely existent numerals. We never say 'all'. That is, we never make the idealized assumption that all numerals exist, which will go beyond what really exist.

The Axioms of SF:

(1) The axiom schemes of classical propositional logic.

(2) Identity Axioms: for terms t, s, r of the type o,

$$t = t, \quad t = r \wedge s = r \to t = s,$$
$$t = s \to r\,[t/n] = r\,[s/n]\,.$$

(3) Arithmetic Axioms: axioms characterizing S:

$$\neg St = 0, \; St = Sr \to t = r,$$

and the primitive recursive definition axioms for the base functions $+$, $\cdot$, and *pow*:

$$r+0=r,\ r+St=S(r+t),$$
$$r\cdot 0=0,\ r\cdot St=r\cdot t+r,$$
$$r^0=1,\ r^{St}=r^t\cdot r,$$

and the axioms characterizing $I_<$:

$$I_<(s,t)=0\vee I_<(s,t)=1,$$
$$\neg r<0,\ r<St\leftrightarrow r<t\vee r=t,$$

and the axioms characterizing finite sum and finite product:

$$\sum\nolimits_{i\leq 0}t[i]=t[0],\ \sum\nolimits_{i\leq Sr}t[i]=\sum\nolimits_{i\leq r}t[i]+t[Sr],$$
$$\prod\nolimits_{i\leq 0}t[i]=t[0],\ \prod\nolimits_{i\leq Sr}t[i]=\prod\nolimits_{i\leq r}t[i]\cdot t[Sr].$$

(4) Selection Axioms:

$$s\{J(0,t_1,t_2)\}=s\{t_1\},$$
$$s\{J(St,t_1,t_2)\}=s\{t_2\}.$$

(5) Reduction Axioms:

$$s\{Ap(J(t,t_1,t_2),\mathbf{s})\}=s\{J(t,Ap(t_1,\mathbf{s}),Ap(t_2,\mathbf{s}))\},$$
$$s\{Ap(\lambda\mathbf{x}.t,\mathbf{r})\}=s\{t[\mathbf{r}/\mathbf{x}]\}.$$

(6) Recursion Axioms:

$$\mathrm{Re}ij(0,r,b,t[i,j])=J(I_<(r,b(0)),r,b(0)),$$
$$\mathrm{Re}ij(Ss,r,b,t[i,j])=J\left(I_<(t',b(Ss)),t',b(Ss)\right),$$
$$\text{where } t'\equiv t[s,\mathrm{Re}ij(s,r,b,t[i,j])].$$

Rules:

(1) Modus Ponens: $\varphi\rightarrow\psi,\ \varphi\Longrightarrow\psi$.
(2) Induction Rule: $\varphi[0],\ \varphi[n]\rightarrow\varphi[Sn]\Longrightarrow\varphi[t]$.

A *proof* of a formula φ in **SF** is a sequence of formulas of **SF** ending with φ, such that for each formula in the sequence, either it is an axiom or it is derived from the previous formulas by one of the rules. We say that φ is a *theorem* of **SF**, if we *can* construct such a proof, and we denote that fact as **SF**$\vdash\varphi$. Note that we will keep the finitistic spirit in our meta-language claims about **SF** as a formal system. We won't consider 'the existence of a proof' in the abstract, non-constructive sense, and we won't consider whether a formula is 'non-provable'. When we claim that we 'can' construct a proof, either we actually give the proof or we present an elementary recursive procedure for constructing the proof.

Note that the axioms (4) and (5) are stated in schematic formats, because we do not have equalities between terms of higher types and we cannot state Selection

Axioms as $J(0,t_1,t_2)=t_1$, $J(St,t_1,t_2)=t_2$, for instance. Moreover, note that **SF** has an induction rule. Since **SF** is quantifier-free, this is quantifier-free induction. In an application of Induction Rule, the variable n above is called the inductive variable. It usually disappears in the conclusion.

Substitution Rule is obviously derivable since all axioms are in schematic formats:

Corollary 2.1. *If SF$\vdash \varphi[x]$, then SF$\vdash \varphi[t]$ for any t of the same type as x.*

A deduction from premises, $\{\varphi_1,...,\varphi_k\} \vdash_{\mathbf{SF}} \psi$, can be defined as usual, except for the requirement that when Induction Rule is applied, the inductive variable should not occur free in $\varphi_1,...,\varphi_k$. Then, it is easy to see that we have Deduction Theorem:

Corollary 2.2. *If $\{\varphi_1,...,\varphi_k,\varphi\} \vdash_{SF} \psi$, then $\{\varphi_1,...,\varphi_k\} \vdash_{SF} \varphi \to \psi$.*

Since we will focus on developing ordinary mathematics in **SF**, not on the meta-properties of the formal system **SF**, whenever we simply assert a formula φ, we always mean **SF**$\vdash \varphi$. Similarly, when we claim that ψ is derivable from φ, we always mean $\varphi \vdash_{\mathbf{SF}} \psi$.

2.1.2 Arithmetic in SF

We will first develop some basic arithmetic in **SF**. We have to be a little careful here, because we cannot use quantifiers in proving those arithmetic theorems. Moreover, sufficiently rich arithmetic must be developed before we can use bounded primitive recursion to construct useful terms, because in using bounded primitive recursion, we usually must first prove that the recursive construction does respect the bound, so that we can ignore the bound and show that the constructed term satisfies the recursive equation. Therefore, this is a little different from developing arithmetic in **PRA**. In developing arithmetic in **SF**, we will define many new constants. Sometimes we will say something like 'define a function f, $f(x) \equiv_{df} t[x]$', by which we really mean 'let f be a new constant symbol defined by $f \equiv_{df} \lambda x.t[x]$'.

First, by using Arithmetic Axioms and Induction Rule, we can easily prove simple arithmetic laws for S, $+$, $\cdot$, and pow, such as the commutative and associative laws for $+$ and $\cdot$, the distributive laws for $+$ and $\cdot$, and various laws for the power function, for instance,

$$m^{n+k} = m^n \cdot m^k, \ m^{n \cdot k} = (m^n)^k, \ (m \cdot n)^k = m^k \cdot n^k.$$

We will omit the details here.

Then, we have some basic theorems for inequality:

$$0 < Sn, \ m < n \leftrightarrow Sm < Sn,$$
$$k < m < n \to k < n, \ m < n \to \neg n \le m,$$

$$Sm < n \leftrightarrow m < n \wedge Sm \neq n,$$
$$n < m \vee n = m \vee m < n.$$

These can be proved by inductions on n, in the given order. We will omit the details here. Then, familiar laws of inequality with respect to addition, multiplication, and exponentiation can be derived, for instance,

$$m \leq n \leftrightarrow m + k \leq n + k,$$
$$m \leq n \rightarrow m \cdot k \leq n \cdot k \wedge m^k \leq n^k \wedge k^m \leq k^n.$$

These can be proved by inductions on k. We will again omit the details here.

To use bounded primitive recursion to construct terms satisfying recursive equations, we can use the following lemma:

Lemma 2.3. *Suppose that* $t[i,j]$, r *are terms of the type* o, *and* b *is be a term of the type* $(o \rightarrow o)$, *and variables* i, j, n *are not free in* b, r. *Suppose that we have*

$$r < b(0), \;\; j < b(i) \rightarrow t[i,j] < b(Si).$$

Then, we can construct a term $q[n]$, *such that*

$$q[0] = r, \;\; q[Sn] = t[n, q[n]].$$

Proof. Let $q[n] \equiv \mathrm{Re}\,ij(n, r, b, t[i,j])$. By $r < b(0)$ and Recursion Axioms and Selection Axioms, it directly follows that $q[0] = r$. Note that by Recursion Axioms, we generally have

$$t[n, q[n]] < b(Sn) \rightarrow q[Sn] = t[n, q[n]].$$

Moreover, by the assumption $j < b(i) \rightarrow t[i,j] < b(Si)$, we have

$$q[Sn] < b(Sn) \rightarrow t[Sn, q[Sn]] < b(SSn).$$

Then, it is easy to see that we can use an induction on n to show that

$$t[n, q[n]] < b(Sn) \wedge q[Sn] = t[n, q[n]].$$

□

When b satisfies the condition of the lemma, we say that iterating $t[i,j]$ (for j) starting from r is bounded by b. We say that $t[i,j]$ is iteratively bounded for j if we can find such b for any r. Note that $f = \lambda n.q[n]$ will be a type $(o \rightarrow o)$ term satisfying

$$f(0) = r, \;\; f(Sn) = t[n, f(n)].$$

Sometimes we will write a construction of a term by bounded primitive recursion in this format, and say that it defines f as a function, or more accurately, a type $(o \rightarrow o)$ term.

Then, we can define the predecessor function *pred* and define subtraction for natural numbers by bounded primitive recursion. First, we can define *pred* by the following recursive equations:

$$pred\,(0) = 0,\ \ pred\,(Sn) = n.$$

More accurately, let $r \equiv 0$, $t\,[i,j] \equiv i$, $b \equiv \lambda i.Si$. Then, we have $r < b\,(0)$ and $t\,[i,j] < b\,(Si)$. Therefore, by the lemma above, we can construct a term $q\,[n]$, such that $q\,[0] = r = 0$, $q\,[Sn] = t\,[n, q\,[n]] = n$. Then, we can let

$$pred \equiv_{df} \lambda n.q\,[n]\,.$$

Similarly, we can define substraction $-$ so that

$$s - 0 = s,\ \ s - Sn = pred\,(s - n)\,.$$

Let $r \equiv s$, $t\,[i,j] \equiv pred\,(j)$, and $b \equiv \lambda i.\,(s+1)$. It is easy to see that iterating $t\,[i,j]$ starting from r is bounded by b. Therefore, we can define $-$ similarly. Recursive equations for $-$ then follow from the lemma above. Familiar properties of *pred* and $-$ can then be derived, for instance,

$$\begin{gathered}
m,n > 0 \to (m \le n \leftrightarrow pred\,(m) \le pred\,(n))\,,\\
m \le k \to m - n \le k - n,\\
m - n = Sm - Sn,\\
n < m \leftrightarrow 0 < m - n,\\
n < m \to (m-n) + n = m,\\
m \le n \leftrightarrow m - n = 0.
\end{gathered}$$

These can be proved by inductions on n, in the given order. Other common arithmetic properties of S, *pred*, $+$, $-$, $\cdot$, *pow*, and $<$ are also provable. We omit the details here.

For any formula φ, by an induction on the construction of φ, we can construct a corresponding term t_φ representing it, so that

$$(t_\varphi = 0 \vee t_\varphi = 1) \wedge (\varphi \leftrightarrow t_\varphi = 0)\,.$$

For instance, for $\varphi \equiv (t = s)$, we can let $t_\varphi \equiv sg\,((t-s)+(s-t))$, where sg is defined by

$$sg\,(0) = 0,\ \ sg\,(Sn) = 1.$$

Similarly, we can let $t_{\varphi\wedge\psi} \equiv sg\,(t_\varphi + t_\psi)$. Note that it is critical here that the language of **SF** is quantifier-free and allows only equalities between numerical terms. With representing terms available, we sometimes write $J\,(t_\varphi, t_1, t_2)$ as $J\,(\varphi, t_1, t_2)$.

We can easily prove some basic properties of finite sum and finite product, including the associative laws, distributive laws, and common inequalities for them. For instance, we have

$$s \cdot \sum_{i \le m} t[i] = \sum_{i \le m} s \cdot t[i],$$
$$\sum_{i \le n} t[i] + \sum_{i \le m} t[i + Sn] = \sum_{i \le Sn+m} t[i],$$
$$\sum_{i \le m} t[i] + \sum_{i \le m} s[i] = \sum_{i \le m} (t[i] + s[i]),$$
$$k \le m \to t[k] \le \sum_{i \le m} t[i] \wedge \sum_{i \le k} t[i] \le \sum_{i \le m} t[i],$$

and we have similar formulas for finite product. These can be proved by inductions on m.

Bounded quantifiers are defined as:

$$(\forall i \le m)\, \varphi[i] \equiv_{df} \sum_{i \le m} t_\varphi[i] = 0,$$
$$(\exists i \le m)\, \varphi[i] \equiv_{df} \neg \forall i \le m \neg \varphi[i].$$

We sometimes simply write $(\forall i \le m)\, \varphi[i]$, $(\exists i \le m)\, \varphi[i]$ as $\forall i \le m \varphi[i]$, $\exists i \le m \varphi[i]$. We will sometimes use notations like $(\forall i < m)$ and $(\exists i < m)$. Their definitions are obvious. Basic properties characterizing bounded quantifiers can be easily proved. For instance, we have

$$(\forall i \le Sm)\, \varphi[i] \leftrightarrow (\forall i \le m)\, \varphi[i] \wedge \varphi[Sm],$$
$$(\forall i \le m)\, \varphi[i] \wedge n \le m \to \varphi[n] \wedge (\forall i \le n)\, \varphi[i],$$
$$(\forall i \le m)\, (\varphi[i] \to \psi[i]) \to ((\forall i \le m)\, \varphi[i] \to (\forall i \le m)\, \psi[i]),$$
$$\varphi \to (\forall i \le m)\, \varphi, \text{ where } i \text{ is not free in } \varphi,$$
$$(\forall i \le m)\, \varphi[i] \leftrightarrow \prod_{i \le m} t_{\neg\varphi}[i] = 1,$$
$$(\forall i \le \overline{m})\, \varphi[i] \leftrightarrow \varphi[0] \wedge \ldots \wedge \varphi[\overline{m}],$$

for the bounded universal quantifier, and we have similar formulas for the bounded existential quantifier. The first formula above directly follows from the definition. The second follows from corresponding properties for finite sum. The next three can be proved by an induction on m. The last one also directly follows from the definition.

Moreover, if k does not occur in $\psi[m]$, we have

$$\text{if } \mathbf{SF} \vdash \psi[m] \to (k \le m \to \varphi[k]),$$
$$\text{then } \mathbf{SF} \vdash \psi[m] \to (\forall i \le m)\, \varphi[i].$$

To see this, assuming that we already have a proof of $\psi[m] \to (k \le m \to \varphi[k])$, we can first use an induction on k to prove

$$\psi[m] \to k \le m \to \sum_{i \le k} t_\varphi[i] = 0.$$

Then, the required formula follows as an instance when $k = m$. As a consequence, we have

$$\text{if } \mathbf{SF} \vdash \varphi[k], \text{ then } \mathbf{SF} \vdash (\forall i \le k)\, \varphi[i].$$

We have similar results for the bounded existential quantifier $(\exists i \leq k)$. These mean that $(\forall i \leq m)$ and $(\exists i \leq m)$ do behave like quantifiers.

Basic arithmetic theorems involving bounded quantifiers, finite sum and finite product are then straightforward consequences of these definitions and the properties of finite sum, finite product and bounded quantifiers. For instance, we have

$$(\forall i \leq m)(t[i] \leq s[i]) \rightarrow \sum\nolimits_{i \leq m} t[i] \leq \sum\nolimits_{i \leq m} s[i],$$
$$(\forall i \leq m)(t[i] = 0) \leftrightarrow \sum\nolimits_{i \leq m} t[i] = 0,$$
$$(\exists i \leq m)(t[i] > 0) \leftrightarrow \sum\nolimits_{i \leq m} t[i] > 0,$$

and we have similar formulas for finite product.

The term $\max_{i \leq m} t[i]$ can then be constructed to satisfy these recursive equations:

$$\max_{i \leq 0} t[i] = t[0], \quad \max_{i \leq Sm} t[i] = \max\left(\max_{i \leq m} t[i], t[Sm]\right),$$

where the function max is defined as $\max(m,n) = J(m < n, n, m)$. We can use $\sum_{i \leq m} t[i]$ as the bound in the bounded recursive construction of $\max_{i \leq m} t[i]$. The basic properties of $\max_{i \leq m} t[i]$ are easily proved by inductions. For instance, we have

$$k \leq m \rightarrow t[k] \leq \max_{i \leq m} t[i],$$
$$(\forall i \leq m)(t[i] \leq n) \rightarrow \max_{i \leq m} t[i] \leq n.$$

Bounded minimalization can be defined as:

$$(\mu i \leq m)\,\varphi[i] \equiv_{df} \sum\nolimits_{i \leq m} \prod\nolimits_{j \leq i} t_\varphi[j].$$

We sometimes simply write $(\mu i \leq m)\,\varphi[i]$ as $\mu i \leq m \varphi[i]$. Basic properties characterizing it are also easy to prove. For instance, we have

$$(\forall i \leq m) \neg\varphi[i] \rightarrow (\mu i \leq m)\,\varphi[i] = Sm,$$
$$n < (\mu i \leq m)\,\varphi[i] \rightarrow \neg\varphi[n],$$
$$(\mu i \leq m)\,\varphi[i] < m \rightarrow \varphi[(\mu i \leq m)\,\varphi[i]],$$
$$(\mu i \leq m)\,\varphi[i] \leq m \leftrightarrow (\exists i \leq m)\,\varphi[i].$$

The first can be easily proved by an induction on m. To prove the second, we can first prove

$$\prod\nolimits_{j \leq k} t_\varphi[j] \leq 1,$$
$$\varphi[n] \wedge n \leq k \rightarrow \prod\nolimits_{j \leq k} t_\varphi[j] = 0.$$

Now, in general, we have

$$(\forall i \le m)(s[i] \le 1 \wedge (i \ge n \to s[i] = 0)) \to \sum_{i \le m} s[i] \le n.$$

Therefore, it follows that $\varphi[n] \to (\mu i \le m)\varphi[i] \le n$. The third can be proved by an induction on m. The last again follows from the above formula.

These allow us to state Induction Rule in the following format:

$$\text{if } \mathbf{SF} \vdash \varphi[0], \forall i \le k \varphi[i] \to \varphi[Sk],$$
$$\text{then } \mathbf{SF} \vdash \varphi[k].$$

To prove it, we can use the original Induction Rule on k to show that $\forall i \le k \varphi[i]$.

With these, we can encode sequences of numbers. This can be done in a few ways. For instance, to take the well-known Gödel β-function approach (Murawski [24], pp. 30–34), first note that the predicates $Div(a,b)$ (i.e., 'a is divisible by b'), $\Pr(a)$ (i.e., 'a is a prime number'), and $RP(a,b)$ (i.e., 'a and b are mutually prime') are easily defined using bounded quantifiers. Similarly, using bounded recursion, we can construct the function $m \div n$, giving the integer quotient of m divided by n, and the function $rm(m,n)$, giving the remainder of m divided by n. (We can make the convention that $m \div 0 = 0$.)

To prove the Basic Theorem of Arithmetic, we first construct the function lpf such that $lpf(m)$ is the least prime factor of m, if any, and it is 1, otherwise. Then, we can construct a function pf such that

$$pf(m,0) = m,$$
$$pf(m,k+1) = pf(m,k) \div lpf(pf(m,k)).$$

That is, $pf(m,0)$, $pf(m,1)$, $pf(m,2)$, ... is the sequence obtained by dividing the least prime factors repeatedly, until it reaches 1 (or 0 if it starts with 0), and then it stays constant from there. Then, it is easy to see that

$$prf(m,k) \equiv pf(m,k) \div pf(m,k+1)$$

gives the $(k+1)$th prime factor of m. Moreover,

$$0 < m \to m = \prod_{k \le m} prf(m,k), \tag{2.1}$$

which is a prime factorization of m. This can be proved by an induction on m, using the fact that

$$1 < m \to pf(m,1) < m \wedge pf(m,k+1) = pf(pf(m,1),k),$$

which can be proved by an induction on k.

To prove the uniqueness of prime factorization, first note that

$$RP(m,n) \to RP(rem(n,m),m).$$

Then, we can prove that

$$RP(k,m) \wedge Div(m \cdot n,k) \rightarrow Div(n,k). \tag{2.2}$$

To prove this, we can first use an induction on k to prove this formula $\varphi[k]$:

$$(\forall m,n \leq N)(m \cdot n \leq N \wedge RP(k,m) \wedge Div(m \cdot n,k) \rightarrow Div(n,k)).$$

For this, assume that $(\forall i < k)\,\varphi[i]$, and then assume that $m \cdot n \leq N$, $RP(k,m)$, $Div(m \cdot n,k)$. We have $RP(rem(m,k),k)$ and $Div(rem(m,k) \cdot n,k)$. Therefore, there exists $l \leq N$ such that $rem(m,k) \cdot n = k \cdot l$. Now, $rem(m,k) < k$ and $Div(k \cdot l, rem(m,k))$. Therefore, the inductive assumption is applicable and we get $Div(l, rem(m,k))$, from which we have $Div(n,k)$. This proves (2.2). It then easily follows that

$$\Pr(p) \wedge (\forall i \leq m)\Pr(t[i]) \rightarrow \left(Div\left(\prod_{i \leq m} t[i], p\right) \leftrightarrow (\exists i \leq m)(t[i] = p)\right).$$

Then, the uniqueness of the factorization in (2.1) follows as usual. This proves the Basic Theorem of Arithmetic.

We give a rather detailed presentation here because classical proofs are usually stated with inductions on quantified statements. We see here that only inductions on quantifier-free formulas are needed. Since we have bounded quantifiers available, quantifier-free inductions are actually sufficient for many ordinary proofs, as long as the proofs do not require generating values that cannot be bounded by elementary recursive functions.

These provide sufficient arithmetic basics for encoding sequences by the Gödel β-function. $\beta(m,i)$ can be constructed as a term using bounded quantifiers and bounded minimalization. Following Murawski [24], pp. 30–34, we let

$$OP(m,n) \equiv_{df} (m+n)(m+n)+m+1.$$

Then,

$$\beta(m,i) \equiv_{df} \mu k \leq m \left(\exists n,l \leq m \begin{bmatrix} m = OP(n,l) \wedge \\ Div(n, 1 + (Op(k,i)+1) \cdot l) \end{bmatrix} \right).$$

The code of the sequence $k_0, \ldots, k_{n-1}$ can then be defined by bounded minimalization

$$\langle k_0, \ldots, k_{n-1} \rangle \equiv_{df} \mu m \leq s \begin{pmatrix} \beta(m,0) = \overline{n} \wedge \\ \beta(m,1) = k_0 \wedge \ldots \wedge \beta(m,\overline{n}) = k_{n-1} \end{pmatrix},$$

where the bound $s \equiv s[k_0, \ldots, k_{n-1}]$ can be constructed using elementary recursive functions. More specifically, first note that the factorial function $m!$ can be defined by bounded primitive recursion. Let

$$c \equiv_{df} \max\left(OP(k_0,0)+1, \ldots, OP\left(k_{n-1}, \overline{n-1}\right)+1\right)!,$$
$$d \equiv_{df} (1 + (Op(k_0,1)+1) \cdot c) \cdot \ldots \cdot \left(1 + \left(Op\left(k_{n-1}, \overline{n-1}\right)+1\right) \cdot c\right).$$

Then, we can let $s[k_0,...,k_{n-1}] \equiv_{df} OP(d,c)$. Length and decoding functions are similarly defined:

$$lh(m) \equiv_{df} \beta(m,0), \quad (m)_i \equiv_{df} \beta(m,i+1).$$

Basic properties stating that this is a coding function can be proved:

$$lh(\langle k_0,...,k_{n-1}\rangle) = n, \quad (\langle k_0,...,k_{n-1}\rangle)_{\bar{i}} = k_i.$$

The predicate *Seq* for sequence numbers is defined by bounded quantifiers

$$Seq(m) \equiv_{df} (\forall k < m)(lh(k) \neq lh(m) \vee (\exists i < lh(m))((m)_i \neq (k)_i)).$$

Concatenation of sequences is also defined by bounded minimalization

$$m * n \equiv_{df} (\mu k \leq r[m,n]) \begin{pmatrix} lh(k) = lh(m) + lh(n) \wedge \\ (\forall i < lh(m))((k)_i = (m)_i) \wedge \\ (\forall i < lh(n))\left((k)_{lh(m)+i} = (n)_i\right) \end{pmatrix}.$$

The bound $r[m,n]$ can be constructed in a similar way as the construction of $s[k_0,...,k_{n-1}]$ above. Then, the basic properties of these can be proved.

Similarly, given a numerical term $t[n]$, we can construct a term $\bar{t}[n]$ that intuitively encodes the sequence $t[0],...,t[n-1]$, that is,

$$\bar{t}[n] \equiv_{df} (\mu m \leq s[n])(\beta(m,0) = n \wedge \forall i \leq n-1(\beta(m,i+1) = t[i])),$$

where the bound $s[n]$ can be constructed from $t[n]$ using the finite product operator. Then, we have

$$lh(\bar{t}[n]) = n, \quad i < n \rightarrow (\bar{t}[n])_i = t[i].$$

With this, we can use bounded course-of-value recursion to construct terms. That is, given numerical terms $r, s[m,n], b[m]$, we can construct a term $t[m]$ such that

$$t[0] = J(r \leq b[0], r, b[0]),$$
$$t[Sm] = J(s[m,\bar{t}[m]] \leq b[Sm], s[m,\bar{t}[m]], b[Sm]).$$

2.1.3 A Finitistic Interpretation of SF

Note that all elementary recursive functions can be expressed by terms in **SF**. We will first show that the reverse is also true. We will need some standard concepts and results from the theory of typed lambda calculus (see, for instance, Barendregt [3]). A *one-step reduction* is a pair of terms of one of the following forms:

$$\langle s\{Ap(J(t,t_1,t_2),\mathbf{s})\}, s\{J(t,Ap(t_1,\mathbf{s}),Ap(t_2,\mathbf{s}))\}\rangle,$$
$$\langle s\{Ap(\lambda\mathbf{x}.t,\mathbf{r})\}, s\{t[\mathbf{r}/\mathbf{x}]\}\rangle.$$

Subterms of the format $Ap(J(t,t_1,t_2),\mathbf{s})$ or $Ap(\lambda\mathbf{x}.t,\mathbf{s})$ are called *redices*, and the indicated redices above are the *reduced redices* in each reduction. A term is *normal* if it contains no redex. A reduction sequence is a sequence of terms such that each pair of adjacent terms is a one-step reduction. We say that t is *reducible to* t' if we can construct a reduction sequence $t \equiv t_0, ..., t_n \equiv t'$. If t' is further normal, t' is called a *normal form* of t. Adapting the proof from the standard typed lambda calculus, we have:

Theorem 2.4. *(Normal Form Theorem) Each term is reducible to a normal form.*

Proof. First we need some notions. We define the *height* $|\sigma|$ of a type σ by

$$|o| = 0,$$
$$|(\sigma_1,...,\sigma_n \to \sigma)| = \max(|\sigma_1|,...,|\sigma_n|,|\sigma|)+1.$$

The *height of a redex* $Ap(J(t,t_1,t_2),\mathbf{s})$ or $Ap(\lambda\mathbf{x}.t,\mathbf{r})$ is the height of the type of $J(t,t_1,t_2)$ or $\lambda\mathbf{x}.t$. A *one-step strict reduction* is a one-step reduction of a term t such that the reduced redex is of the maximum height among the redices in t, and either it is the rightmost occurrence of a J-redex of that height in t if there is such a J-redex, or it is the rightmost occurrence of a λ-redex of that height if there is no J-redex of that height. (We agree that if s_1 occurs in s_2 and s_2 occurs in t, then the occurrence of s_1 in t is at the right of the occurrence of s_2 in t.) For a term t, let $|t|$ denote the length of t as a string of symbols.

If t is not normal, there is a unique t_1 such that $\langle t,t_1\rangle$ is a strict reduction. Clearly, $t_1 \equiv sr(t)$ for some function sr, where we agree that $sr(t) \equiv t$ in case t is already normal. Moreover, note that $|sr(t)| \le |t|^2$. Beginning with any term t we can then construct a sequence of strict reductions $sr^0(t) \equiv t, sr^1(t) \equiv sr(t), sr^2(t) \equiv sr(sr(t)),....$

For each term t, let $m(t)$ be the maximum height of the redices in t, which is set to 0 if t is normal, and let $n(t)$ be the number of occurrences of J-redices of the height $m(t)$, and let $l(t)$ be the number of occurrences of λ-redices of that height. Now suppose that t is an arbitrary term and $m(t) > 0$.

If $n(t) > 0$, in the one-step strict reduction from t, the reduced redex is a J-redex $Ap(J(t_0,t_1,t_2),\mathbf{s})$ and it is reduced to $J(t_0,Ap(t_1,\mathbf{s}),Ap(t_2,\mathbf{s}))$. In case t_1 or t_2 or both are still J-terms, one or two more J-redices of the same height will be created in this one-step strict reduction, but no other new redex of the same or any higher height will be created. If t_2 is a J-term, $Ap(t_2,\mathbf{s})$ must be the reduced redex in the next one-step strict reduction. Continuing such strict reductions, we will reach a term in the position of t_2 which is not a J-term, and then for the next strict reduction we will go back to the corresponding $Ap(t_1,\mathbf{s})$ and see if t_1 is still a J-term. Continuing the procedure we will finally reduce all J-redices of the same height created by the original reduction and decrease the number of J-redices in t of that height by one. Repeating this process, we can reduce all J-redices in t of that height. Note that the number of reduction steps needed here is less than the number of J-subterms in the original term t, which is less than $|t|$. Therefore, either $m\left(sr^{|t|}(t)\right) < m(t)$, or $m\left(sr^{|t|}(t)\right) = m(t)$ and $n\left(sr^{|t|}(t)\right) = 0$. That is, after $|t|$ steps, either all redices of the original maximum height are reduced and the maximum height of all redices is strictly decreased, or all J-redices of the original maximum height are reduced. Each reduction step at most doubles the length of the term. Therefore, $\left|sr^{|t|}(t)\right| \leq 2^{|t|}|t|$.

Next, let $t' \equiv sr^{|t|}(t)$. If $m(t') = m(t)$ and $n(t') = 0$, the next reduced redex in the strict reduction from t' is a λ-redex $Ap(\lambda\mathbf{x}.t_0, \mathbf{r})$ of the height $m(t)$, and it is reduced to $t_0[\mathbf{r}/\mathbf{x}]$. Note that the heights of the terms in $\mathbf{r}$ are all strictly less than $m(t)$. It is routine to check that the substitution $t_0[\mathbf{r}/\mathbf{x}]$ will not generate new redices of the height $m(t)$ (or any larger height), and the number $l(t')$ of λ-redices of the height $m(t)$ in t' decreases by one. $l(t') < |t'|$. It means that after $|t'|$ strict reductions, we must reach a term with a strictly less maximum height of redices, that is, $m\left(sr^{|t'|}(t')\right) < m(t)$. Since $t' \equiv sr^{|t|}(t)$ and $|t'| \leq 2^{|t|}|t|$, our rough estimate is that $m\left(sr^{2^{|t|}|t|}(t)\right) < m(t)$.

This shows that all redices will finally be reduced within a number of steps bounded by $m(t)$ iterations of the power function 2^n. □

Note that the number of reduction steps needed for reducing t cannot be uniformly bounded by any fixed elementary recursive function of $|t|$. This reduction process is essentially a diagonalization over all elementary recursive functions.

Normal terms in **SF** have some special properties, which guarantee that definable numerical functions in **SF** are only elementary recursive functions. We define a proper subtype of a type $(\sigma_1, ..., \sigma_n \to \sigma)$ to be one of $\sigma_1, ..., \sigma_n, \sigma$ or one of their proper subtypes. Then, we have

Lemma 2.5. *If $t[x_1, ..., x_n]$ is a normal term of the type σ, and $x_1, ..., x_n$ are all free variables in t, and their types are $\sigma_1, ..., \sigma_n$ respectively, then any subterm s of t must satisfy one of the following:*

(i) s is one of $x_1, ..., x_n$, or one of the constants 0, $S, +, \cdot$, pow, $I_<$, or

(ii) the type of s is o, or σ, or a proper subtype of one of $\sigma_1, ..., \sigma_n, \sigma$.

Proof. Consider the subterms of t that are of the largest height among the subterms satisfying none of these conditions. Suppose that s is maximum (with respect to the subterm relation) among these subterms. s has to be a proper subterm of t. Let s' be the smallest subterm of t containing s as a proper subterm. Routine verifications will show that s' will not satisfy any of the conditions either, but it will have a larger height. That will be a contradiction. □

Corollary 2.6. *If $t[m_1, ..., m_l]$ is a numerical normal term whose free variables are all among the numerical variables $m_1, ..., m_l$, then $t[m_1, ..., m_l]$ contains no occurrence of λ and all its subterms of a type other than o are among the constants S, $+$, $\cdot$, pow, $I_<$.*

This means that a numerical term $t[m_1, ..., m_l]$ with only numerical free variables represents an elementary recursive function. Because of Reduction Axioms, each numerical term is **SF**-provably equal to its normal form. Then, by the Corollary, such a numerical term represents a function composed from the base functions by composition, bounded recursion, finite sum and finite product. It is thus an elementary recursive function. In particular, a closed numerical term is provably equal to a numeral. More generally, numerical terms with free variables of higher types are schemas for elementary recursive functions. They become elementary recursive functions when free variables of higher types are instantiated by closed terms of appropriate types. Similarly, an arbitrary term t of some higher type can be viewed as a schematic representation of elementary recursive functions, in the sense that for any sequences of terms $\mathbf{s}_1, ..., \mathbf{s}_n$ of appropriate types, $t(\mathbf{s}_1)...(\mathbf{s}_n)$ becomes a numerical term and thus represents (schematically) elementary recursive functions.

Next, consider mathematical proofs in **SF**. First we have

Lemma 2.7. *If ψ is a closed formula of **SF** such that **SF**$\vdash \psi$, then there is a proof of ψ in **SF** consisting of only closed formulas, and therefore it does not use any induction.*

Proof. Consider the last application of Induction Rule

$$\frac{\varphi[0], \varphi[n] \to \varphi[Sn]}{\varphi[t]}$$

in the original proof of ψ. We can assume that free variables in $\varphi[t]$ have been instantiated by closed terms. Then, t is provably equal to a numeral $\overline{m}$. Therefore, this application of Induction Rule can be replaced by repeated applications of modus ponens of the format

$$\frac{\varphi\left[\overline{i}\right], \varphi\left[\overline{i}\right] \to \varphi\left[\overline{i+1}\right]}{\varphi\left[\overline{i+1}\right]},$$

for $i = 0, ..., m-1$. Each instance $\varphi\left[\overline{i}\right] \to \varphi\left[\overline{i+1}\right]$ has a proof in **SF** with one less application of Induction Rule. Therefore, all such applications of Induction Rule can be eventually eliminated, and the resulted proof will consist of only closed formulas. □

We will call a proof in **SF** consisting of only closed formulas and using no induction a '*closed proof*'. Other proofs with free variables can be viewed as schematic presentations of closed proofs obtained by instantiating free variables and eliminating inductions as in the lemma above. Closed proofs use only propositional inference rules and equation substitutions. These are valid inference rules when applied to realistic sentences. Note that eliminating an application of Induction Rule will introduce new closed instances of axioms as premises, that is, the premises for deriving the instances $\varphi\left[\overline{i}\right] \to \varphi\left[\overline{i+1}\right]$ above. The quantifier-free feature of the language of **SF** is essential for allowing the elimination of inductions.

Now, consider how to interpret **SF** as a realistic theory about concrete computational devices. Terms in **SF** can be treated as expressions referring to programs in a concrete computer. In particular, a numeral is a program that outputs itself. A closed numerical term in normal form is a composition of numerals, the base functions S,

$+$, $\cdot$, *pow*, $I_<$, and the operators bounded primitive recursion, finite sum and finite product. It is a program producing a concrete numeral output when executed according to the primitive recursive equations defining base functions, bounded primitive recursion, finite sum and finite product. An arbitrary closed numerical term is also a program, since it can be transformed into a normal term. A closed term of an arbitrary type is a program that transforms any sequences of terms $\mathbf{s}_1, ..., \mathbf{s}_n$ of appropriate types $\sigma_1, ..., \sigma_n$ (as programs) into another term $t(\mathbf{s}_1)...(\mathbf{s}_n)$ of the type o (as a program).

Then, a closed atomic formula $t = s$ is a realistic assertion about two such programs, saying that they output the same numeral. Note that when interpreted as realistic assertions about a concrete computer, not all closed instances of axioms are literally true of a concrete computer, because for realistic assertions about computers, we have to consider the physical limitations of a concrete computer. For instance, the function symbol S is interpreted as the computer operation of adding 1 to a numeral. Since to some point this will cause memory overflow, some instances of the axiom $St = Sr \to t = r$ may not be literally true when so interpreted. For example, suppose that the way a computer handles overflow is such that adding 1 to the maximum numeral $\overline{N}$ that the computer can handle will not change its value. Then, interpreted for that computer, $S\overline{N} = S\overline{N-1} = \overline{N}$, but $\overline{N} \neq \overline{N-1}$. However, as long as the numerals involved are not too large, an axiom instance *can* be interpreted as a literally true assertion about a concrete computer.

Recall that a closed proof in **SF** consists of propositional inferences and equation substitutions. It can become a series of sound logical deductions on realistic assertions about a concrete computer, as long as the numerals involved are not too large. That is how **SF** can be interpreted as a realistic theory about concrete programs. Again, the quantifier-free nature of the language of **SF** is essential here. It means that a closed formula never refers to 'all numerals'. Therefore, it may have a chance of being interpreted into a true assertion about a finite computer. A term with free variables is essentially a schematic presentation of multiple closed terms, that is, closed terms resulted from instantiating free variables by closed terms of appropriate types. When interpreting proofs in **SF** as deductions on realistic sentences, we can also treat a proof with free variables as a schematic presentation of many closed proofs obtained by instantiating free variables and eliminating inductions, as in the lemma above.

Before closing this section, we want to note that there is another way of formalizing strict finitism and it is perhaps more faithful to the idea of strict finitism. We can add a constant symbol N of the type o to the language of **SF**. N intuitively means the largest numeral accepted by the system. All numerical functions take the same values as before as long as they do not exceed N. Otherwise, they stay at N. That is, $SN = N$, $N + x = N$, and so on. We add the axiom schemas $SN = N$, $t \leq N$, and we replace the axiom schema $St = Sr \to t = r$ and $r < St \leftrightarrow r < t \vee r = t$ by

$$t < N \wedge r < N \wedge St = Sr \to t = r,$$
$$t < N \to (r < St \leftrightarrow r < t \vee r = t).$$

Note that other axioms do not need to be revised. The resulted system $\mathbf{SF}^N$ can have any finite initial segment of natural numbers as a model. Then, if we translate N into the largest numeral that a concrete computer can handle and if we assume that the computer handles overflows in the way corresponding to $SN = N$, then all axiom instances can be interpreted as literally true assertions about the computer (ignoring other physical limitations). This is a formalization of strict finitism different from **SF**. We did not choose this formalization $\mathbf{SF}^N$, because **SF** is a little closer to ordinary mathematics and is more convenient. In $\mathbf{SF}^N$, many ordinary theorems will have to be preceded by assumptions like $t < N \wedge r < N$ above. Otherwise, there is no essential difference between the two. Each closed proof in **SF** involves only finitely many numerals, which either explicitly appear in the proof, or are implicitly referred to as values of closed numerical subterms. As long as these numerals are not too large, extra assumptions like $t < N \wedge r < N$ are always literally true when interpreted as realistic assertions about concrete finite things. Then, **SF** and $\mathbf{SF}^N$ do not have any essential difference.

2.2 Doing Mathematics in Strict Finitism

This section will introduce some semi-formal notations to help us express mathematics in strict finitism in a simpler and more familiar manner. They will allow us to state mathematical claims in strict finitism in a format much like stating theorems in classical mathematics. In particular, we will be able to use quantifiers and talk about sets and functions, although the uses of quantifiers and the apparent references to sets and functions can always be eliminated, and speaking in this manner never really goes beyond the formal system **SF**. These notations will allow us to use the logical laws of intuitionistic predicate logic and the axiom of choice in proving theorems in strict finitism, and they also make most of the constructions in constructive mathematics by Errett Bishop [5] available to strict finitism. These show that strict finitism is not as weak as it might appear to be.

2.2.1 Mathematical Claims in Strict Finitism

Developing mathematics in strict finitism means constructing terms (of any types) in **SF** and proving that those terms satisfy some desired conditions, which means proving some (quantifier-free) formulas containing those terms in **SF**. Therefore, a claim in strict finitism reports what terms have been constructed and which condition about the terms has been verified. This is much like a computer programmer's job, namely, designing programs and demonstrating that the programs meet a given specification. These programs are then resources for simulating other natural phenomena in applications. Our task is to show that applied mathematics can be construed in this manner.

The constructed terms and the desired condition about the terms can contain free variables. That is, we can do the job in a schematic manner, allowing a single construction to produce multiple concrete terms and realize multiple conditions in the same format. When free variables are instantiated by any closed terms of appropriate types, we get the construction of concrete terms and concrete conditions about those terms.

Moreover, we want to allow presenting the constructions of terms and verifications of conditions in some informal and abbreviated manner, so that routine details can be omitted and the presentations of mathematical work in strict finitism will not be unnecessarily lengthy. For instance, sometimes we do not need to give the constructed terms explicitly. We may be satisfied with the recognition that some terms can be (really can be) constructed. Similarly, terms can be constructed hypothetically, that is, assuming that some other terms satisfying some conditions are already available. For instance, we may want to construct a term t, depending on another given term s, so that as long as s satisfies $\varphi[s]$, t will satisfy $\psi[s,t]$. (A real computer programmer frequently has to do something similar.) This means constructing a term T (of a higher type) and verifying that $\varphi[x] \to \psi[x,T(x)]$. That is, T operates on any term satisfying φ to produce a term satisfying ψ. However, frequently, we want to state this informally and more conveniently, for instance, as 'for any x such that $\varphi[x]$, there exists y such that $\psi[x,y]$'.

To allow all this, we first use quantifiers to state what terms have been constructed and which condition about the constructed terms has been verified. Recall that every positive claim we make in strict finitism can be eventually stated in the format 'We have constructed terms $\mathbf{t}$ and derive the formula $\varphi[\mathbf{t}]$ (in **SF**)'. We will state this in a symbolic format as follows:

Definition 2.8. Suppose that $\varphi[\mathbf{x},\mathbf{y},\mathbf{p}]$ is a formula of **SF**, and suppose that $\mathbf{x},\mathbf{y},\mathbf{p}$ are all and different free variables in φ. A **claim** *in strict finitism* is a symbolic formula

$$\exists\mathbf{x}\forall\mathbf{y}\varphi[\mathbf{x},\mathbf{y},\mathbf{p}], \tag{FinC}$$

which means that we have constructed some terms $\mathbf{t}$ of appropriate types and prove that

$$\mathbf{SF} \vdash \varphi[\mathbf{t},\mathbf{y},\mathbf{p}].$$

$\mathbf{t}$ may contain variables in $\mathbf{p}$ but not those in $\mathbf{y}$. The variables in $\mathbf{p}$ are free variables (as parameters) in the claim. A **proof** *of the claim* in strict finitism consists of the required terms $\mathbf{t}$ and a proof of $\varphi[\mathbf{t},\mathbf{y},\mathbf{p}]$ in **SF**. The constructed terms $\mathbf{t}$ are **witnesses** for the claim.

Informally, we will also state (FinC) as '*there exist* $\mathbf{x}$ *such that for all* $\mathbf{y}$, φ'. However, quantifiers here are not understood as they are in classical mathematics. The existential quantifier only means that relevant terms have been constructed, and the universal quantifier is only to indicate free variables independent of the constructed terms in the condition to be verified within **SF**. The symbols $\exists$ and $\forall$ will occur *only in such contexts* and other ways of nesting them are meaningless. We use existential quantifiers only because we do not want to mention the details of those constructed terms in the claim. Our interest is only to communicate the fact that they can (really can) be constructed. A proof of the claim must explicitly contain the terms required. We will accept informal arguments demonstrating that such terms *can* be constructed, but the informal arguments must allow extracting such terms from the arguments, and this 'allow extracting' must itself be understood in the strictly finitistic sense (e.g., by an elementary recursive procedure). In particular, we do not consider abstract mathematical proofs in classical mathematics concluding that so and so terms exist.

2.2.2 Defined Logical Constants on Claims

Then, we introduce some *new* and *defined* logical constants $\neg^*$, $\vee^*$, $\wedge^*$, $\rightarrow^*$, $\exists^*$, and $\forall^*$ in our informal language, to allow expressing claims like (FinC) in strict finitism in more readable formats and to allow more familiar informal arguments for proving claims like (FinC) in strict finitism. These logical constants are explicitly defined. They may not be equivalent to their corresponding classical logical constants.

Definition 2.9. Suppose that $\varphi \equiv \exists \mathbf{x} \forall \mathbf{y} \varphi_1 [\mathbf{x}, \mathbf{y}]$ and $\psi \equiv \exists \mathbf{u} \forall \mathbf{v} \psi_1 [\mathbf{u}, \mathbf{v}]$ are claims in strict finitism, where $\mathbf{x}, \mathbf{y}, \mathbf{u}, \mathbf{v}$ are distinct variables. (We suppress the parameters here.) Define

(1) $(\varphi \wedge^* \psi) \equiv_{df} \exists \mathbf{xu} \forall \mathbf{yv} (\varphi_1 \wedge \psi_1)$;
(2) $(\varphi \vee^* \psi) \equiv_{df} (\varphi_1 \vee \psi_1)$ if $\mathbf{x}, \mathbf{y}, \mathbf{u}, \mathbf{v}$ are all empty, otherwise,

$$(\varphi \vee^* \psi) \equiv_{df} \exists n \mathbf{xu} \forall \mathbf{yv} ((n = 0 \wedge \varphi_1) \vee (n \neq 0 \wedge \psi_1));$$

(3) $(\exists^* z \varphi) \equiv_{df} \exists z \mathbf{x} \forall \mathbf{y} \varphi_1$ if z does not occur in $\mathbf{x}, \mathbf{y}$, otherwise $(\exists^* z \varphi) \equiv_{df} \varphi$;
(4) $(\forall^* z \varphi) \equiv_{df} \exists \mathbf{X} \forall z \mathbf{y} \varphi_1 [\mathbf{X}(z), \mathbf{y}]$ if z does not occur in $\mathbf{x}, \mathbf{y}$, otherwise $(\forall^* z \varphi) \equiv_{df} \varphi$;
(5) $(\varphi \rightarrow^* \psi) \equiv_{df} \exists \mathbf{UY} \forall \mathbf{xv} (\varphi_1 [\mathbf{x}, \mathbf{Y}(\mathbf{x}, \mathbf{v})] \rightarrow \psi_1 [\mathbf{U}(\mathbf{x}), \mathbf{v}])$;
(6) $(\neg^* \varphi) \equiv_{df} (\varphi \rightarrow^* S0 = 0) \equiv \exists \mathbf{Y} \forall \mathbf{x} (\neg \varphi_1 [\mathbf{x}, \mathbf{Y}(\mathbf{x})])$;
(7) $(\varphi \leftrightarrow^* \psi) \equiv_{df} (\varphi \rightarrow^* \psi) \wedge^* (\psi \rightarrow^* \varphi)$.

Note that as defined symbols, $\forall^*$ and $\exists^*$ are not $\forall, \exists$ used in (FinC) for stating claims in strict finitism, and the rest are not logical constants $\neg$, $\vee$, $\wedge$, $\rightarrow$ and $\leftrightarrow$ in **SF**. We can use these defined logical constants to construct new claims in strict finitism from given claims, as in a first-order language. For a formula φ constructed using these defined logical constants from formulas in the format (FinC), after the defined logical constants $\neg^*$, $\vee^*$, $\wedge^*$, $\rightarrow^*$,$\leftrightarrow^*$, $\exists^*$, and $\forall^*$ are eliminated, it will eventually reduce to a claim in the format (FinC) again. The intuitive meanings of these defined logical constants are obvious, except perhaps for $\rightarrow^*$.

The definition of $(\varphi \rightarrow^* \psi)$ above is Bishop's numerical implication in [4]. Intuitively, it means that to claim that $\exists \mathbf{u} \forall \mathbf{v} \psi_1 [\mathbf{u}, \mathbf{v}]$ follows from the assumption $\exists \mathbf{x} \forall \mathbf{y} \varphi_1 [\mathbf{x}, \mathbf{y}]$ in strict finitism, one must take an arbitrary $\mathbf{x}$ and derive $\exists \mathbf{u} \forall \mathbf{v} \psi_1 [\mathbf{u}, \mathbf{v}]$

from $\forall \mathbf{y} \varphi_1 [\mathbf{x}, \mathbf{y}]$, which means that one must construct a term $\mathbf{U}$ that operates on arbitrary $\mathbf{x}$ and derive $\psi_1 [\mathbf{U}(\mathbf{x}), \mathbf{v}]$ from $\forall \mathbf{y} \varphi_1 [\mathbf{x}, \mathbf{y}]$, which in turn means that one must construct $\mathbf{Y}$, operating on $\mathbf{x}, \mathbf{v}$, and derive $\psi_1 [\mathbf{U}(\mathbf{x}), \mathbf{v}]$ from $\varphi_1 [\mathbf{x}, \mathbf{Y}(\mathbf{x}, \mathbf{v})]$. The first step implies that the construction of $\mathbf{u}$ to satisfy $\psi_1 [\mathbf{u}, \mathbf{v}]$ can depend on any hypothetical $\mathbf{x}$ satisfying $\forall \mathbf{y} \varphi_1 [\mathbf{x}, \mathbf{y}]$. The second step means that to use the hypothesis $\forall \mathbf{y} \varphi_1 [\mathbf{x}, \mathbf{y}]$, we can only use a constructed instance $\varphi_1 [\mathbf{x}, \mathbf{Y}(\mathbf{x}, \mathbf{v})]$ of it.

This is at least as strong as the intuitionistic implication, in the sense that if φ and ψ have the format as in the definitions above and $\varphi \to^* \psi$ is provable in strict finitism, then 'if φ, then ψ' is provable in intuitionism. However, it is different from the intuitionistic implication. The standard intuitionistic interpretation of 'if φ, then ψ' refers to an 'arbitrary proof' of φ, that is, a proof of 'if φ, then ψ' will operate on an 'arbitrary proof' of φ and produce a proof of ψ. This requires 'an arbitrary proof' as a primitive notion. Numerical implication amounts to restricting that 'arbitrary proof'. First, the construction of $\mathbf{u}$ to satisfy $\psi_1 [\mathbf{u}, \mathbf{v}]$ for any $\mathbf{v}$ does not have to start from an 'arbitrary proof' of $\exists \mathbf{x} \forall \mathbf{y} \varphi_1 [\mathbf{x}, \mathbf{y}]$. Instead, it can start from an arbitrary $\mathbf{x}$ that hypothetically witnesses $\forall \mathbf{y} \varphi_1 [\mathbf{x}, \mathbf{y}]$. Second, the derivation of $\psi_1 [\mathbf{u}, \mathbf{v}]$ for an arbitrary $\mathbf{v}$ cannot start from a proof of $\forall \mathbf{y} \varphi_1 [\mathbf{x}, \mathbf{y}]$. It can only start from a constructed instance $\varphi_1 [\mathbf{x}, \mathbf{Y}(\mathbf{x}, \mathbf{v})]$.

A natural question is, 'is this numerical interpretation of implication too strong?' In particular, in deriving $\psi_1 [\mathbf{U}(\mathbf{x}), \mathbf{v}]$, it allows using only an instance $\varphi_1 [\mathbf{x}, \mathbf{Y}(\mathbf{x}, \mathbf{v})]$, not the full universal claim $\forall \mathbf{y} \varphi_1 [\mathbf{x}, \mathbf{y}]$. The lemma below shows that this numerical interpretation actually allows using finitely many instances of $\forall \mathbf{y} \varphi_1 [\mathbf{x}, \mathbf{y}]$.

Lemma 2.10. *To prove the claim*

$$\exists UY \forall xv (\varphi_1 [x, Y(x, v)] \to \psi_1 [U(x), v])$$

in strict finitism, it is sufficient to prove

$$\exists UY' M \forall xv ((\forall n \leq M(x, v)) \varphi_1 [x, Y'(n, x, v)] \to \psi_1 [U(x), v]),$$

or prove the following for some fixed N,

$$\exists UY_0 ... Y_N \forall xv (\varphi_1 [x, Y_0(x, v)] \wedge ... \wedge \varphi_1 [x, Y_N(x, v)] \to \psi_1 [U(x), v]).$$

Proof. Suppose that we have constructed the terms U, Y', M and derive

$$(\forall n \leq M(x, v)) \varphi_1 [x, Y'(n, x, v)] \to \psi_1 [U(x), v]$$

in **SF**. Let

$$T[x, v] \equiv_{df} (\mu n \leq M(x, v)) \neg \varphi_1 [x, Y'(n, x, v)].$$

Then,

$$\varphi_1 [x, Y'(T[x, v], x, v)] \to \forall n \leq M(x, v) \varphi_1 [x, Y'(n, x, v)].$$

Then, we can let Y be $\lambda xv.Y'(T[x, v], x, v)$ in the first part of the conclusion. The second part follows from the first part if we let $M \equiv \overline{N}$ and let

$$Y' \equiv \lambda nxv. (J(n=0, Y_0(x,v), J(n=1, Y_1(x,v), \ldots))).$$

□

This means that for deriving $\psi_1[U(x),v]$, one can actually use finitely many instances of $\forall y \varphi_1[x,y]$, that is, instances $\varphi_1[x, Y'(n,x,v)]$ for all $n \leq M(x,v)$, where the number of instances can depend on x, v. Still, it does not allow operating on an arbitrary proof of $\forall y \varphi_1[x,y]$ and it does not depend on infinitely many instances of $\forall y \varphi_1[x,y]$. This, we believe, captures the finitistic numerical content of implication. To justify this, we first note that in realistic mathematics, we never refer to an arbitrary *proof* of our hypotheses in deriving some conclusion from the hypotheses. Secondly, the lemma actually says that for finitistic implication, the universal quantifier in the antecedent should not be taken too literally. A proof of the implication must derive the consequent $\psi_1[U(x),v]$ from *finitely* many instances of the antecedent $\forall y \varphi_1[x,y]$. This perhaps captures the spirit of finitism. Thirdly, we will see that the logical constant $\rightarrow^*$ so defined does have the common features of logical implication.

In fact, the following theorem shows that all defined logical constants $\neg^*$, $\vee^*$, $\wedge^*$, $\rightarrow^*$, $\leftrightarrow^*$, $\exists^*$, and $\forall^*$ follow the intuitionistic logical laws. The definitions of starred logical constants in (1) – (6) above are essentially Gödel's *Dialectica* interpretation of intuitionistic logic. Therefore, the proof of this theorem is similar to the proof of *Dialectica* interpretation (see, for instance, Troelstra [34], p. 234).

Theorem 2.11. $\neg^*$, $\vee^*$, $\wedge^*$, $\rightarrow^*$, $\exists^*$, *and* $\forall^*$ *follow the laws of intuitionistic predicate logic, as well as the axiom of choice. That is, for any claims* φ, ψ, χ, *the following claims are provable in strict finitism:*

(1) $\varphi \rightarrow^* \psi \rightarrow^* \varphi$;
(2) $(\varphi \rightarrow^* \chi \rightarrow^* \psi) \rightarrow^* (\varphi \rightarrow^* \chi) \rightarrow^* (\varphi \rightarrow^* \psi)$;
(3) $0 = 1 \rightarrow^* \varphi$;
(4) $\varphi \wedge^* \psi \rightarrow^* \varphi$;
(5) $\varphi \wedge^* \psi \rightarrow^* \psi$;
(6) $\varphi \rightarrow^* \psi \rightarrow^* \varphi \wedge^* \psi$;
(7) $\varphi \rightarrow^* \varphi \vee^* \psi$;
(8) $\psi \rightarrow^* \varphi \vee^* \psi$;
(9) $(\varphi \rightarrow^* \chi) \rightarrow^* (\psi \rightarrow^* \chi) \rightarrow^* \varphi \vee^* \psi \rightarrow^* \chi$;
(10) $\forall^* z(\varphi \rightarrow^* \psi) \rightarrow^* \forall^* z\varphi \rightarrow^* \forall^* z\psi$;
(11) $\varphi \rightarrow^* \forall^* z\varphi$, *where* x *is not free in* φ;
(12) $\forall^* z\varphi \rightarrow^* \varphi[t/z]$;
(13) $\forall^* z(\varphi \rightarrow^* \psi) \rightarrow^* \exists^* z\varphi \rightarrow^* \exists^* z\psi$;
(14) $\exists^* z\varphi \rightarrow^* \varphi$, *where* x *is not free in* φ;
(15) $\varphi[t/z] \rightarrow^* \exists^* z\varphi$;
(16) $\forall^* z \exists^* w \varphi[z,w] \rightarrow \exists^* W \forall^* z \varphi[z, W(z)]$.

Moreover, for any claims φ, ψ, *if the claims* φ *and* $\varphi \rightarrow^* \psi$ *are provable, then* ψ *is also provable; and if the claim* φ *is provable, then* $\forall^* z\varphi$ *is also provable.*

Proof. Suppose that $\varphi \equiv \exists x' \forall y \varphi_1 [x', y]$, $\psi \equiv \exists u' \forall v \psi_1 [u', v]$, and $\chi \equiv \exists z' \forall w \chi_1 [z', w]$. To simplify our notations, we will write these claims as $\varphi_1 [x', y]$, $\psi_1 [u', v]$, and $\chi_1 [z', w]$, with the convention that variables with a prime are bound by $\exists$ and other variables are bound by $\forall$. For simplicity, we will consider only a single $\exists$ and a single $\forall$ in the prefix. The case for multiple quantifiers is similar.

(1) $\psi \to^* \varphi$ is $\psi_1 [u, v'(u,y)] \to \varphi_1 [x'(u), y]$. Therefore, $\varphi \to^* \psi \to^* \varphi$ is

$$\varphi_1 \left[x_1, y_1'(x_1, u, y)\right] \to \psi_1 \left[u, v'(x_1)(u,y)\right] \to \varphi_1 \left[x'(x_1)(u), y\right].$$

(We change x', y in the first occurrence of φ_1 to x_1', y_1, to avoid confusing with x', y in the last occurrence of φ_1 in the formula $\varphi \to^* \psi \to^* \varphi$. Moreover, here we use the same variable names x' and v', ignoring the fact that their types should be different from their types in the previous formulas.) Then, to prove this formula, we have to construct terms for y_1', v', and x'. This is trivial. We can just choose y_1' and x' such that $y_1'(x_1, u, y) = y$ and $x'(x_1)(u) = x_1$. This proves (1).

(2) $\chi \to^* \psi$ is $\chi_1 [z, w'(z,v)] \to \psi_1 [u'(z), v]$. Therefore, $\varphi \to^* \chi \to^* \psi$ is

$$\varphi_1 \left[x, y'(x,z,v)\right] \to \chi_1 \left[z, w'(x)(z,v)\right] \to \psi_1 \left[u'(x)(z), v\right].$$

$\varphi \to^* \chi$ is

$$\varphi_1 \left[x_1, y_1'(x_1, w_1)\right] \to \chi_1 \left[z_1'(x_1), w_1\right],$$

and $\varphi \to^* \psi$ is

$$\varphi_1 \left[x_2, y_2'(x_2, v_2)\right] \to \psi_1 \left[u_2'(x_2), v_2\right].$$

Therefore, $(\varphi \to^* \chi) \to^* (\varphi \to^* \psi)$ is

$$(A_1 \to C_1) \to (A_2 \to B_2),$$

where

$$\begin{aligned}
A_1 &\equiv \varphi_1 \left[x_1'(y_1, z_1, x_2, v_2), y_1\left(x_1'(y_1, z_1, x_2, v_2), w_1'(y_1, z_1, x_2, v_2)\right)\right],\\
C_1 &\equiv \chi_1 \left[z_1\left(x_1'(y_1, z_1, x_2, v_2)\right), w_1'(y_1, z_1, x_2, v_2)\right],\\
A_2 &\equiv \varphi_1 \left[x_2, y_2'(y_1, z_1)(x_2, v_2)\right],\\
B_2 &\equiv \psi_1 \left[u_2'(y_1, z_1)(x_2), v_2\right].
\end{aligned}$$

Let $\mathbf{x}_3 \equiv (y, w, u, y_1, z_1, x_2, v_2)$, $\mathbf{y}_3 \equiv (y_1, z_1, x_2, v_2)$. Then, finally,

$$(\varphi \to^* \chi \to^* \psi) \to^* ((\varphi \to^* \chi) \to^* (\varphi \to^* \psi))$$

is

$$(A_3 \to C_3 \to B_3) \to ((A_4 \to C_4) \to (A_5 \to B_5)), \tag{2.3}$$

where

$$\begin{aligned}
A_3 &\equiv \varphi_1\left[x'(\mathbf{x}_3), y\left(x'(\mathbf{x}_3), z'(\mathbf{x}_3), v'(\mathbf{x}_3)\right)\right],\\
C_3 &\equiv \chi_1\left[z'(\mathbf{x}_3), w\left(x'(\mathbf{x}_3)\right)\left(z'(\mathbf{x}_3), v'(\mathbf{x}_3)\right)\right],\\
B_3 &\equiv \psi_1\left[u\left(x'(\mathbf{x}_3)\right)\left(z'(\mathbf{x}_3)\right), v'(\mathbf{x}_3)\right],\\
A_4 &\equiv \varphi_1\left[x_1'(y,w,u)(\mathbf{y}_3), y_1\left(x_1'(y,w,u)(\mathbf{y}_3), w_1'(y,w,u)(\mathbf{y}_3)\right)\right],\\
C_4 &\equiv \chi_1\left[z_1\left(x_1'(y,w,u)(\mathbf{y}_3)\right), w_1'(y,w,u)(\mathbf{y}_3)\right],\\
A_5 &\equiv \varphi_1\left[x_2, y_2'(y,w,u)(y_1,z_1)(x_2,v_2)\right],\\
B_5 &\equiv \psi_1\left[u_2'(y,w,u)(y_1,z_1)(x_2), v_2\right].
\end{aligned}$$

By Lemma 2.10, to prove (2.3), we can instead prove

$$(A_3 \to C_3 \to B_3) \wedge (A_4 \to C_4) \wedge A_5 \wedge A_5' \to B_5,$$

where A_5' is obtained from A_5 by replacing y_2' by y_2''. The idea is to match A_5 with A_4, A_5' with A_3, C_4 with C_3, and B_5 with B_3. We need to construct x', z', v', x_1', w_1', y_2', y_2'', and u_2' to satisfy the above. To match B_5 with B_3, we need

$$\begin{aligned}
v'(\mathbf{x}_3) &= v_2,\\
u_2'(y,w,u)(y_1,z_1)(x_2) &= u\left(x'(\mathbf{x}_3)\right)\left(z'(\mathbf{x}_3)\right).
\end{aligned}$$

To match C_4 with C_3, we need

$$\begin{aligned}
z'(\mathbf{x}_3) &= z_1\left(x_1'(y,w,u)(\mathbf{y}_3)\right),\\
w_1'(y,w,u)(\mathbf{y}_3) &= w\left(x'(\mathbf{x}_3)\right)\left(z'(\mathbf{x}_3), v'(\mathbf{x}_3)\right).
\end{aligned}$$

To match A_5 with A_4, we need

$$\begin{aligned}
x_1'(y,w,u)(\mathbf{y}_3) &= x_2,\\
y_2'(y,w,u)(y_1,z_1)(x_2,v_2) &= y_1\left(x_1'(y,w,u)(\mathbf{y}_3), w_1'(y,w,u)(\mathbf{y}_3)\right).
\end{aligned}$$

Finally, to match A_5' with A_3, we need

$$\begin{aligned}
x'(\mathbf{x}_3) &= x_2,\\
y_2''(y,w,u)(y_1,z_1)(x_2,v_2) &= y\left(x'(\mathbf{x}_3), z'(\mathbf{x}_3), v'(\mathbf{x}_3)\right).
\end{aligned}$$

These requirements can be satisfied simultaneously. This proves (2).

(3) is trivial.

(4) $\varphi \wedge^* \psi$ is $\varphi_1[x',y] \wedge \psi_1[u',v]$. Therefore, $\varphi \wedge^* \psi \to^* \varphi$ is

$$\varphi_1\left[x, y'(x,u,y_1)\right] \wedge \psi_1\left[u, v'(x,u,y_1)\right] \to \varphi_1\left[x_1'(x,u), y_1\right].$$

We only need to construct y', x_1' such that $y'(x,u,y_1) = y_1$ and $x_1'(x,u) = x$.

(5) is proved similarly.

(6) Since $\varphi \wedge^* \psi$ is $\varphi_1[x',y] \wedge \psi_1[u',v]$, $\psi \to^* \varphi \wedge^* \psi$ is

$$\psi_1\left[u_1,v_1'(u_1,y,v)\right]\to\varphi_1\left[x'(u_1),y\right]\wedge\psi_1\left[u'(u_1),v\right].$$

Therefore, $\varphi\to^*\psi\to^*\varphi\wedge^*\psi$ is

$$\begin{aligned}&\varphi_1\left[x_1,y_1'(x_1,u_1,y,v)\right]\\ \to\;&\psi_1\left[u_1,v_1'(x_1)(u_1,y,v)\right]\\ \to\;&\varphi_1\left[x'(x_1)(u_1),y\right]\wedge\psi_1\left[u'(x_1)(u_1),v\right].\end{aligned}$$

We can let y_1', v_1', x', u' be such that $y_1'(x_1,u_1,y,v)=y$, $v_1'(x_1)(u_1,y,v)=v$, $x'(x_1)(u_1)=x_1$, and $u'(x_1)(u_1)=u_1$. This proves (6).

(7), (8), (9) are similar.

(10) $\forall^*z\psi$ is $\psi_1\left[u'(z),v,z\right]$ and $\forall^*z\varphi$ is $\varphi_1\left[x'(z_1),y,z_1\right]$. Therefore, $\forall^*z\varphi\to^*\forall^*z\psi$ is

$$\varphi_1\left[x\left(z_1'(x,z,v)\right),y'(x,z,v),z_1'(x,z,v)\right]\to\psi_1\left[u'(x)(z),v,z\right].$$

$\varphi\to^*\psi$ is $\varphi_1\left[x_2,y_2'(x_2,v_2)\right]\to\psi_1\left[u_2'(x_2),v_2\right]$. Therefore, $\forall^*z(\varphi\to^*\psi)$ is

$$\varphi_1\left[x_2,y_2'(z_2)(x_2,v_2),z_2\right]\to\psi_1\left[u_2'(z_2)(x_2),v_2,z_2\right].$$

Let $\mathbf{y}_3=(y_2,u_2,x,z,v)$. Then, $\forall^*z(\varphi\to^*\psi)\to^*\forall^*z\varphi\to^*\forall^*z\psi$ is

$$(A_1\to B_1)\to(A_2\to B_2),$$

where

$$\begin{aligned}A_1&\equiv\varphi_1\left[x_2'(\mathbf{y}_3),y_2\left(z_2'(\mathbf{y}_3)\right)\left(x_2'(\mathbf{y}_3),v_2'(\mathbf{y}_3)\right),z_2'(\mathbf{y}_3)\right],\\ B_1&\equiv\psi_1\left[u_2\left(z_2'(\mathbf{y}_3)\right)\left(x_2'(\mathbf{y}_3)\right),v_2'(\mathbf{y}_3),z_2'(\mathbf{y}_3)\right],\\ A_2&\equiv\varphi_1\left[x\left(z_1'(y_2,u_2)(x,z,v)\right),y'(y_2,u_2)(x,z,v),z_1'(y_2,u_2)(x,z,v)\right],\\ B_2&\equiv\psi_1\left[u'(y_2,u_2)(x)(z),v,z\right].\end{aligned}$$

Therefore, to match A_1 with A_2 and B_1 with B_2, we can construct u', v_2', z_2', y', z_1', x_2' such that

$$\begin{aligned}z_2'(\mathbf{y}_3)&=z;\\ v_2'(\mathbf{y}_3)&=v;\\ z_1'(y_2,u_2)(x,z,v)&=z_2'(\mathbf{y}_3);\\ x_2'(\mathbf{y}_3)&=x\left(z_1'(y_2,u_2)(x,z,v)\right);\\ y'(y_2,u_2)(x,z,v)&=y_2\left(z_2'(\mathbf{y}_3)\right)\left(x_2'(\mathbf{y}_3),v_2'(\mathbf{y}_3)\right);\\ u'(y_2,u_2)(x)(z)&=u_2\left(z_2'(\mathbf{y}_3)\right)\left(x_2'(\mathbf{y}_3)\right).\end{aligned}$$

This proves (10).

(11) and (12) are trivial.

(13) $\exists^*z\varphi$ is $\varphi_1\left[x',y,z'\right]$ and $\exists^*z\psi$ is $\psi_1\left[u',v,z_1'\right]$. Therefore, $\exists^*z\varphi\to^*\exists^*z\psi$ is

$$\varphi_1\left[x,y'(x,z,v),z\right] \rightarrow \psi_1\left[u'(x,z),v,z_1'(x,z)\right].$$

Again, $\forall^* z(\varphi \rightarrow^* \psi)$ is

$$\varphi_1\left[x_2,y_2'(z_2)(x_2,v_2),z_2\right] \rightarrow \psi_1\left[u_2'(z_2)(x_2),v_2,z_2\right].$$

Let $\mathbf{y}_3 = (y_2,u_2,x,z,v)$. Then, $\forall^* z(\varphi \rightarrow^* \psi) \rightarrow^* \exists^* z\varphi \rightarrow^* \exists^* z\psi$ is

$$(A_1 \rightarrow B_1) \rightarrow (A_2 \rightarrow B_2),$$

where

$$A_1 \equiv \varphi_1\left[x_2'(\mathbf{y}_3),y_2\left(z_2'(\mathbf{y}_3)\right)\left(x_2'(\mathbf{y}_3),v_2'(\mathbf{y}_3)\right),z_2'(\mathbf{y}_3)\right],$$

$$B_1 \equiv \psi_1\left[u_2\left(z_2'(\mathbf{y}_3)\right)\left(x_2'(\mathbf{y}_3)\right),v_2'(\mathbf{y}_3),z_2'(\mathbf{y}_3)\right],$$
$$A_2 \equiv \varphi_1\left[x,y'(y_2,u_2)(x,z,v),z\right],$$
$$B_2 \equiv \psi_1\left[u'(y_2,u_2)(x,z),v,z_1'(y_2,u_2)(x,z)\right].$$

Then, we can let

$$x_2'(\mathbf{y}_3) = x;$$
$$z_2'(\mathbf{y}_3) = z;$$
$$v_2'(\mathbf{y}_3) = v;$$
$$z_1'(y_2,u_2)(x,z) = z_2'(\mathbf{y}_3);$$
$$y'(y_2,u_2)(x,z,v) = y_2\left(z_2'(\mathbf{y}_3)\right)\left(x_2'(\mathbf{y}_3),v_2'(\mathbf{y}_3)\right);$$
$$u'(y_2,u_2)(x,z) = u_2\left(z_2'(\mathbf{y}_3)\right)\left(x_2'(\mathbf{y}_3)\right).$$

(14) and (15) are again trivial.
(16) Note that $\forall^* z\exists^* w\varphi[z,w]$ is $\varphi_1[x'(z),y,z,w'(z)]$, and $\exists^* W\forall^* z\varphi[z,W(z)]$ is $\varphi_1[x'(z),y,z,W'(z)]$. Therefore, (16) is trivial.

Now, suppose that the claims φ and $\varphi \rightarrow^* \psi$ are provable in strict finitism. That is, we have constructed terms x', y', u' and derive $\varphi_1[x',y]$ and $\varphi_1[x,y'(x,v)] \rightarrow \psi_1[u'(x),v]$. Substitute x' for x in the second, we have a proof of $\varphi_1[x',y'(x',v)] \rightarrow \psi_1[u'(x'),v]$. Substitute $y'(x',v)$ for y in the first, we have a proof of $\varphi_1[x',y'(x',v)]$. Therefore, we have a proof of $\psi_1[u'(x'),v]$, which is a proof of the claim ψ.

Finally, suppose that φ is provable in strict finitism. That is, we have constructed a term $t[z]$, which may depend on the parameter z, and derive $\varphi_1[t[z],y,z]$. Now, $\forall^* z\varphi$ is $\varphi_1[x'(z),y,z]$. Therefore, $\forall^* z\varphi$ is also provable. □

This theorem shows that in proving claims in strict finitism, we can use (1)–(16) in the theorem as axioms and use the rules Modus Ponens for $\rightarrow^*$and Generalization for $\forall^*$. The latter means that we can assume that free variables are all bound by $\forall^*$. The theorem also implies that we can use skills such as Deduction Theorem and other natural deduction rules if the deduction consists only of the axioms

(1)–(16), the axioms of **SF**, Modus Ponens, and Generalization, that is, if it does not use Induction Rule, which we will discuss later. For instance, suppose that we start from a premise φ and use the axioms (1)–(16), the axioms of **SF**, Modus Ponens, and Generalization to derive a conclusion ψ. Generalization should not be applied to free variables in φ, of course. Then, we have a proof of $\varphi \rightarrow^* \psi$ using the axioms (1)–(16), the axioms of **SF**, Modus Ponens, and Generalization. This is Deduction Theorem.

Now, suppose that $\varphi_1, ..., \varphi_n$ is a proof using the axioms (1)–(16), the axioms of **SF**, Modus Ponens, and Generalization. Eliminating defined logical symbols in each statement in the proof, we get a sequence of claims $\psi_1, ..., \psi_n$ in strict finitism. Each ψ_i is in the format $\exists \mathbf{x} \forall \mathbf{y} \chi$, stating that some terms are constructed and some condition about the terms (expressed as a quantifier-free formula of **SF**) is verified. The theorem above implies that if φ_i is an axiom among (1)–(16), the corresponding claim ψ_i is provable in strict finitism and the required terms witnessing it can be constructed routinely from φ_i. Similarly, if φ_i is derived from some previous statements by Modus Ponens or Generalization, assuming that the required witness terms for those previous statements as claims are already constructed, the required witness terms for the claim ψ_i can also be constructed routinely. Therefore, we can automatically extract the terms witnessing the last claim ψ_n in strict finitism. If $\varphi \equiv \exists \mathbf{x} \forall \mathbf{y} \varphi_1 [\mathbf{x}, \mathbf{y}]$, $\psi \equiv \exists \mathbf{u} \forall \mathbf{v} \psi_1 [\mathbf{u}, \mathbf{v}]$, and we derive ψ by such a proof using φ as a premise, then this means that a mechanical procedure can automatically extract the terms $\mathbf{U}, \mathbf{Y}$ from the proof and generate a proof in **SF** of the formula

$$\varphi_1 [\mathbf{x}, \mathbf{Y}(\mathbf{x}, \mathbf{v})] \rightarrow \psi_1 [\mathbf{U}(\mathbf{x}), \mathbf{v}].$$

This is what we really get when we derive the conclusion $\exists \mathbf{u} \forall \mathbf{v} \psi_1 [\mathbf{u}, \mathbf{v}]$ from the premise $\exists \mathbf{x} \forall \mathbf{y} \varphi_1 [\mathbf{x}, \mathbf{y}]$ using the axioms and rules in the theorem above.

On the other side, since $\rightarrow^*$is not the intuitionistic implication, we have some provable claims that do not hold in the intuitionistic logic. For instance,

Lemma 2.12. *If $\varphi [x, y]$ is quantifier-free, then*

$$\neg^* \exists^* x \forall^* y \varphi [x, y] \leftrightarrow^* \forall^* x \exists^* y \neg \varphi [x, y].$$

In particular, if $\psi [y]$, $\psi' [y]$ are quantifier-free, then

$$\neg^* \forall y \psi [y] \leftrightarrow^* \exists y \neg \psi [y], \quad \neg^* \exists y \psi [y] \leftrightarrow^* \forall y \neg \psi [y];$$
$$\neg^* \neg^* \forall y \psi [y] \leftrightarrow^* \forall y \psi [y], \quad \neg^* \neg^* \exists y \psi [y] \leftrightarrow^* \exists y \psi [y];$$
$$\neg^* \left(\forall y \psi [y] \wedge \forall y \psi' [y] \right) \leftrightarrow^* \neg^* \forall y \psi [y] \vee \neg^* \forall y \psi' [y].$$

Proof. Using the convention in the proof of the last theorem, $\exists^* x \forall^* y \varphi [x, y]$ is $\varphi [x', y]$ and then $\neg^* \exists^* x \forall^* y \varphi [x, y]$ is $\neg \varphi [x, y' (x)]$, which is just $\forall^* x \exists^* y \neg \varphi [x, y]$. The rest follows easily. □

However, note that we do not have a general proof of

$$\neg^* \forall^* x \exists^* y \varphi [x, y] \rightarrow^* \exists^* x \forall^* y \neg \varphi [x, y].$$

$\forall^* x \exists^* y \varphi[x,y]$ is $\varphi[x, y'(x)]$. Therefore, $\neg^* \forall^* x \exists^* y \varphi[x,y]$ is $\neg\varphi[x'(y), y(x'(y))]$. In other words, $\forall^* x \exists^* y \varphi[x,y]$ means that we can construct y' such that $\varphi[x, y'(x)]$. Therefore, to positively deny this, we have to construct x' that gives a counter-example $x'(y)$ for each y as the potential candidate for y' in $\varphi[x, y'(x)]$. That is $\neg\varphi[x'(y), y(x'(y))]$. However, this does not require us to construct a fixed x' as a counter-example for all potential candidate for y' in $\varphi[x, y'(x)]$. That is, it does not imply $\neg\varphi[x', y]$, which is $\exists^* x \forall^* y \neg\varphi[x,y]$. Similarly, for arbitrary claims φ, ψ, we do not generally have

$$\neg^*\neg^*\varphi \to^* \varphi, \quad \neg^*(\varphi \wedge \psi) \to \neg^*\varphi \vee \neg^*\psi, \text{ or } \neg^*\forall^* x\varphi \to \exists^* x \neg^* \varphi.$$

Therefore, these starred logical constants are not classical logical constants either.

It is easy to see that defined logical constants $\neg^*$, $\vee^*$, $\wedge^*$, $\to^*$, $\leftrightarrow^*$, $\exists^*$, $\forall^*$ and the logical constants $\neg$, $\vee$, $\wedge$, $\to$, $\leftrightarrow$ in **SF** and $\exists$, $\forall$ used in (FinC) are consistent whenever both are syntactically appropriate in a relevant context. For instance, if φ, ψ are (quantifier-free) formulas in **SF**, then $\varphi \wedge^* \psi$ is just $\varphi \wedge \psi$, after the defined logical constant $\wedge^*$is eliminated. Similarly, $\exists^* x \forall^* y \varphi$ is just $\exists x \forall y \varphi$ in the format (FinC). Moreover, it can be easily proved that $(\varphi \vee^* \psi) \leftrightarrow^* (\varphi \vee \psi)$. Therefore, *from now on, we will omit the stars on these symbols.* Furthermore, we will simply use natural language terms 'and', 'or', 'there exists', 'for all', 'if ... then ...', and 'not' to say these defined logical constants when stating a formula in natural language. We will never use these terminologies in their 'classical meaning'. In this manner, any statement constructed using these logical constants eventually says again that some terms are constructed to satisfy some condition expressed by a formula in the language of **SF**. For instance, 'for any x, there exists y, such that $\varphi[x,y]$' always means 'can construct Y that operates on any x such that $\varphi[x, Y(x)]$'.

When developing mathematics in strict finitism, we translate a theorem in classical mathematics (or a variant of it) into a claim in strict finitism, with logical constants in classical mathematics translated into $\neg^*$, $\vee^*$, $\wedge^*$, $\to^*$, $\leftrightarrow^*$, $\exists^*$, and $\forall^*$. Therefore, every mathematical theorem that we can prove in strict finitism is eventually a claim in the format (FinC) above, stating that some terms in **SF** have been constructed and some condition has been verified within **SF**. This assigns numerical content to a classical theorem. Proving the theorem in strict finitism means constructing the relevant terms and deriving the relevant formula in **SF**.

To see how this is related to the applicability of mathematics, suppose that $\forall n \forall \mathbf{x} \varphi[n, \mathbf{x}]$ with a quantifier-free φ is a mathematical premise expressing an idealized assumption about some finite and discrete physical quantity. For instance, the continuity assumption about the population growth on the Earth can be expressed by a statement of this format. It is our mathematical premise corresponding to some literally true realistic premise about a finite and discrete physical quantity. We will see that in such cases, usually $\varphi[\overline{m}, \mathbf{t}]$ can be translated into a literally true assertion about that discrete physical quantity as long as m is not too large. In the case of population growth, this is due to the fact that while population growth is not literally continuous, it is sufficiently smooth at the macro-scale. We will see that this 'sufficiently smooth at the macro-scale', together with bridging postulations, will imply $\varphi[\overline{m},\mathbf{t}]$ for not too large m. In other words, the magnitude of m in $\varphi[\overline{m},\mathbf{t}]$ corresponds to the degree of smoothness of population growth at the macro-scale.

Then, suppose that we prove a theorem in strict finitism in the format $\forall n \forall \mathbf{x} \varphi[n,\mathbf{x}] \rightarrow \forall k \psi[k]$, with ψ a quantifier-free formula of **SF**. That is, we derive $\forall k \psi[k]$ with $\forall n \forall \mathbf{x} \varphi[n,\mathbf{x}]$ as an assumption. The numerical interpretation of implication means that we actually construct terms $N, \mathbf{X}$ and derive $\varphi[N(k), \mathbf{X}(k)] \rightarrow \psi[k]$ within **SF**, with k a free variable. Therefore, for each numeral instance $\overline{m_0}$, we can get a closed proof of

$$\varphi[N(\overline{m_0}), \mathbf{X}(\overline{m_0})] \rightarrow \psi[\overline{m_0}]$$

in **SF**. This means that for deriving a particular instance $\psi[\overline{m_0}]$, we do not really need the premise $\varphi[\overline{m},\mathbf{x}]$ for arbitrarily large m. If $N(\overline{m_0})$ is not too large, so that $\varphi[N(\overline{m_0}), \mathbf{X}(\overline{m_0})]$ is implied by our realistic premises about discrete population growth and our bridging postulations, then the instance $\psi[\overline{m_0}]$ will follow from our realistic premises and bridging postulations (together with the axioms of strict finitism as true realistic assertions about concrete programs). This will be our basis for explaining applicability. See Sect. 3.7 for the details.

Note that N is an elementary recursive function constructed in the proof. Therefore, we can examine the value $N(\overline{m_0})$ to see if it *is* too large. On the other hand, in classical mathematics, if $\forall n \forall \mathbf{x} \varphi[n,\mathbf{x}] \rightarrow \forall k \psi[k]$ is true, then there also 'exist' functions $N, \mathbf{X}$ such that $\varphi[N(k), \mathbf{X}(k)] \rightarrow \psi[k]$ is true for all k. Actually, they can be constant functions. That is, we can define them as follows: let $N(k) \equiv_{df} c$ and $\mathbf{X}(k) \equiv_{df} \mathbf{d}$ if there exist a number c and entities $\mathbf{d}$ such that $\neg\varphi[c,\mathbf{d}]$, and otherwise arbitrarily assign values to $N(k)$ and $\mathbf{X}(k)$. However, such a function N is useless for examining if the proof preserves truth about that discrete physical quantity, because it does not give any hint about how large $N(\overline{m_0})$ is for a particular m_0.

2.2.3 Recursive Constructions and Inductions

The real restriction for strict finitism (compared with intuitionistic mathematics or Bishop's constructive mathematics, for instance) is on available recursive constructions and inductions. From the axioms and rules of **SF** we have only bounded primitive recursion on numerical terms and induction on quantifier-free formulas in **SF**. Lemma 2.3 shows how to use bounded primitive recursion to construct numerical terms.

Recursions can also be used to construct sequences of higher types. A sequence of the type σ items is a term of the type $(o \rightarrow \sigma)$. If t is of the type $(o \rightarrow \sigma)$, we frequently write $t(n)$ as t_n, or write t as $(t_n)_n$ or simply (t_n), to indicate that we consider it a sequence. We also call a term $T[n]$ of the type σ a sequence, by which we mean $\lambda n.T[n]$. To use numerical recursions to construct sequences of higher types, we generally proceed as follows: For a term T of any type, $T(\mathbf{x}_1)...(\mathbf{x}_l)$ becomes a numerical term for some variables $\mathbf{x}_1,...,\mathbf{x}_l$ of appropriate types. Instead of constructing a term $T[n]$ to satisfy some recursive equation directly, we construct a numerical term $q[n, \mathbf{x}_1, ..., \mathbf{x}_l]$, which is to be $T[n](\mathbf{x}_1)...(\mathbf{x}_l)$, to satisfy some appropriate recursive equation, and then we let

$$T[n] \equiv \lambda \mathbf{x}_1 .....\lambda \mathbf{x}_l . q[n, \mathbf{x}_1, ..., \mathbf{x}_l].$$

For this strategy to work, we usually need some *extensionality conditions*. The *extensional equality* of two terms T, R of the same type is defined as

$$(T \simeq R) \equiv_{df} \forall \mathbf{x}_1 ... \mathbf{x}_l (T(\mathbf{x}_1)...(\mathbf{x}_l) = R(\mathbf{x}_1)...(\mathbf{x}_l)),$$

where $\mathbf{x}_1, ..., \mathbf{x}_l$ are appropriate variables not free in T, R. Extensional equality is a substitute for equality for terms of higher types.

For example, to encode a finite sequence of terms $t_0, ..., t_{l-1}$ of some type σ, let

$$\langle t_0, ..., t_{l-1} \rangle \equiv_{df} \lambda \mathbf{x}_1 .....\lambda \mathbf{x}_n . \langle t_0(\mathbf{x}_1)...(\mathbf{x}_n), ..., t_{l-1}(\mathbf{x}_1)...(\mathbf{x}_n) \rangle,$$

where $\mathbf{x}_1, ..., \mathbf{x}_n$ are new variables of some uniquely determined types such that $t_0(\mathbf{x}_1)...(\mathbf{x}_n)$ is of the type o, and $<>$ on the right hand side is the coding function for sequences of numerical terms (see Subsection 2.1.2). Similarly, define

$$\begin{aligned} (t)_i &\equiv_{df} \lambda \mathbf{x}_1 .....\lambda \mathbf{x}_n . (t(\mathbf{x}_1)...(\mathbf{x}_n))_i, \\ lh(t) &\equiv_{df} lh(t(\mathbf{0}^{\sigma_1})...(\mathbf{0}^{\sigma_n})), \\ Seq(t) &\equiv_{df} \forall \mathbf{x}_1 ... \forall \mathbf{x}_n (Seq(t(\mathbf{x}_1)...(\mathbf{x}_n))). \end{aligned}$$

Here, $()_i$, lh, Seq on the right hand side are the corresponding decoding function, length function, and sequence number predicate for sequences of natural numbers. $\sigma_1, ..., \sigma_n$ are the sequences of the types of $\mathbf{x}_1, ..., \mathbf{x}_n$ respectively. For a sequence of types $\sigma = (\rho_1, ..., \rho_m)$, $\mathbf{0}^\sigma$ is the sequence $(0^{\rho_1}, ..., 0^{\rho_m})$ of terms, and for each type σ, 0^σ is defined as follows:

$$\begin{aligned} 0^o &\equiv_{df} 0, \\ 0^{(\rho_1, ..., \rho_m \to \rho)} &\equiv_{df} \lambda x^{\rho_1} ... x^{\rho_m} . 0^\rho. \end{aligned}$$

Therefore, $0^\sigma(\mathbf{x}_1)...(\mathbf{x}_n) = 0$ for variables $\mathbf{x}_1, ..., \mathbf{x}_n$ of appropriate types. Then, concatenation can be defined by

$$t * s \equiv_{df} \lambda \mathbf{x}_1 .....\lambda \mathbf{x}_n . (t(\mathbf{x}_1)...(\mathbf{x}_n) * s(\mathbf{x}_1)...(\mathbf{x}_n)).$$

We can prove

$$\begin{aligned} \langle t_0, ..., t_{l-1} \rangle * \langle t_l \rangle &\simeq \langle t_0, ..., t_{l-1}, t_l \rangle, \\ (t * s) * r &\simeq t * (s * r) \end{aligned}$$

and so on. Note that we use extensional equalities here.

Now, consider inductions. Suppose that we can prove a statement of the format

$$\exists n\varphi[0,n] \wedge (\exists n\varphi[m,n] \rightarrow \exists n\varphi[Sm,n]),$$

where $\varphi[m,n]$ is a quantifier-free formula. This means that we have constructed terms r and N and derived

$$\varphi[0,r], \ \varphi[m,n] \rightarrow \varphi[Sm,N(m,n)]$$

in **SF**. Suppose that the term $N(m,n)$ is iteratively bounded in n. Then, we can construct a term $q[m]$ so that $q[0] = r \wedge q[Sm] = N(m,q[m])$. Therefore, we have

$$\varphi[0,q[0]], \ \varphi[m,q[m]] \rightarrow \varphi[Sm,q[Sm]].$$

Then, a quantifier-free induction in **SF** derives $\varphi[m,q[m]]$, which then implies $\exists n\varphi[m,n]$. It means that we can actually have Σ_1^0-induction if relevant terms witnessing our proof of the inductive step are iteratively bounded. By similar strategies, sometimes we can get Π_1^0-induction, or inductions that are even more complex on apparent.

Inductions on formulas with parameters ranging over a domain, or inductions with assumptions, will be very useful in applications. In the simple case, we have

Lemma 2.13. *Suppose that $\varphi[n,\mathbf{x}]$ is a quantifier-free formula whose free variables are all in n, $\mathbf{x}$, where n and $\mathbf{x}$ are distinct variables, and suppose that $\chi[\mathbf{x}]$ is any claim (which may contain quantifiers) whose free variables are all in $\mathbf{x}$, and suppose that*

$$\chi[\mathbf{x}] \rightarrow \varphi[0,\mathbf{x}],$$
$$\chi[\mathbf{x}] \rightarrow (\varphi[n,\mathbf{x}] \rightarrow \varphi[Sn,\mathbf{x}]).$$

Then

$$\chi[\mathbf{x}] \rightarrow \varphi[n,\mathbf{x}].$$

Proof. Suppose that $\chi[\mathbf{x}] \equiv \exists\mathbf{y}\forall\mathbf{z}\chi_1[\mathbf{y},\mathbf{z},\mathbf{x}]$ with χ_1 quantifier-free. By the assumptions, after eliminating the defined symbol $\rightarrow^*$, we have

$$\exists\mathbf{Z}_0\forall\mathbf{y}(\chi_1[\mathbf{y},\mathbf{Z}_0(\mathbf{y}),\mathbf{x}] \rightarrow \varphi[0,\mathbf{x}]),$$
$$\exists\mathbf{Z}\forall\mathbf{y}(\chi_1[\mathbf{y},\mathbf{Z}(\mathbf{y}),\mathbf{x}] \rightarrow (\varphi[n,\mathbf{x}] \rightarrow \varphi[Sn,\mathbf{x}])).$$

$\mathbf{x}, n$ are free variables in this statement. They are implicitly quantified by $\forall^*$. Therefore, these are finally transformed into the following claims in strict finitism:

$$\exists\mathbf{Z}_0\forall\mathbf{x}\forall\mathbf{y}(\chi_1[\mathbf{y},\mathbf{Z}_0(\mathbf{x})(\mathbf{y}),\mathbf{x}] \rightarrow \varphi[0,\mathbf{x}]),$$
$$\exists\mathbf{Z}\forall\mathbf{n}\forall\mathbf{x}\forall\mathbf{y}(\chi_1[\mathbf{y},\mathbf{Z}(n,\mathbf{x})(\mathbf{y}),\mathbf{x}] \rightarrow (\varphi[n,\mathbf{x}] \rightarrow \varphi[Sn,\mathbf{x}])).$$

Proving these means that we can construct closed terms $\mathbf{Z}_0$, $\mathbf{Z}$ such that

$$\chi_1[\mathbf{y},\mathbf{Z}_0(\mathbf{x},\mathbf{y}),\mathbf{x}] \rightarrow \varphi[0,\mathbf{x}],$$
$$\chi_1[\mathbf{y},\mathbf{Z}(n,\mathbf{x},\mathbf{y}),\mathbf{x}] \rightarrow (\varphi[n,\mathbf{x}] \rightarrow \varphi[Sn,\mathbf{x}]).$$

Let $\mathbf{Z}' \equiv \lambda n\mathbf{xy}.J(n, \mathbf{Z}_0(\mathbf{x},\mathbf{y}), \mathbf{Z}(n-1,\mathbf{x},\mathbf{y}))$. Then

$$\chi_1[\mathbf{y}, \mathbf{Z}'(0,\mathbf{x},\mathbf{y}), \mathbf{x}] \leftrightarrow \chi_1[\mathbf{y}, \mathbf{Z}_0(\mathbf{x},\mathbf{y}), \mathbf{x}],$$
$$\chi_1[\mathbf{y}, \mathbf{Z}'(Sn,\mathbf{x},\mathbf{y}), \mathbf{x}] \leftrightarrow \chi_1[\mathbf{y}, \mathbf{Z}(n,\mathbf{x},\mathbf{y}), \mathbf{x}].$$

Let $\psi[n] \equiv \forall i \leq n\chi_1[\mathbf{y}, \mathbf{Z}'(i,\mathbf{y},\mathbf{x}), \mathbf{x}] \to \varphi[n,\mathbf{x}]$. Then, we will have

$$\psi[0] \wedge (\psi[n] \to \psi[Sn]).$$

ψ is quantifier-free. By an induction in **SF** we obtain

$$\forall i \leq n\chi_1[\mathbf{y}, \mathbf{Z}'(i,\mathbf{y},\mathbf{x}), \mathbf{x}] \to \varphi[n,\mathbf{x}].$$

Let $s[n,\mathbf{y},\mathbf{x}] \equiv \mu i \leq n \neg\chi_1[\mathbf{y}, \mathbf{Z}'(i,\mathbf{y},\mathbf{x}), \mathbf{x}]$. Then,

$$\chi_1[\mathbf{y}, \mathbf{Z}'(s[n,\mathbf{y},\mathbf{x}], \mathbf{y},\mathbf{x}), \mathbf{x}] \to \varphi[n,\mathbf{x}],$$

Therefore, $\chi[\mathbf{x}] \to \varphi[n,\mathbf{x}]$. □

By similarly examining if some relevant terms are iteratively bounded, sometimes we can have Σ_1^0 or more complex inductions with assumptions.

This means that Deduction Theorem and other natural deduction rules still hold when we use Induction Rule in a deduction from premises.

Theorem 2.14. *Suppose that we can derive a claim ψ, from a claim φ as the premise, using Definition 2.9, the axioms and rules in Theorem 2.11 and Lemma 2.12 (where Generalization is not applied to variables free in φ), and the axioms and rules of* **SF** *(including Induction Rule on quantifier-free formulas). Then, we have a proof of the claim $\varphi \to^* \psi$ in strict finitism. Similarly, if we can derive ψ from $\varphi[x]$ as the premise in the same manner and x does not occur free in ψ, then we have a proof of the claim $\exists^* x\varphi[x] \to^* \psi$ in strict finitism.*

Proof. This can be proved by an induction on the length of a derivation from φ to ψ. □

In summary, this is what we will do in developing mathematics within strict finitism. We translate a theorem in classical mathematics into a claim in strict finitism, using logical constants $\neg^*$, $\vee^*$, $\wedge^*$, $\to^*$, $\leftrightarrow^*$, $\exists^*$, and $\forall^*$ to replace classical logical constants. Sometimes we have to modify the classical theorem into a (classically) logically equivalent format before doing the translation, because two classically equivalent statements may have different finitistic content. The claim eventually says that some terms can be constructed to satisfy some condition that is a quantifier-free formula in **SF**, as in (FinC). We then prove the claim informally, using the axioms and rules of **SF**, Definition 2.9, the axioms and rules in Theorem 2.11 and Lemma 2.12, plus some forms of induction (e.g. those above) that can be reduced to the quantifier-free induction in **SF**, plus the techniques in natural deduction, including Deduction Theorem, $\exists$-Introduction Rule, and so on. Theorem 2.11 and Theorem 2.14 guarantee that relevant terms in **SF** demanded by the final claim can be automatically extracted from the informal proof, and a derivation in **SF** of the condition for those terms implied in the final claim can also be automatically generated.

In particular, this means that most of the proofs in constructive mathematics (i.e., Bishop and Bridges [6]) are actually available to us. We only need to examine the recursive constructions and inductions used in a constructive proof, to see if they are reducible to bounded primitive recursion and quantifier-free induction.

2.3 Sets and Functions

Defined logical constants on mathematical claims in strict finitism, including defined quantifiers, allow us to express claims in strict finitism in some simplified format, very close to the statements in classical mathematics. In order to develop advanced mathematics in strict finitism, we need more ways of simplifying the presentations of claims in strict finitism. Sets and functions are also meta-language notions to allow us to do this. Sets are conditions for classifying terms of various types. Functions are terms that apply to terms satisfying some conditions and produce other terms satisfying some other conditions. Sets and functions together allow us to state sophisticated conditional constructions of terms and state complex conditions about terms in simpler, more readable and more familiar formats.

The basic ideas for representing sets and functions are from Bishop and Bridges [6] Chap. 3, but some changes are required to fit into our more restrictive framework here.

2.3.1 Sets

Classification needs equality. A set provides a way to say that a term belongs to a class and that two terms are equal (when considered as members of the class). Therefore, a set will be a pair of statements, defining respectively the *membership condition* and the *equality condition*. More accurately, for a pair of formulas $A \equiv \langle \varphi[a], \psi[a,b] \rangle$, we will call A a *set form of the type* σ, if a is the only free variable in φ, a and b are the only free variables in ψ, a and b are of the type σ. We usually write $\varphi[a]$ as $a \in A$, and write $\psi[a,b]$ as $a =_A b$ (or simply $a = b$ when no real ambiguity will arise). They are the membership condition and the equality condition of the set. In that case, A is a *set of the type* σ if

$$(1)\ \forall a,b,c \begin{pmatrix} a \in A \wedge b \in A \wedge c \in A \to \\ a =_A a \wedge (a =_A b \to b =_A a) \wedge (a =_A b \wedge b =_A c \to a =_A c) \end{pmatrix},$$

$$(2)\ \forall a,b (a \simeq b \wedge a \in A \to b \in A \wedge a =_A b).$$

(2) means that these conditions are extensional. In particular, the equality $=_A$ is a more coarse-grained equivalence relation than the extensional equality. If $(a \in A)$ is the claim $\exists \mathbf{x} \forall \mathbf{y} \varphi_0 [a, \mathbf{x}, \mathbf{y}]$ and $\mathbf{x}$ are of the types $\rho_1, ..., \rho_m$, we call $(\sigma, \rho_1, ..., \rho_m)$ the *signature* of the set A and call $\rho_1, ..., \rho_m$ the *witness types* of A, and we use $a \in_{\mathbf{x}} A$

to denote the formula $\forall \mathbf{y} \varphi_0 [a, \mathbf{x}, \mathbf{y}]$, read as '$a$ belongs to A with $\mathbf{x}$ as the witnesses'. We also use the notation $\{a : \varphi[a]\}$ to denote a set when the equality condition is obvious from the context. Similarly, $\{t[x] : \varphi[x]\}$ denotes a set with the membership condition $\exists x (a \simeq t[x] \wedge \varphi[x])$.

Therefore, we talk about sets in our informal presentations of the work done in strict finitism. They provide a way to identify terms in a more coarse-grained manner for some specific purpose. For instance, we will use type $(o \to o)$ terms to encode real numbers, but we want to say that two terms may encode the same number. This needs a more coarse-grained identity relation between type $(o \to o)$ terms than the syntactical identity or the extensional equality. Moreover, a set provides a condition for specifying terms of some special interest for a specific purpose. For instance, only terms that satisfy some condition (i.e., terms representing Cauchy sequences) can encode real numbers. Obviously, this condition should respect the equality between set members.

Note that these are only convenient ways for presenting a piece of mathematical work done in strict finitism. The statement 'A is a set' is to be understood as the conjunction of (1) and (2) above. Therefore, in the end, apparent references to 'sets' can be eliminated, and we get statements with only logical constants $\neg^*$, $\vee^*$, $\wedge^*$, $\to^*$, $\exists^*$, $\forall^*$, $\exists$, and $\forall$ introduced in the last section. These statements are eventually claims in strict finitism, in the format (FinC) in the last section.

This also means that we must be careful in quantifying over sets in our informal presentations. A quantification like 'for all sets A, ...' will actually be 'for all formulas φ, ψ satisfying the conditions for sets, ...'. If this quantification is not nested, it can be understood as a schematic claim of the format 'if (1) and (2), then ...' involving two arbitrary formulas φ, ψ, where (1) and (2) are the formulas above. It is not unlike the general claim 'for any formula φ, we have $\varphi \to \varphi$'. We can certainly prove such schematic claims in strict finitism. Similarly, a quantification like 'for some set A, ...' will actually be 'for some formulas φ, ψ, we have (1) and (2), and ...'. Again, if this is not nested and the context explicitly shows that the formulas φ, ψ can be constructed, then this is acceptable, and it is also reduced to a claim in strict finitism. However, if we want to nest these quantifiers and other logical constants, we must be very careful. We could try using the tricks for defining the logical constants $\neg^*$, $\vee^*$, $\wedge^*$, $\to^*$, $\exists^*$, $\forall^*$, $\exists$, and $\forall$ in the last section, to allow the nested quantifiers and logical constants to be eliminated, but we will not do that in this monograph, because we will see that we can always avoid quantifying over arbitrary sets. (The trick is to quantify over parameters in the formulas defining sets. See below.)

There are some generalizations of the notion of set. First, a set form for an *n-place multiple set* of the type $(\sigma_1, ..., \sigma_n)$ will be a pair $A \equiv \langle \varphi[\mathbf{a}], \psi[\mathbf{a}, \mathbf{b}] \rangle$, where $\mathbf{a}$ and $\mathbf{b}$ are each of the type $(\sigma_1, ..., \sigma_n)$. Then, A is a set when some conditions similar to (1) and (2) above hold. Second, a set form $A[\mathbf{w}]$ for a *parameterized set* with the parameters $\mathbf{w}$ is a pair of formulas $\langle \varphi[a, \mathbf{w}], \psi[a, b, \mathbf{w}] \rangle$, where φ, ψ may contain free variables $\mathbf{w}$ other than a, b. Then, $A[\mathbf{w}]$ is a set if (1) and (2) above hold with $\varphi[a], \psi[a, b]$ replaced by $\varphi[a, \mathbf{w}], \psi[a, b, \mathbf{w}]$. A form for a *family of sets* is a pair $\langle A[\mathbf{w}], \chi[\mathbf{w}] \rangle$ consisting of a set form $A[\mathbf{w}]$ for a parameterized set and a formula $\chi[\mathbf{w}]$ with $\mathbf{w}$ as the only free variables. Then '$\langle A[\mathbf{w}], \chi[\mathbf{w}]\rangle$ is a family of sets' is to mean 'For all $\mathbf{w}$, if $\chi[\mathbf{w}]$, then $A[\mathbf{w}]$ is a set'. We will also say '$\{A_{\mathbf{w}} : \chi[\mathbf{w}]\}$ is a family of sets'. A special case of these is a family $\{A_{\mathbf{i}} : \mathbf{i} \in I\}$ of sets indexed by another set I. We will also use the notation $\{A_{\mathbf{i}}\}_{\mathbf{i}\in I}$. Parameterized sets and families of sets allow us to quantify over *some* sets by quantifying over parameters or indices. That is, we can say 'For all $\mathbf{i}$, ... $A_{\mathbf{i}}$ ...', instead of using quantifiers on sets (or actually, formulas). Clearly, we can also combine these generalizations. So, we have parameterized multiple sets, families of parameterized sets or multiple sets and so on.

We use $\forall x \in A$ to mean $\forall x\,(x \in A \to ...)$ and use $\exists x \in A$ to mean $\exists x\,(x \in A \wedge ...)$. When sets A and B are of the same type, A is a subset of B, or $A \subseteq B$, means

$$\forall x \in A\,(x \in B) \wedge \forall x,y \in A\,(x =_A y \leftrightarrow x =_B y) \wedge$$
$$\forall x,y \in B\,(x \in A \wedge x =_B y \to y \in A)\,.$$

The last condition says that the membership condition for A is extensional relative to the equality relation of B. We will make this convention: *when we mention a subset of a set, unless otherwise stated, we always assume that it has the same equality condition as the super-set.* So, a subset is determined when its membership condition (extensional relative to the equality for its super-set) is given. Further, $A = B$ is $(A \subseteq B) \wedge (B \subseteq A)$. We also need stronger notions of containment and equality between sets. For A, B of the same signature, $A \prec B$ means

$$A \subseteq B \wedge \forall a, \mathbf{x}\,(a \in_{\mathbf{x}} A \to a \in_{\mathbf{x}} B)\,,$$

and $A \cong B$ means $A \prec B \wedge B \prec A$. When using $\{A_{\mathbf{i}}\}_{\mathbf{i}\in I}$ to denote a family of sets, we always assume that the following extensionality condition relative to the indices holds:

$$\forall \mathbf{i}_1, \mathbf{i}_2 \in I\,(\mathbf{i}_1 =_I \mathbf{i}_2 \to A_{\mathbf{i}_1} \cong A_{\mathbf{i}_2})\,.$$

These notions of subset apply to multiple sets, parameterized sets and other generalizations as well. The same holds for other notions related to sets to be introduced below. We will omit the details.

The intersection and union of two sets A and B are defined only in case they have equivalent equality conditions. In that case, the equality condition of the intersection and union is clear, while the membership conditions are respectively

$$(x \in A \cap B) \equiv_{df} x \in A \wedge x \in B,$$
$$(x \in A \cup B) \equiv_{df} x \in A \vee x \in B.$$

The complement $A - B$ is a subset of A with the membership condition

$$x \in A - B \equiv_{df} x \in A \wedge \neg x \in B.$$

A sequence $(A_n)_n$ of sets is a parameterized set with n as (one of) the parameter(s). In case the equality condition does not depend on the parameter n, especially when A_n is a subset of a set B for all n, the union and intersection of the sequence are sets defined as usual:

$$(x \in \cup_{n=0}^{\infty} A_n) \equiv_{df} \exists n \, (x \in A_n);$$
$$(x \in \cap_{n=0}^{\infty} A_n) \equiv_{df} \forall n \, (x \in A_n).$$

Similarly, for any family $\{A_{\mathbf{i}}\}_{\mathbf{i} \in I}$ of subsets of a set B, we can define the set $\cup_{\mathbf{i} \in I} A_{\mathbf{i}}$ and $\cap_{\mathbf{i} \in I} A_{\mathbf{i}}$, if the equality conditions for $\{A_{\mathbf{i}}\}_{\mathbf{i} \in I}$ do not depend on the parameters $\mathbf{i}$. Clearly, these are still subsets of B.

Products of sets can be constructed in a few ways, as multiple sets, as sets of sequences, or as sets of codes of finite sequences. First, let $A_1, ..., A_n$ be n sets of the types $\sigma_1, ..., \sigma_n$ respectively. The product $A_1 \times ... \times A_n$ is the multiple set of the type $(\sigma_1, ..., \sigma_n)$ with the membership and equality conditions:

$$((x_1, ..., x_n) \in A_1 \times ... \times A_n) \equiv_{df} (x_1 \in A_1 \wedge ... \wedge x_n \in A_n),$$
$$((x_1, ..., x_n) =_{A_1 \times ... \times A_n} (y_1, ..., y_n)) \equiv_{df} (x_1 =_{A_1} y_1 \wedge ... \wedge x_n =_{A_n} y_n).$$

When A_is are multiple sets or parameterized sets, the products are defined similarly.

Suppose that (A_n) is a parameterized set of the type σ with the parameter n. Then, $\prod_{n=0}^{\infty} A_n$ is a set of the type $(o \to \sigma)$:

$$\left(x \in \prod_{n=0}^{\infty} A_n \right) \equiv_{df} \forall n \, (x_n \in A_n),$$
$$\left(x =_{\prod_{n=0}^{\infty} A_n} y \right) \equiv_{df} \forall n \, (x_n =_{A_n} y_n).$$

It is easy to prove that if A_n is a set for all n, then this defines a set.

We can use the same format of definition to define $\prod_{n=0}^{N} A_n$:

$$\left(x \in \prod_{n=0}^{N} A_n \right) \equiv_{df} \forall n \le N \, (x_n \in A_n),$$
$$\left(x =_{\prod_{n=0}^{N} A_n} y \right) \equiv_{df} \forall n \le N \, (x_n =_{A_n} y_n).$$

We can also define $\prod_{n=0}^{N} A_n$ as a set of the type σ again, with the codes of finite sequences as its members:

$$\left(x \in \prod_{n=0}^{N} A_n \right) \equiv_{df} Seq(x) \wedge lh(x) = N + 1 \wedge \forall n \le N \, ((x)_n \in A_n),$$
$$\left(x =_{\prod_{n=0}^{N} A_n} y \right) \equiv_{df} \forall n \le N \, ((x)_n =_{A_n} (y)_n).$$

In actual applications, we will choose the one that is more convenient for us. Note that these are parameterized sets with N as a parameter. Given a set A, we can define A^N as $\prod_{n=0}^{N} A$.

Moreover, given a set A, we can also define the set $A^{<\infty}$ of all finite sequences of elements of A:

$$(x \in A^{<\infty}) \equiv_{df} Seq(x) \wedge \forall k < lh(x)\,((x)_k \in A)\,,$$
$$(x =_{A^{<\infty}} y) \equiv_{df} lh(x) = lh(y) \wedge \forall k < lh(x)\,((x)_k =_A (y)_k)\,.$$

This means that given a set A, we can quantify over all finite sequences of the members of A as in $(\forall x \in A^{<\infty})$, $(\exists x \in A^{<\infty})$.

Basic properties about subsets, unions, intersections and products of sets are easily proved. However, note that we do not have the notion of power set, and we cannot quantify over all subsets of a set.

Sometimes we consider a set with an inequality relation, which means a set A together with a formula defining a relation $\neq$. We need an inequality $x \neq y$ to be symmetric and stronger than $\neg x = y$, and we need it to imply distinguishability. That is, '$\neq$ is an inequality relation on A' is the formula

$$\forall x, y \in A\,(x \neq y \rightarrow \neg x = y \wedge y \neq x \wedge \forall z\,(x \neq z \vee y \neq z))\,.$$

Usually there is a natural inequality relation on a set.

2.3.2 Examples: Sets of Numbers

Here are some examples of sets. First, the set $\mathbb{N}$ of natural numbers is a type o set:

$$(n \in \mathbb{N}) \equiv_{df} n = n,$$
$$(n =_{\mathbb{N}} m) \equiv_{df} n = m.$$

Then, we assume a fixed coding for integers, for instance, coding a positive integer n as $2n$ and coding a negative integer $-n$ as $2n-1$. Then, we can define the set of integers, $\mathbb{Z}$, as a set of the type o:

$$(a \in \mathbb{Z}) \equiv_{df} a = a,$$
$$(a =_{\mathbb{Z}} b) \equiv_{df} a = b.$$

Common functions or predicates of integers, for instance, $a+b$, $a-b$, ab, $|a|$, $0 <$, $0 \leq a$, $a \leq b$, $a < b$ and so on, can be constructed as terms or formulas. For example,

$$(0 \leq a) \equiv_{df} Div\,(a, 2)\,,$$
$$|a| \equiv_{df} J\,(0 \leq a, a, a+1)\,.$$

We will use the symbols $+$, $-$, $<$, $\leq$ ambiguously. Contexts should be able to make clear what they mean.

Then, we assume a fixed coding of rational numbers as pairs of integers into natural numbers, for instance, coding $r = \frac{a}{b}$ as

$$OP(a,b) = (a+b)(a+b)+a+1.$$

Let π_1, π_2 be the corresponding decoding function, namely, $\pi_1(OP(a,b)) = a$, $\pi_2(OP(a,b)) = b$. Then, we can define the set of rational numbers, $\mathbb{Q}$, as a set of the type o:

$$(r \in \mathbb{Q}) \equiv_{df} \exists a \leq r \exists b \leq r\,(b \neq 0 \wedge r = OP(a,b)),$$
$$(r_1 =_{\mathbb{Q}} r_2) \equiv_{df} \pi_1(r_1) \cdot \pi_2(r_2) = \pi_2(r_1) \cdot \pi_1(r_2).$$

We will again use the common notations for the common functions or predicates of rational numbers, constructed as terms or formulas, for instance, $p+q$, $p-q$, p/q, pq, $|p|$, p^{-1}, $\lceil p \rceil$ (the least integer greater than p), $p \geq q$, $p > q$ and so on. (A default value for 0^{-1} can be set.) For example,

$$p+q \equiv_{df} OP(\pi_1(p)\pi_2(q) + \pi_2(p)\pi_1(q), \pi_2(p)\pi_2(q)).$$

The basic arithmetic properties of these functions and predicates of rational numbers are easy to prove. Exponentiation with integer exponents p^n and p^{-n} can also be constructed, for example,

$$p^n \equiv_{df} OP(\pi_1(p)^n, \pi_2(p)^n),$$

where the exponentiation on the right hand side is the exponentiation function of integers. Contexts can always resolve any ambiguity regarding whether a numerical term should be treated as a natural number, or an integer, or a rational number.

We will frequently use bounded primitive recursions to construct sequences of rational numbers. Here, we must note that the bound in a bounded primitive recursion should be the bound of the codes of relevant rational numbers, not the bound of those rational numbers themselves. For a rational number $\frac{p}{q}$, its code is bounded by $4(p+q)^2 + 2p + 1$. Therefore, ignoring details, we can take it that the code of a rational number is bounded by its numerator and denominator. Then, when a numerical term $t[n]$ is iterated for n to construct a sequence of rational numbers, the primitive recursion pattern is bounded as long as both the numerator sequence and the denominator sequence are bounded by some elementary recursive function. In particular, this is the case when we iterate ordinary operations such as addition, substraction, multiplication, division, taking absolute value, max, min and so on for rational numbers.

For instance, suppose that $r[n] \equiv \frac{p[n]}{q[n]}$ is a sequence of rational numbers and we want to construct a term $s[n]$ by recursion such that $s[n] = r[0] \cdots r[n]$, that is

$$s[0] = r[0], \quad s[n+1] = s[n] \cdot r[n+1].$$

The numerator and denominator of $s[n]$ are bounded by $\prod_{i \leq n} p[i]$ and $\prod_{i \leq n} q[i]$ respectively. Therefore, $s[n]$ (i.e. the code of $s[n]$ as a rational number) is bounded by a term $b[n]$ constructed from p and q. Then, $s[n]$ can be constructed by bounded primitive recursion.

We will call operations such as addition, substraction, multiplication, division, taking absolute value, max, min and so on '*iteratively bounded operations*' for rational numbers. These operations can be iterated for recursively constructing sequences of rational numbers (usually for approximating real numbers) by bounded recursion. Note that the exponentiation p^n is iteratively bounded for p but not iteratively bounded for n.

The set of real numbers, $\mathbb{R}$, is then a set of the type $\sigma = (o \to o)$ and the signature (σ) ([6], p. 18):

$$(x \in \mathbb{R}) \equiv_{df} \forall m, n > 0\,(x(n) \in \mathbb{Q} \wedge |x(m) - x(n)| \leq 1/m + 1/n)\,,$$
$$(x =_{\mathbb{R}} y) \equiv_{df} \forall n > 0\,(|x(n) - y(n)| \leq 2/n)\,.$$

Therefore, we essentially take elementary recursive Cauchy sequences of rational numbers as real numbers. Note that $t \in \mathbb{R}$ is a Π^0_1 formula. A proof of $t \in \mathbb{R}$ is then simply a proof within **SF** of a quantifier-free formula (with free variables).

We can define

$$(x \leq y) \equiv_{df} \forall n > 0\,(x(n) \leq y(n) + 2/n)$$
$$(x < y) \equiv_{df} (\exists n > 0)\,(x(n) < y(n) - 2/n)\,.$$

Note that $x \leq y$ is a Π^0_1 formula and $x < y$ is a Σ^0_1 formula. That is, proving $x < y$ requires finding a witness n such that $x(n) < y(n) - 2/n$. It is easy to verify that these definitions are consistent with the corresponding classical notions. It follows from Lemma 2.12 that

$$\neg x \leq y \leftrightarrow x > y,\ x \leq y \leftrightarrow \neg x > y$$

(which is different from the intuitionistic case). However, note that $x \leq y$ does not generally imply $x < y \vee x = y$.

Here are some common subsets and parameterized subsets of $\mathbb{R}$. Their equality conditions are all the same as that of $\mathbb{R}$, so we give only their membership conditions:

The set of positive real numbers $\mathbb{R}^+$: signature: $((o \to o), o)$,

$$x \in \mathbb{R}^+ \equiv_{df} x \in \mathbb{R} \wedge x > 0.$$

$\mathbb{R}^-$ is defined similarly. Note that $x > 0$ requires a witness.

The set of non-negative real numbers $\mathbb{R}^{+0}$: signature: $((o \to o))$,

$$x \in \mathbb{R}^{+0} \equiv_{df} x \in \mathbb{R} \wedge x \geq 0.$$

Closed interval $[a, b]$: parameters: a, b of the type $(o \to o)$, signature: $((o \to o))$,

$$x \in [a, b] \equiv_{df} x \in \mathbb{R} \wedge a \leq x \wedge x \leq b.$$

Open interval (a, b): parameters: a, b of the type $(o \to o)$, signature: $((o \to o), o, o)$,

$$x \in (a,b) \equiv_{df} x \in \mathbb{R} \wedge a < x \wedge x < b.$$

Note that each of $a < x$ and $x < b$ needs a type o witness.

The intervals $[a,b)$, $(a,b]$, $(\infty,b]$, $(a,\infty]$ and so on are defined similarly. $[a,b]$ is also called a compact interval.

For subsets of real numbers, we always assume this standard inequality relation

$$(x \neq y) \equiv_{df} (x > y \vee x < y).$$

It is easy to show that $x \neq y \rightarrow z \neq x \vee z \neq y$. That is, it satisfies the condition for an inequality relation.

2.3.3 Functions

A function is a term that applies to terms belonging to the domain set of the function, *together with their witnesses*, and results in terms belonging to the range set. Suppose that A and B are sets of the signatures $(\sigma_0, \sigma_1, ..., \sigma_n)$ and $(\rho_0, \rho_1, ..., \rho_m)$ respectively, and f is a term of the type $(\sigma_0, \sigma_1, ..., \sigma_n \rightarrow \rho_0)$. f is a function from A to B, or $f : A \rightarrow B$, if

$$\begin{aligned}&\forall x_0 \mathbf{x}_1 \left(x_0 \in_{\mathbf{x}_1} A \rightarrow f(x_0, \mathbf{x}_1) \in B\right) \wedge \\ &\forall x_0 \mathbf{x}_1 y_0 \mathbf{y}_1 \left(x_0 \in_{\mathbf{x}_1} A \wedge y_0 \in_{\mathbf{y}_1} A \wedge x_0 =_A y_0 \rightarrow f(x_0, \mathbf{x}_1) =_B f(y_0, \mathbf{y}_1)\right).\end{aligned}$$

Note that a function operates on the witnesses for an element belonging to its domain. From $f : A \rightarrow B$ it follows that

$$x_0 \in_{\mathbf{x}_1} A \wedge x_0 \in_{\mathbf{x}_2} A \rightarrow f(x_0, \mathbf{x}_1) =_B f(x_0, \mathbf{x}_2).$$

So we frequently simply write $f(x_0)$ instead of $f(x_0, \mathbf{x}_1)$, as equal members of a set can be treated as the same in most contexts. Similarly, sometimes we use notations like $\forall x_0 \in A(....f(x_0)....)$, while literally it should be

$$\forall x_0 \mathbf{x}_1 \left(x_0 \in_{\mathbf{x}_1} A \rightarrow ...f(x_0, \mathbf{x}_1)...\right).$$

Such *simplified notations* are more readable, and contexts can always determine how to complete them, as long as we always keep in mind that a function operates on the witnesses as well as the elements belonging to its domain.

For example, the reciprocal function x^{-1} on the set $\mathbb{R}^+$ operates on an arbitrary sequence $x = (x_n)_n$ of rational numbers *and* an arbitrary natural number m as the putative witness for $x \in \mathbb{R}^+$. That is, it is a witness when $0 < x_m - 2/m$, namely, $x_m > 2/m$. It's easy to see that for x and m such that $x \in \mathbb{R}^+$ and $x_m > 2/m$, we can find N, M, such that for all $n,k \geq N$, we have $\left|x_{nM}^{-1} - x_{kM}^{-1}\right| \leq 1/n + 1/k$. Then, we can construct a term t, such that $t(x,m)(k) = x_{NM}^{-1}$ for $k \leq N$ and $t(x,m)(k) = x_{kM}^{-1}$ for

$k > N$. Then, it is easy to see that $t(x,m) \in \mathbb{R}$ if $x \in_m \mathbb{R}^+$. $\lambda x.t$ is then the reciprocal function $\cdot^{-1}$ on the set $\mathbb{R}^+$.

When A is an n-place multiple set, σ_0, x_0, and y_0 above are actually sequences of types and variables σ_0, $\mathbf{x}_0$, $\mathbf{y}_0$. In that case, we will call f an n-place function. Similarly, when B is a multiple set, ρ_0 is a sequence ρ_0 of types and f is a sequence $\mathbf{f}$ of terms of the type $(\sigma_0, \sigma_1, ..., \sigma_n \to \rho_0)$. To emphasize that, we will call $\mathbf{f}$ a multiple function. In particular, suppose that A_1, ..., A_n and B are sets, and A_i has the signature (σ_i, ρ_i) and B has the type ρ. Recall that $A_1 \times ... \times A_n$ is a multiple set. Therefore, a function $f : A_1 \times ... \times A_n \to B$ is an n-place function. It will be a term of the type

$$(\sigma_1, \rho_1, ..., \sigma_n, \rho_n \to \rho).$$

Note that constant terms S, $+$, $\cdot$, and *pow* are functions from $\mathbb{N}$ to $\mathbb{N}$, or 2-place functions from $\mathbb{N} \times \mathbb{N}$ to $\mathbb{N}$. Therefore, the terminology 'function' used here is consistent with our previous uses of it.

Given the term f and sets A and B as pairs of formulas, the statement 'f is a function from A to B' is a statement using logical constants $\neg^*$, $\vee^*$, $\wedge^*$, $\to^*$, $\exists^*$, and $\forall^*$ and so on, and is therefore a claim in strict finitism. That is, all apparent references to functions can in principle be eliminated. A statement referring to a function actually expresses a condition about a term in some simplified format. For instance,

$$(f : \mathbb{R}^+ \to \mathbb{R}) \wedge \forall x \in \mathbb{R}^+ (x \cdot f(x) = 1)$$

gives the condition for a term f as the reciprocal function on $\mathbb{R}^+$. Spelling it out, we will get a quite complex statement using the symbols $\neg^*$, $\vee^*$, $\wedge^*$, $\to^*$, $\exists^*$, $\forall^*$, $\exists$, and $\forall$, as well as the symbols $\neg$, $\vee$, $\wedge$, and $\to$ in **SF**. Then, after these defined logical constants are eliminated, it eventually becomes a claim in strict finitism in the format (FinC) in Sect. 2.2.1, stating that some terms of **SF** can be constructed, together with f, to satisfy some condition expressed as a quantifier-free formula in **SF**.

The set of functions from A to B, $F(A,B)$, is defined as follows:

$$(f \in F(A,B)) \equiv_{df} f : A \to B,$$
$$(f =_{F(A,B)} g) \equiv_{df} \forall x_0 \mathbf{x}_1 (x_0 \in_{\mathbf{x}_1} A \to f(x_0, \mathbf{x}_1) =_B g(x_0, \mathbf{x}_1)).$$

This equality condition, which can be expressed as $\forall x \in A (f(x) = g(x))$ in a simplified manner, is called *extensional equality for functions*. We agree that when defining sets of functions, for instance, the set of continuous functions on $\mathbb{R}$, *the extensional equality for functions is always tacitly assumed unless it is explicitly stated otherwise.*

Notions like '$f : A \to B$ is onto', '$f : A \to B$ is an inverse of $g : B \to A$', '$f : A \to B$ is one-one' and so on are defined as usual. If A, B are sets with inequality relations, '$f : A \to B$ respects inequalities' is to mean 'if $f(x) \neq f(y)$ then $x \neq y$ for all $x, y \in A$'. Given terms f, g, we can construct a term $g \circ f$ such that if $f : A \to B$ and $g : B \to C$, then $g \circ f : A \to C$. The basic properties of these notions are easily proved. In particular, if $f : A \to B$ is onto, then for $y \in B$, there exists $x \in A$ such that $f(x) = y$. Since the axiom of choice holds, it means that there exists g such that $g : B \to A$ and $f \circ g$ is the identity function on B.

A set A is countable if

$$\exists f\,((f : \mathbb{N} \to A) \wedge (f \text{ is onto})),$$

and A is finite if

$$\exists x \exists m\,((\forall n < m)\,(x(n) \in A) \wedge (\forall z \in A)\,(\exists n < m)\,(x(n) =_A z)).$$

The extensionality condition enables us to encode a finite set into a single object of the same type. Suppose that x, m witness A's being a finite set and let $a \equiv \overline{x}(m) = \langle x(0), \ldots, x(m-1) \rangle$. Then

$$(\forall n < m)\,((a)_n \in A) \wedge (\forall z \in A)\,(\exists n < m)\,((a)_n =_A z).$$

So a encodes A. Note that the extensionality of set conditions is needed here, because we have only $(a)_n \simeq x(n)$ and we need extensionality to infer $(a)_n \in A$.

If $f : A_1 \to B$ and $A_1 \cong A_2$ then $f : A_2 \to B$. The stronger equality is necessary here, for otherwise A_1 and A_2 may even have different signatures. On the other hand, if in a context we have $f : A \to B$ and $C \subseteq A$, then there is a natural way to construct a function from C to B as the restriction of f to C. First, $C \subseteq A$ implies that there exist $\mathbf{W}$ such that for any $a, \mathbf{u}$, $a \in_{\mathbf{u}} C$ implies $a \in_{\mathbf{W}(a,\mathbf{u})} A$. $\mathbf{W}$ are the witnesses for $C \subseteq A$. Then, let f' be defined by $f'(a, \mathbf{u}) \equiv f(a, \mathbf{W}(a, \mathbf{u}))$. Clearly, $f' : C \to B$. We will denote f' by $f|_C$, or simply by f ambiguously, while remembering that it actually contains the witnesses for $C \subseteq A$.

We will frequently use such *simplified notations* without stating so explicitly. For example, after defining a function on $\mathbb{R}$, we always use the same notation for its restrictions to the subsets of $\mathbb{R}$, such as various intervals. Further, if in a context we have $f_1 : A_1 \to B$ and $f_2 : A_2 \to B$, and $C \subseteq A_1 \wedge C \subseteq A_2$, then '$f_1 = f_2$ on C' is to mean

$$\forall x \mathbf{u}_1 \mathbf{u}_2\,(x \in C \wedge x \in_{\mathbf{u}_1} A_1 \wedge x \in_{\mathbf{u}_2} A_2 \to f_1(x, \mathbf{u}_1) = f_2(x, \mathbf{u}_2)),$$

or $\forall x \in C\,(f_1(x) = f_2(x))$ in simplified notations. The assertion '$f_1 = f_2$ on C' actually involves A_1, A_2, which do not appear in the simplified expression.

Sets and functions together provide a way to make sophisticated claims on the constructions of terms and on conditions about the constructed terms. For instance, we will study the set $C(\mathbb{R}, \mathbb{R})$ of continuous functions from $\mathbb{R}$ to $\mathbb{R}$ and make claims about such arbitrary functions. A function from $\mathbb{R}$ to $\mathbb{R}$ is a term of the type $((o \to o) \to (o \to o))$ satisfying some condition, and the set $C(\mathbb{R}, \mathbb{R})$ of continuous functions from $\mathbb{R}$ to $\mathbb{R}$ actually expresses an extra condition on these terms. The condition $f \in C(\mathbb{R}, \mathbb{R})$ is itself a claim in strict finitism in the format (FinC). We will see that it requires witnesses. When we make a claim about an arbitrary continuous function f from $\mathbb{R}$ to $\mathbb{R}$, we are making a conditional claim in the format

$$\forall^* f\,(f \in C(\mathbb{R}, \mathbb{R}) \to^* \varphi[f]) \tag{2.4}$$

where the implication is understood as the numerical implication, and the universal quantification is also to be eliminated by the Axiom of Choice. Therefore, in the end, we are still claiming that some terms can be constructed to satisfy some condition expressed as a quantifier-free formula in **SF**.

Intuitively, the constructed terms will operate on an arbitrary term f of the type $((o \to o) \to (o \to o))$, together with witnessing terms for its belonging to $C(\mathbb{R}, \mathbb{R})$, and the resulted terms will satisfy some quantifier-free but very complex condition. Referring to functions and sets allows us to state these in a simple and more familiar format. In particular, it allows us to state sophisticated nested conditional constructions, while spelling out the numerical implications and quantifications will result in very complex quantifier-free formulas. For instance, the claim 'f is a function from $\mathbb{R}$ to $\mathbb{R}$' already contains nested universal quantifications and numerical implications, because it is $\forall x (x \in \mathbb{R} \to f(x) \in \mathbb{R})$, and $x \in \mathbb{R}$ itself contains universal quantifications and numerical implications. We will see that the extra condition of continuity demands a witness for continuity and also contains universal quantifications and numerical implications referring to arbitrary type $(o \to o)$ terms that are real numbers. Then, these nested quantifications and numerical implications are again nested in (2.4). It is foreseeable that spelling out all these will result in an extremely complex formula in the format (FinC). Sets and functions allow us to state these in a highly abstract and simplified manner.

2.3.4 Partial Functions

In the theory of integration we will need partial functions from $\mathbb{R}$ to $\mathbb{R}$. The domain of a partial function on $\mathbb{R}$ is supposed to be a subset of $\mathbb{R}$. Subsets of $\mathbb{R}$ may have different witness types and hence they cannot all be put into a single family of subsets. Therefore, we cannot quantify over all such partial functions in our formulas and cannot define the set of all partial functions. Here, we will introduce a strategy to allow us to treat some very broad families of partial functions. In particular, we will see that we will be able to treat all Lebesgue integrable functions $\mathbb{R}$ to $\mathbb{R}$ in a single family of partial functions $\mathbb{R}$ to $\mathbb{R}$.

Given a family of parameterized subsets $\mathscr{D} \equiv \{D_{\mathbf{i}} : \mathbf{i} \in I\}$ of a set X indexed by an n-place multiple set I, we can define the set $\mathscr{F}(\mathscr{D}, Y)$ of partial functions from X to a set Y with domains in the family $\mathscr{D}$. This is an $(n+1)$-place multiple set (or $(n+m)$-place multiple set if Y is an m-place multiple set):

$$((\mathbf{i}, f) \in \mathscr{F}(\mathscr{D}, Y)) \equiv_{df} (\mathbf{i} \in I \wedge f : D_{\mathbf{i}} \to Y),$$
$$((\mathbf{i}_1, f_1) =_{\mathscr{F}(\mathscr{D},Y)} (\mathbf{i}_2, f_2)) \equiv_{df} \left(D_{\mathbf{i}_1} = D_{\mathbf{i}_2} \wedge \forall x \in D_{\mathbf{i}_1} \left(f_1(x) = f_2(x)\ \right)\right).$$

Note that $f_1(x) = f_2(x)$ in the above actually has a format like $f_1(x, a) = f_2(x, u(a))$, where a is a hypothetical witness for $x \in D_{\mathbf{i}_1}$, and u is a term witnessing $D_{\mathbf{i}_1} \subseteq D_{\mathbf{i}_2}$, that is, a term such that $x \in_a D_{\mathbf{i}_1} \to x \in_{u(a)} D_{\mathbf{i}_2}$. Now, suppose that $D_{\mathbf{i}_1} = D_{\mathbf{i}_2}$ and $\forall x \in D_{\mathbf{i}_1} (f_1(x) = f_2(x)\ )$. Then, from $x \in_b D_{\mathbf{i}_2}$ we also have $x \in_{v(u)} D_{\mathbf{i}_1}$ for some v, and then $x \in_{u(v(b))} D_{\mathbf{i}_2}$ again. From $f_2 : D_{\mathbf{i}_2} \to Y$ we have $f_2(x,b) = f_2(x,u(v(b)))$. Let a above be $v(b)$ here, we have $f_1(x,v(b)) = f_2(x,u(v(b)))$. Therefore, $f_2(x,b) = f_1(x,v(b))$. That is, we also have $\forall x \in D_{\mathbf{i}_2}(f_2(x) = f_1(x))$, and the definition of the equality above is symmetric. Other conditions for defining a set are also easy to verify.

Then, we can quantify over all partial functions in $\mathscr{F}(\mathscr{D},Y)$ in our formulas, by which we mean a quantification like

$$\forall \mathbf{i} \forall f\,((\mathbf{i},f) \in \mathscr{F}(\mathscr{D},Y) \to \ldots\ldots).$$

We will say that $D_{\mathbf{i}}$ is the domain of $(\mathbf{i},f)$. Therefore, the domain (actually the parameter of the domain) of a partial function is uniquely determined by the partial function. $(\mathbf{i}_1,f_1)$ is a restriction of $(\mathbf{i}_2,f_2)$ if

$$D_{\mathbf{i}_1} \subseteq D_{\mathbf{i}_2} \wedge \left(f_1 = f_2 \text{ on } D_{\mathbf{i}_1}\right).$$

If $(\mathbf{i}_1,f_1)$ is a restriction of $(\mathbf{i}_2,f_2)$, then $(\mathbf{i}_1,f_1) = (\mathbf{i}_1,f_2)$. When no real ambiguities will occur, we simply call f a partial function and call $D_{\mathbf{i}}$ the domain of f, denoted by $Dmn(f)$. So, for example, $(\mathbf{i},f) \in \mathscr{F}(\mathscr{D},Y)$ will be simplified as $f \in \mathscr{F}(\mathscr{D},Y)$. Similarly, we will simply write the equality as $f_1 = f_2$.

Notions such as 'onto', 'one-one' can apply to a partial function $(\mathbf{i},f)$ viewed as a function $f : D_{\mathbf{i}} \to Y$.

We say that the family $\mathscr{D}$ is *closed under finite intersection*, if for any finite sequence $\mathbf{i}_1,\ldots,\mathbf{i}_n$ of indices there exists $\mathbf{j}$ such that $D_{\mathbf{j}} = \cap_{k=1}^{n} D_{\mathbf{i}_k}$. We say that $\mathscr{D}$ is closed under countable intersection, if for any sequence $(\mathbf{i}_n)$ of indices there exists $\mathbf{j}$ such that $D_{\mathbf{j}} \subseteq \cap_{n=1}^{\infty} D_{\mathbf{i}_n}$. Note that the latter needs only inclusion but not equality.

Chapter 3
Calculus

This chapter develops the basics of calculus in strict finitism. Notions such as limit, convergence, continuity, differentiability, and Riemann integration are introduced, and their basic properties are proved. A case study of demonstrating applicability by reducing to strict finitism is also presented in the last section of this chapter.

Since the laws of intuitionistic logic are available for our informal arguments in strict finitism, we will follow the techniques for developing calculus in Bishop's constructive mathematics (see Chap. 2 of Bishop and Bridges [6]). Recall that the essential difference between strict finitism and Bishop's constructive mathematics is on the inductions and recursive constructions available. Therefore, our critical job here is to make sure that various recursive constructions employed are bounded by elementary recursive functions and that the inductions used are reducible to the quantifier-free induction.

3.1 The Real Number System

The set $\mathbb{R}$ of real numbers is already defined. Note that the set of rational numbers $\mathbb{Q}$ is a set of the type o, which is different from the type of $\mathbb{R}$. However, clearly $x \in \mathbb{Q} \to \lambda m.x \in \mathbb{R}$. For convenience, we frequently ignore the type difference between rational numbers and real numbers and simply write $0 \in \mathbb{R}$, $x \in \mathbb{Q} \to x \in \mathbb{R}$ and so on, which will actually mean $\lambda m.0 \in \mathbb{R}$ and $x \in \mathbb{Q} \to \lambda m.x \in \mathbb{R}$ respectively. Similarly, note that as a rational number, 2 is $\frac{+2}{+1}$ and is encoded as the numeral $OP(2 \cdot 2, 2 \cdot 1)$. However, we will simply treat the numerical terms $0, 1, 2, \ldots$ (i.e. 0, $S0$, $SS0$, ...) as if they represent the rational numbers 0, 1, 2, ..., and we will also treat them as real numbers. These are not essential issues. Readers should be able to rectify such details and we will simply adopt a manner of speech that is more convenient.

Common functions for real numbers such as $+$, $-$, $\cdot$, $/$, $|\cdot|$, max, min, x^n and so on are constructed as closed terms of appropriate types. For instance,

F. Ye, *Strict Finitism and the Logic of Mathematical Applications*, Synthese Library 355,
DOI 10.1007/978-94-007-1347-5_3, © Springer Science+Business Media B.V. 2011

$$x + y \equiv_{df} \lambda n.\,(x(2n) + y(2n))\,.$$

It is straightforward to verify that

$$x \in \mathbb{R} \wedge y \in \mathbb{R} \to x + y \in \mathbb{R}. \tag{3.1}$$

Let

$$\varphi[x,m,n] \equiv m > 0 \wedge n > 0 \to |x(m) - x(n)| \le 1/m + 1/n.$$

Then, by the definition of $x \in \mathbb{R}$, (3.1) means

$$\forall m \forall n \forall m' \forall n' \left(\varphi[x,m,n] \wedge \varphi\left[y,m',n'\right]\right) \to \forall m \forall n\,(\varphi[x+y,m,n])\,.$$

By the numerical interpretation of implication, proving this requires constructing terms $s \equiv s[x,y,m,n]$, $t \equiv t[x,y,m,n]$, $s' \equiv s'[x,y,m,n]$, $t' \equiv t'[x,y,m,n]$ such that

$$\varphi[x,s,t] \wedge \varphi\left[y,s',t'\right] \to \varphi[x+y,m,n]\,.$$

This is trivial, for we can let $s \equiv 2m$, $t \equiv 2n$, $s' \equiv 2m$, $t' \equiv 2n$. To prove that $+ : \mathbb{R} \times \mathbb{R} \to \mathbb{R}$, we still need to prove that it respects the equality for $\mathbb{R}$, that is,

$$x =_{\mathbb{R}} x' \wedge y =_{\mathbb{R}} y' \to x + y =_{\mathbb{R}} x' + y'.$$

This is also trivial.

The above is a rather detailed presentation of some constructions in strict finitism. Such details are usually straightforward but tedious. We will omit such details in the rest of this monograph, as long as we consider them trivial. For instance, similar constructions and proofs can be easily supplemented for the functions $-$, $|\cdot|$, max, and min. Readers should keep in mind that every claim we make in the end states that some terms are constructed and some conditions about the terms expressed as quantifier-free formulas of **SF** are verified.

Similarly, note that if $x \equiv (x_n)$ is a real number, then $x_n < x_1 + 2$ for all $n > 0$. Therefore, we can define $x \cdot y$ as $\lambda n.\,(x(2kn)\,y(2kn))$, where k is $\max\{\lceil x_1 + 2 \rceil, \lceil y_1 + 2 \rceil\}$. Similarly, let $k \equiv m\,(\lceil x_1 + 2 \rceil)^m$. Then, the sequence $\left(x^m_{kn}\right)_n$ is a real number and we can define x^m as $\lambda n.x^m_{kn}$. When r is a rational number, for $x \equiv (x_n)$ a real number, we can define $r \cdot x \equiv_{df} \left(r x_{\max(1,\lceil |r| \rceil)n}\right)_n$. In particular, when $|r| \le 1$, $r \cdot x \equiv_{df} (r x_n)$. This will simplify some constructions.

Recall that $<$, $\le$, and $\neq$ are already defined for real numbers. $>$ and $\ge$ can be defined in obvious ways. Note that $x < y$, $x > y$, $x \neq y$ are Σ^0_1 formulas, that is, they require a witness, while $x \ge y$, $x \le y$, $x = y$ are Π^0_1 formulas. Ignoring the type difference, we will also use notations like $2 + x$, $x > 1$ and so on when x is a variable for real numbers. Note that for a real number $x = (x_n)$, $x_1 - 1 \le x \le x_1 + 1$. Moreover, given that x, y are real numbers, we have more useful characterizations of the equality and inequalities: if $x, y \in \mathbb{R}$, then

$$(x = y) \leftrightarrow \forall k > 0 \exists M \forall m \geq M (|x(m) - y(m)| \leq 1/k),$$
$$(x \leq y) \leftrightarrow \forall k > 0 \exists M \forall m \geq M (x(m) \leq y(m) + 1/k),$$
$$(x < y) \leftrightarrow \exists k > 0 \exists M \forall m \geq M (x(m) < y(m) - 1/k).$$

Furthermore, given x, y such that $x < y$, for any z, it is decidable if $z < y$ or $z > x$. This fact will be very useful in various constructions.

The reciprocal function $\cdot^{-1}$ is defined on $\mathbb{R}^+ \cup \mathbb{R}^-$. Note that this function has the signature $((o \to o), o, o, o)$, because the defining formula is $x \in \mathbb{R}^+ \vee x \in \mathbb{R}^-$, and $x \in \mathbb{R}^+$ and $x \in \mathbb{R}^-$ each contains one existential quantifier and the disjunction $\vee$ generates one more existential quantifier. Therefore, there are three witnesses i, n, m for $x \in \mathbb{R}^+ \cup \mathbb{R}^-$. $i = 0$ or 1 indicates that the element belongs to $\mathbb{R}^+$ or $\mathbb{R}^-$; n and m witness that the element belongs to $\mathbb{R}^+$ or $\mathbb{R}^-$ respectively, that is, $x(n) > 2/n$ or $x(m) < -2/m$. Then, the function $\cdot^{-1}$ actually operates on (x, i, n, m). It is easy to construct a term $t[x, n, m]$ (depending on $x(n) - \frac{2}{n}, -\frac{2}{m} - x(m)$) such that $t[x, n, m] > 2$ and for all $k > t[x, n, m]$, $|x(k)| > 1/t[x, n, m]$, if n and m are the above witnesses. Then, we can let

$$\cdot^{-1} \equiv_{df} \lambda xinm.\lambda j.x\left(\max(j, n, m, 2)\, t[x, n, m]^2\right)^{-1}.$$

It can be verified that this is a function from $\mathbb{R}^+ \cup \mathbb{R}^-$ to $\mathbb{R}$ and hence we can simply write $\cdot^{-1}(x, i, n, m)$ as x^{-1} when it is verified that $x \in \mathbb{R}^+ \cup \mathbb{R}^-$.

Then, the division function $/$ is a 2-place function from $\mathbb{R} \times (\mathbb{R}^+ \cup \mathbb{R}^-)$ to $\mathbb{R}$, and it can be defined as

$$x/y \equiv_{df} x \cdot y^{-1}.$$

The basic properties of these functions are easily proved within strict finitism. For instance, we can prove that these operations satisfy the conditions for a field in algebra, and we can prove the basic properties characterizing $|\cdot|$, max and min, and we can prove some basic inequalities involving these operations. Some of these are straightforward from the definitions and some will require constructing relevant witnesses.

For instance, to prove that $x > y > 0 \to x^{-1} < y^{-1}$, we must construct a witness for $x^{-1} < y^{-1}$ from the witnesses for $x > y$ and $y > 0$. That is, given m, n such that $x(m) > y(m) + 2/m$ and $y(n) > 2/n$, we must construct k such that $x^{-1}(k) < y^{-1}(k) - 2/k$. This is obvious. We will omit the details here.

Note that $x = y$ and $x \leq y$ are Π_1^0 formula and we have only quantifier-free inductions. However, the finite transitivity of equalities and inequalities always hold:

Lemma 3.1. *If we can prove*

$$\varphi \to \forall n (t[n] \in \mathbb{R} \wedge t[n] = t[n+1]),$$

then we have

$$\varphi \to \forall n (t[n] \in \mathbb{R} \wedge t[n] = t[0]).$$

The same holds with $=$ *replaced by* $\leq$.

Proof. From the assumption, it follows that for $m > 0$,

$$|t[n](m) - t[n+1](m)| \leq 2/m.$$

Then, by a quantifier-free induction, we have $|t[n](m) - t[0](m)| \leq \frac{2n}{m}$. Since this holds for arbitrary m, we have $t[n] = t[0]$. □

We also have the following Cantor's Theorem ([6], p. 27):

Theorem 3.2. *For any sequence (a_n) of real numbers and any two real numbers x, y with $x < y$, there exists a real number z such that $x \leq z \leq y$ and $z \neq a_n$ for each n.*

Proof. It suffices to show that for any sequence (a_n) of real numbers, we can construct a real number x such that $0 \leq x \leq 1$ and $x \neq a_n$ for all $n > 0$. The idea is to construct a sequence (k_n), such that for any n, (i) $a_n < \frac{k_{n+1}}{2^{2(n+1)}}$ or $a_n > \frac{k_{n+1}+1}{2^{2(n+1)}}$, and (ii) $k_{n+1} = 4k_n$ or $4k_n + 3$, which means that $\frac{k_{n+1}}{2^{2(n+1)}} = \frac{k_n}{2^{2n}}$ or $\frac{k_n + \frac{3}{4}}{2^{2n}}$. Then, we can let $x \equiv \left(\frac{k_n}{2^{2n}}\right)$. x will be a real number and $\frac{k_{n+1}}{2^{2(n+1)}} \leq x \leq \frac{k_{n+1}+1}{2^{2(n+1)}}$, which implies that $x \neq a_n$. To find k_{n+1}, we divide the interval $\left[\frac{k_n}{2^{2n}}, \frac{k_n+1}{2^{2n}}\right]$ into 4 equal subintervals, and then compare $a_n\left(2^{2n+3}\right)$ with the middle point $\frac{4k_n+2}{2^{2(n+1)}}$ of the interval. If $a_n\left(2^{2n+3}\right) \leq \frac{4k_n+2}{2^{2(n+1)}}$, then we are sure that $a_n < \frac{4k_n+3}{2^{2(n+1)}}$, and we let $k_{n+1} = 4k_n + 3$. Otherwise, $a_n > \frac{4k_n+1}{2^{2(n+1)}}$, and we let $k_{n+1} = 4k_n$. The recursive construction is obviously bounded by an elementary recursive function. □

This recursive construction is more explicit than the one on [6], p. 27, which is not very clearly available to **SF**, that is, not very clearly bounded by elementary recursive functions. Cases like this are common in developing mathematics within strict finitism. That is, we may need recursive constructions that are more straightforward and more explicit than the recursive constructions in Bishop's constructive mathematics. As a consequence, our constructions are closer to realistic computer programs.

A real number a is an upper bound of a set $A \subseteq \mathbb{R}$, if $\forall x (x \in A \rightarrow a \geq x)$, and a is the supremum $\sup A$ of A, if a is an upper bound of A and for any $k > 0$ there exists $x \in A$ such that $x > a - k^{-1}$. Lower bound and infimum are defined similarly. Note that verifying that a is the supremum of A involves constructing a term witnessing the condition. It is well known that several classical theorems about the topology of the real number system cannot be constructivized, including the theorem that any set with an upper bound has a supremum. However, we still have some finitistic substitutes ([6], p. 37).

Theorem 3.3. *Suppose that there exists $a \in A$ and there exists an upper bound b of A, and suppose that for any $x, y \in \mathbb{R}$ such that $x < y$, either there exists $z \in A$ such that $x < z$, or y is an upper bound of A. Then,* $\sup A$ *exists.*

Proof. To estimate $c = \sup A$ up to the precision $1/k$, we can proceed as follows: Take two rational numbers $p < a$ and $q > b$. Choose a constant number $k_0 > 0$ so that $(q-p)/2^{k_0} < \frac{1}{2}$. For each k, divide the rational interval $[p, q]$ evenly into 2^{k+k_0}

sub-intervals $[p_i, p_{i+1}]$, $p_0 = p < \cdots < p_{2^{k+k_0}} = q$. We have $p_{i+1} - p_i < 1/2^k$. From the assumption it follows that we have a numerical term $s[p,q,k,i]$ that equals to 0 or 1 indicating the case where there exists $z \in A$ such that $p_i < z$, or the case where p_{i+1} is an upper bound of A. Let c_k be p_{i+1} with i the largest index so that there exists $z \in A$ such that $p_i < z$. So, c_k is an upper bound of A. It is easy to verify that $c = (c_k)$ is a real number and is $\sup A$. (Note that this is again a simpler construction than the one on [6], p. 37.) □

A set A is *totally bounded* if for each $k > 0$ there exists a finite sequence of real numbers a, such that $(a)_i \in A$, for $i < lh(a)$, and for each $x \in A$, there exists $i < lh(a)$, such that $|x - (a)_i| < k^{-1}$. It is easy to show that if A is totally bounded, then the conditions of the theorem above hold. Therefore, we have

Corollary 3.4. *If A is totally bounded, then* $\sup A$ *and* $\inf A$ *exist.*

In applications, rational numbers are in principle sufficient for representing real physical quantities, since our scientific theories are accurate only above the Planck scale. We want to use real numbers because we want to use some general procedures for computing physical quantities to represent physical quantities, and we want mathematical operations on physical quantities to be closed and thus simpler. For instance, given that the microscopic space-time structure is unknown and could be discrete or not 4-dimensional, when calculating the ratio of a circumference to its diameter for an ordinary circular physical object, it is physically meaningless to consider the precision of the ratio after a few decimal digits (in common physics units). However, we have a general procedure for computing π up to an arbitrary precision, for instance, by an approximation using polygons, and that same procedure appears in many places in mathematics. It makes our life easier if we simply use the procedure, an elementary recursive function computing the rational approximations of π, as our representation of the ratio. Similarly, we have a procedure for calculating the sequence (p_n) of rational numbers so that p_n^2 approaches 2 closer and closer. We want to consider such procedures and arithmetic operations on them. If we use such procedures to represent physical quantities, the operation of taking square root on physical quantities will be generally available, although in real life we always need only approximations to $\sqrt{2}$ up to some finite precision (certainly quite above 10^{-100} in common physics units). Therefore, using real numbers to represent physical quantities simplifies our theories.

In strict finitism, we actually use terms encoding real numbers to represent physical quantities, such as temporal or spatial distance, mass, energy and so on. Recall that the atomic formulas of **SF** are only equations between numerical terms. Suppose that x is a variable of the type $(o \to o)$ (for a real number) and $t[x]$ is a numerical term in normal form with x as the only free variable. From Lemma 2.5 it follows that all subterms of $t[x]$ other than x itself and the constant function symbols S, $+$, $\cdot$, *pow*, and $I_<$ are numerical terms. Therefore, in $t[x]$, x appears only in contexts like $x(r)$ for some numerical term r. If x encodes a real number, $x(r)$ is intuitively a rational approximation to x. This means that in an atomic formula $t[x] = s[x]$, only rational approximations to x as a real number are really mentioned, not x itself. Then, when this formula is translated into a realistic assertion about real physical quantities, the term $x(r)$ is translated into a term expressing physical properties like 'p seconds', 'p meters', 'p kilograms' and so on, with p a rational number. In other words, in realistic applications, bridging postulations will only translate sentences in **SF** into realistic assertions about physical quantities represented by rational numbers.

For instance, suppose that f, m, a are terms encoding real numbers and consider the statement $f = m \cdot a$. This may be an abstract representation of the relation between force, mass and acceleration for a physical object in the Newtonian mechanics. Recall that in our informal presentation of strict finitism, this statement is a Π_1^0 sentence:

$$\forall n \left(\left| f(n) - m(2kn) \cdot a(2kn) \right| \leq 2/n \right).$$

where k is another term constructed from m and a. Deriving this in strict finitism means deriving the atomic formula

$$\left| f(n) - m(2kn) \cdot a(2kn) \right| \leq 2/n$$

with a free variable n in **SF**. Similarly, using it as a premise to derive another formula $\psi(l)$ means using an instance

$$\left| f(N(l)) - m(2kN(l)) \cdot a(2kN(l)) \right| \leq 2/N(l)$$

of it, according to the numerical interpretation of implication. Either way, only rational approximations to the real numbers f, m, a are involved. Replacing free variables n, l by any numerals, we actually get estimates on how force is approximately close to mass times acceleration, depending on how large n or $N(l)$ is. Some of these estimates can be literally true for real physical quantities, with sufficient but finite precision. (See Sect. 3.7 for a more detailed case study of an example of application.)

3.2 Limit and Continuity

The set of sequences in $\mathbb{R}$, $F(\mathbb{N}, \mathbb{R})$, will also be denoted as $\mathbb{R}^{\mathbb{N}}$. A sequence (a_n) *converges* to y, or $\lim_{n\to\infty} a_n = y$, if

$$\forall k > 0 \exists n \forall m \geq n \left(|a_m - y| < 1/k \right),$$

or equivalently,

$$\exists N \forall k > 0 \forall m \geq N(k) \left(|a_m - y| < 1/k \right).$$

N is a witness for convergence, also called a *modulus of convergence*. ([6], p. 28.) If N is a modulus of convergence for the sequence (a_n), then it is easy to verify that $\left(a_{N(2n)}(2n) \right)_n$ is a real number and is the limit of (a_n). Therefore, $\lim_{n\to\infty} a_n$ can be seen as a term containing N:

$$\lim_{n\to\infty} a_n \equiv_{df} \lambda n.a_{N(2n)}(2n).$$

Note that as a term, $\lim_{n\to\infty} a_n$ contains the modulus of convergence as a subterm, although it is not explicitly shown in the notation.

Similarly, (a_n) is a *Cauchy sequence* if

$$\forall k > 0 \exists n \forall i, j \geq n \left(\left|a_i - a_j\right| < 1/k\right),$$

or equivalently,

$$\exists N \forall k > 0 \forall i, j \geq N(k) \left(\left|a_i - a_j\right| < 1/k\right),$$

and N is a modulus of Cauchyness for (a_n). Then, if N is a modulus of Cauchyness for (a_n), the sequence $(b_k) \equiv \left(\left(a_{N(3k)}\right)_{3k}\right)$ of rational numbers is a real number and is the limit of (a_n). Therefore, a sequence has a limit if and only if it is a Cauchy sequence.

The basic properties of limit can be easily proved. For instance, if $\lim_{n\to\infty} a_n = x$, $\lim_{n\to\infty} b_n = y$, then

$$\lim_{n\to\infty}(a_n + b_n) = x + y, \ \lim_{n\to\infty}(a_n b_n) = xy, \ \lim_{n\to\infty}(|a_n|) = |x|,$$
$$\lim_{n\to\infty} \max(a_n, b_n) = \max(x, y), \ \forall n (a_n \leq b) \to \lim_{n\to\infty} a_n \leq b.$$

Similarly, if $\lim_{n\to\infty} a_n = x$, $x \neq 0$, and $a_n \neq 0$ for all n, then $\lim_{n\to\infty} a_n^{-1} = x^{-1}$. To prove $\lim_{n\to\infty}(a_n + b_n) = x + y$, for instance, we need to construct a modulus of convergence for $(a_n + b_n)$ from any given modulus of convergence for (a_n) and that for (b_n). We omit the details here.

Moreover, consider the last formula above. To derive that $\lim_{n\to\infty} a_n \leq b$, we must show that for any $k > 0$, a sufficient approximation to $\lim_{n\to\infty} a_n$ is less than $b + \frac{1}{k}$. Now, by the construction of the term $\lim_{n\to\infty} a_n$ above, an approximation to $\lim_{n\to\infty} a_n$ is an approximation to some a_n for some sufficiently large n. Therefore, we only need an instance $a_n \leq b$ of the premise $\forall n (a_n \leq b)$ in order to show that the approximation to $\lim_{n\to\infty} a_n$ is less than $b + 1/k$. That is exactly what the numerical implication says.

Suppose that (a_n) is a sequence of real numbers. We want to construct the partial sum $\sum_{i=0}^{n} a_i$. Clearly, to approximate this sum up to $\pm 1/m$ degree of precision, it suffices to use the approximations $a_i((n+1)m)$. So, we let

$$\sum_{i=0}^{n} a_i \equiv_{df} \lambda m. \sum_{i=0}^{n} a_i((n+1)m),$$

where the sum on the right hand side is for rational numbers and can be easily constructed by bounded primitive recursion. Then, some straightforward calculations will verify the basic properties of partial sum. For instance, suppose that $(a_n) \in \mathbb{R}^{\mathbb{N}}$ and $(b_n) \in \mathbb{R}^{\mathbb{N}}$, then

$$\sum_{i=0}^{n} a_i + a_{n+1} = \sum_{i=0}^{n+1} a_i, \ \sum_{i=0}^{n} a_i + \sum_{i=n+1}^{m} a_i = \sum_{i=0}^{m} a_i,$$

$$b\sum_{i=0}^{n} a_i = \sum_{i=0}^{n} ba_i, \ \sum_{i=0}^{n} a_i \pm \sum_{i=0}^{n} b_i = \sum_{i=0}^{n} (a_i \pm b_i),$$

$$\sum_{i=0}^{n} a^i = \frac{a^{n+1}-1}{a-1} \text{ (for } a \neq 1), \quad (\forall i \leq n\,(a_i \leq b_i)) \rightarrow \sum_{i=0}^{n} a_i \leq \sum_{i=0}^{n} b_i.$$

For instance, to prove the first formula, we only need to show that

$$\left(\sum_{i=0}^{n} a_i + a_{n+1}\right)(m) = \sum_{i=0}^{n} a_i\,((n+1)\,2m) + a_{n+1}\,(2m)$$

and

$$\left(\sum_{i=0}^{n+1} a_i\right)(m) = \sum_{i=0}^{n+1} a_i\,((n+2)\,m) = \sum_{i=0}^{n} a_i\,((n+2)\,m) + a_{n+1}\,((n+2)\,m)$$

are arbitrarily close to each other for sufficiently large m. This is trivial.

Note that a finite sum of real numbers is defined directly, not by any recursion on the number of real numbers added. We do not have recursive constructions on real numbers directly, since they are of the type $(o \rightarrow o)$. Similarly, conclusions about finite sum must be proved with the quantifier-free induction. For instance, we have

Lemma 3.5. *If* (a_n), (b_n) *are sequences of real numbers such that* $b_0 = a_0$ *and for all n,* $b_{n+1} = b_n + a_{n+1}$, *then for all n,* $b_n = \sum_{i=0}^{n} a_i$. *Similar conclusions hold when* $=$ *is replaced by* $\leq$ *or* $\geq$.

Proof. This would be trivial if we could use an induction on the equation $b_n = \sum_{i=0}^{n} a_i$ directly, but this is a Π_1^0 formula. However, from $b_{n+1} = b_n + a_{n+1}$ we have for $m > 0$,

$$|b_{n+1}\,(m) - b_n\,(2m) + a_{n+1}\,(2m)| \leq 2/m.$$

Therefore, for $m > 0$,

$$|b_{n+1}\,(2m) - b_n\,(2m) + a_{n+1}\,(2m)| \leq 2/m + 1/m + 1/2m \leq 4/m.$$

These are rational numbers. Then, noting that $|b_0\,(2m) - a_0\,(2m)| \leq 1/m$, a quantifier-free induction (to sum them up) shows that for any $m > 0$,

$$\left| b_n\,(2m) - \sum_{i=0}^{n} a_i\,(2m) \right| \leq \frac{4}{m}n + \frac{1}{m}.$$

From there, it easily follows that $b_n = \sum_{i=0}^{n} a_i$. □

Similarly, the partial product $\prod_{i=0}^{n} a_i$ of a sequence can be defined as

$$\prod_{i=0}^{n} a_i \equiv_{df} \lambda m. \prod_{i=0}^{n} a_i\,(N\,[m,n,a]),$$

where $\prod_{i=0}^{n}$ on the right hand side is the product of rational numbers and can be constructed by bounded primitive recursion, and

$$N[m,n,a] \equiv mn\left(\max_{i\leq n}(|a_i(1)|+1)\right)^n.$$

It can be directly verified that this $N[m,n,a]$ will make $\prod_{i=0}^{n} a_i$ defined above a real number. Basic properties for $\prod_{i=0}^{n} a_i$ similar to the properties for $\sum_{i=0}^{n} a_i$ above can be proved as well. We will omit the details.

Similar to Lemma 3.5 above, we have

Lemma 3.6. *If (a_n), (b_n) are sequences of real numbers such that $b_0 = a_0$ and for all n, $b_{n+1} = rb_n + a_{n+1}$, then for all n, $b_n = \sum_{i=0}^{n} r^{n-i} a_i$. Similar conclusions hold when $=$ is replaced by $\leq$ or $\geq$.*

Proof. Similar to the above, we get an estimate like

$$\left| b_n(tm) - \sum_{i=0}^{n} r(tm)^{n-i} a_i(tm) \right| \leq \frac{s}{m},$$

where t,s are terms depending on r, (a_n), (b_n), n, but not on m. Then, the conclusion follows. □

We will frequently resort to conclusions like these two lemmas implicitly in handling finite sum and finite product of real numbers. Traditionally, they are obtained by simple inductions on equalities or inequalities between real numbers. We see that they can actually be replaced by inductions on similar equalities or inequalities between rational numbers approximating those real numbers. Therefore, they are available to strict finitism. This also means that elementary recursive procedures are essentially enough for these calculations. We do not really need any recursion on functions, which may go beyond elementary recursive procedures.

We define

$$\sum_{i=0}^{\infty} a_i \equiv_{df} \lambda n. \sum_{i=0}^{n} a_i.$$

Therefore, $\sum_{i=0}^{\infty} a_i$ denotes a sequence. We call it a series. The sum of a series $\sum_{i=0}^{\infty} a_i$ can be defined as the limit of the sequence $\lambda n. \sum_{i=0}^{n} a_i$ and is also denoted (ambiguously) by $\sum_{i=0}^{\infty} a_i$:

$$\sum_{i=0}^{\infty} a_i \equiv_{df} \lim_{n\to\infty} \sum_{i=0}^{n} a_i.$$

A modulus of convergence of $\lambda n. \sum_{i=0}^{n} a_i$ is also called a modulus of convergence of the series $\sum_{i=0}^{\infty} a_i$. The equivalence between convergence and Cauchyness implies that the sum exists if and only if

$$\forall k > 0 \exists l \forall m,n > l \left(\left| \sum_{i=m+1}^{n} a_i \right| \leq 1/k \right).$$

Related notions such as the divergence of a series and the absolute convergence of a series are similarly defined.

The basic properties of series can be easily proved. For instance, if $\sum_{i=0}^{\infty} a_i$ and $\sum_{i=0}^{\infty} b_i$ converge, then

$$b\sum_{i=0}^{\infty} a_i = \sum_{i=0}^{\infty} ba_i,\ \sum_{i=0}^{\infty} a_i \pm \sum_{i=0}^{\infty} b_i = \sum_{i=0}^{\infty} (a_i \pm b_i),$$
$$\forall i\,(a_i \leq b_i) \to \sum_{i=0}^{\infty} a_i \leq \sum_{i=0}^{\infty} b_i.$$

These follow from the corresponding properties of finite sum (of real numbers) and limit.

Similarly, we can prove the following: $\sum_{i=1}^{\infty} i^{-1}$ diverges; if $|a| < 1$, then $\sum_{i=0}^{\infty} a^i = \frac{1}{1-a}$; if a series absolutely converges then it converges; if $\sum_{i=0}^{\infty} a_i$ converges and $|b_i| \leq a_i$ for all $i \geq 0$, then $\sum_{i=0}^{\infty} b_i$ converges; if $r < 1$, $N > 0$, and $|a_{n+1}| \leq r\,|a_n|$ for $n \geq N$, then $\sum_{i=0}^{\infty} a_i$ converges. Consider the last one. It suffices to show that $|a_{N+k}| \leq r^k\,|a_N|$ for each k. Now, to derive $|a_{N+k}| \leq r^k\,|a_N|$ from $\forall n \geq N\,(|a_{n+1}| \leq r\,|a_n|)$, we will need an induction. $|a_{N+k}| \leq |a_N|\,r^k$ is a Π_1^0 formula. Therefore, the quantifier-free induction cannot be applied directly. However, we can reduce it to a quantifier-free induction. First, we can choose r to be a rational number. Then, the assumption $\forall n \geq N\,(|a_{n+1}| \leq r\,|a_n|)$ implies that for each k, m,

$$|a_{N+k+1}|\,(m) \leq r\,(|a_{N+k}|\,(m)) + 2/m.$$

This is a quantifier-free formula. Then, by an induction, we have

$$\begin{aligned} |a_{N+k}|\,(m) &\leq r^k\,(|a_N|\,(m)) + \left(r^{k-1} + \ldots + 1\right)\frac{2}{m} \\ &\leq r^k\,(|a_N|\,(m)) + \frac{2}{(1-r)\,m} \end{aligned}$$

for each m. This implies that $|a_{N+k}| \leq r^k\,|a_N|$.

Now, consider continuity. A function $f : [a,b] \to \mathbb{R}$ is *continuous* if

$$\exists\omega\forall x,y \in [a,b]\,\forall n > 0\,(|x-y| \leq \omega\,(n) \to |f\,(x) - f\,(y)| \leq 1/n).$$

ω is a witness for continuity, called a *modulus of continuity*. ([6], p. 38.) Here, ω is assumed to have the type $(o \to o)$, that is, $\omega\,(n)$ is a rational number. We sometimes also assume that ω operates on small rational numbers ε. Then, the condition becomes $|f\,(x) - f\,(y)| \leq \varepsilon$ whenever $|x-y| \leq \omega\,(\varepsilon)$. $C\,([a,b]\,,\mathbb{R})$ denotes the set of such continuous functions. Here are some other sets of continuous functions:

$$(f \in C\,((a,b)\,,\mathbb{R})) \equiv_{df} (f : (a,b) \to \mathbb{R}) \wedge \forall c,d \in (a,b)\,(c \leq d \to f \in C\,([c,d]\,,\mathbb{R})),$$
$$(f \in C\,(\mathbb{R},\mathbb{R})) \equiv_{df} (f : \mathbb{R} \to \mathbb{R}) \wedge \forall c,d \in \mathbb{R}\,(c \leq d \to f \in C\,([c,d]\,,\mathbb{R})).$$

The definition implies that if $f \in C([a,b],\mathbb{R})$, then the set $\{f(x) : x \in [a,b]\}$ is totally bounded. Therefore,

$$\sup_{a \le x \le b}(f) \equiv_{df} \sup\{f(x) : x \in [a,b]\},$$
$$\inf_{a \le x \le b}(f) \equiv_{df} \inf\{f(x) : x \in [a,b]\}$$

exist. It is easy to show that if $f,g \in C([a,b],\mathbb{R})$, then $f+g$, fg, $\max(f,g) \in C([a,b],\mathbb{R})$. For instance, to show that $f+g \in C([a,b],\mathbb{R})$, we need to construct a modulus of continuity for $f+g$ from that for f and that for g. Similarly, if $f \in C([a,b],\mathbb{R})$ and $|f(x)| \ge c > 0$ for all $x \in [a,b]$, then $f^{-1} \in C([a,b],\mathbb{R})$. It is also easy to show that a composition of continuous functions is still continuous.

We have the intermediate value theorem in the following format. ([6], p. 40)

Theorem 3.7. *If $f \in C([a,b],\mathbb{R})$ and $f(a) < f(b)$, then for any y such that $f(a) \le y \le f(b)$, and for any $\varepsilon > 0$, there exists $x \in [a,b]$ such that $|f(x) - y| < \varepsilon$. Moreover, if f is strictly increasing, then there exists $x \in [a,b]$ such that $f(x) = y$.*

Proof. We may assume that ε is a rational number. Divide $[a,b]$ into small intervals $p_0 = a$, ..., $p_N = b$ such that p_1, ..., p_{N-1} are rational numbers and $p_{i+1} - p_i \le \omega(\varepsilon/7)$, where ω is a modulus of continuity for f. Therefore, $|f(p_{i+1}) - f(p_i)| < \varepsilon/6$. For each p_i, we can choose a rational number q_i such that $|f(p_i) - q_i| < \varepsilon/6$, and let q be a rational such that $|y - q| < \varepsilon/6$. Then, $|q_{i+1} - q_i| < \varepsilon/2$. Since $f(a) \le y \le f(b)$, we have $q_0 - \varepsilon/3 \le q \le q_N + \varepsilon/3$. By comparing q with q_0, ..., q_N, we can find q_k such that $|q_k - q| < \varepsilon/2$. Since $|f(p_k) - q_k| < \varepsilon/6$ and $|y - q| < \varepsilon/6$, we have $|f(p_k) - y| < \varepsilon$. For the second half, note that f being strictly increasing implies that we have a term t so that for rational numbers $p,q \in [a,b]$, if $p < q$, then $t(p,q)$ is a positive rational number such that $f(q) - f(p) > t(p,q)$. That is, $t(p,q)$ witnesses that $f(p) < f(q)$ for $p < q$. Then, to estimate the x such that $f(x) = y$ up to the precision $1/k$, we can divide $[a,b]$ into small rational intervals each of length $< 1/2k$, and then for each interval $[p_i, p_{i+1}]$, we can approach y up to the precision of $t(p_i, p_{i+1})/2$ to decide if $f(p_i) < y$ or $y < f(p_{i+1})$. The estimate of x will be p_{i+1} with p_i the last one such that $f(p_i) < y$. □

Corollary 3.8. *If $f \in C([a,b],\mathbb{R})$ and f is strictly increasing, then there exists the inverse function $g \in C([f(a), f(b)],\mathbb{R})$, such that $f(g(y)) = y$ for $y \in [f(a), f(b)]$ and $g(f(x)) = x$ for $x \in [a,b]$.*

Proof. By the theorem and the axiom of choice, we can construct g such that $g : [f(a), f(b)] \to \mathbb{R}$ and $f(g(y)) = y$ for $y \in [f(a), f(b)]$. For $x \in [a,b]$, since f is strictly increasing, $f(x) \in [f(a), f(b)]$. Let $x' = g(f(x))$. Then, $f(x') = f(x)$. Since f is strictly increasing, we have $\neg x < x'$ and $\neg x' < x$. Therefore, $x = x'$. That is, $g(f(x)) = x$. Finally, to see that g is continuous, let t be the term in the proof of the theorem above. That is, for rational numbers p,q, $t(p,q)$ is a rational number such that whenever $p,q \in [a,b]$ and $p < q$, we have $f(q) - f(p) > t(p,q) > 0$. For any $n > 0$, divide $[a,b]$ into small intervals $p_0 = a$, ..., $p_N = b$ such that p_1, ..., p_{N-1} are rational numbers and $p_{i+1} - p_i < 1/4n$ for $i = 0, ..., N-1$. Let ε be a positive rational number such that $\varepsilon < \min_{i=1,...,N-1} t(p_i, p_{i+1})$, $\varepsilon < f(p_1) - f(a)$, and $\varepsilon < f(b) - f(p_{N-1})$. Now, suppose that $y, y' \in [f(a), f(b)]$, $|y - y'| < \varepsilon/2$, $y = f(x)$, and $y' = f(x')$. By the continuity of f, we can choose rational numbers q, q' approximating x, x' sufficiently, so that $|x - q| < 1/4n$, $|x' - q'| < 1/4n$ and $|f(q) - f(q')| < \varepsilon$. This means that q, q' must fall in the same interval $[p_{i-1}, p_{i+1}]$ for some i. That is, $|q - q'| < 1/2n$. Therefore, $|x - x'| < 1/n$. That is, $|g(y) - g(y')| < 1/n$. □

For I an interval, a sequence of functions in $C(I, \mathbb{R})$ is a function from $\mathbb{N}$ to $C(I, \mathbb{R})$. We will use (f_n) to denote such a sequence. (f_n) *converges (uniformly)* on I to another function $g \in C(I, \mathbb{R})$, denoted as $\lim_{n\to\infty} f_n = g$, if

$$\exists N \forall k > 0 \forall m \geq N(k) \forall x \in I \, |f_m(x) - g(x)| < 1/k.$$

Similarly, (f_n) is a *(uniform) Cauchy* sequence on I if

$$\exists N \forall k > 0 \forall m, n \geq N(k) \forall x \in I \, |f_m(x) - f_n(x)| < 1/k.$$

We are only interested in *uniform* convergence or Cauchyness, not point-wise convergence or Cauchyness.

It is easy to see that if (f_n) converges on I, then it is a Cauchy sequence. Now, suppose that (f_n) is a Cauchy sequence on I. Let g be defined such that $g(x)$ is the sequence $\left(\left(f_{N(3k)}(x)\right)_{3k}\right)_k$ where N is a modulus of Cauchyness for (f_n). Then, it is easy to see that $x \in I$ implies that $g(x) \in \mathbb{R}$. Since N does not depend on x, it is also easy to see that (f_n) converges on I to $g(x)$. To see that g is continuous, note that

$$|g(y) - g(x)| \leq |g(y) - f_m(y)| + |f_m(y) - f_m(x)| + |f_m(x) - g(x)|.$$

Therefore, given $k > 0$, we can first use the modulus of convergence for the sequence (f_n) to choose m such that $|g(z) - f_m(z)| < 1/3k$ for all $z \in I$. Then, we can use the modulus of continuity for f_m to get $\omega(k)$ such that $|f_m(y) - f_m(x)| < 1/3k$ and thus $|g(y) - g(x)| < 1/k$ when $|y - x| < \omega(k)$. This also implies that in general, if (f_n) is a sequence of functions in $C(I, \mathbb{R})$ and if (f_n) converges (uniformly) on I to a function g, then g must also be continuous. A modulus of continuity for g can be constructed from a modulus of convergence for (f_n) and a modulus of continuity for f_n.

Given a sequence of functions (f_n) in $C(I, \mathbb{R})$, the corresponding series is defined as the sequence

$$(g_n) \equiv_{df} \lambda n. \lambda x. \sum_{i=0}^{n} f_i(x).$$

We will use $\sum_{i=0}^{\infty} f_i$ to denote both the series and the limit $\lim_{n\to\infty} g_n$. We must prove that $g_n \equiv \lambda x. \sum_{i=0}^{n} f_i(x) \in C(I, \mathbb{R})$. The assumption $\forall n (f_n \in C(I, \mathbb{R}))$ implies that for some ω, $\omega(n)$ is a modulus of continuity for f_n. By a bounded primitive recursion, we can construct a term r such that

$$r[n](m) = \min\{\omega(0)((n+1)m), \ldots, \omega(n)((n+1)m)\}.$$

Then, it is easy to verify that $r[n]$ is a modulus of continuity for g_n.

The comparison test and ratio test for convergence hold for series of functions. That is, if $|f_n(x)| \le g_n(x)$ for all n and $x \in I$, and if $\sum_{n=0}^{\infty} g_n$ converges, then $\sum_{n=0}^{\infty} f_n$ converges; if $|f_{n+1}(x)| \le r|f_n(x)|$ for all n and $x \in I$, where $r < 1$, then $\sum_{n=0}^{\infty} f_n$ converges. Proofs for the same results for series of real numbers can be directly adapted here, since we consider only uniform convergence for series of functions.

Power series $\sum_{i=0}^{\infty} a_n (x - x_0)^i$ are defined naturally. Then, by the ratio test, if there exists $r > 0$ such that $|a_{n+1}| \le r^{-1}|a_n|$ for all n, then the convergence radius of $\sum_{i=0}^{\infty} a_n (x - x_0)^i$ is at least r.

There is no straightforward way of constructing a discontinuous function. Normally, for a function f, the real number $f(x)$ is to be approximated by operations on the estimates of x up to some precision. That is, $f(x)(m)$ normally takes the form $t[x(s[m])]$ for some terms $t[p]$ and $s[m]$. Then, for the sequence $\lambda m.t[x(s[m])]$ to be a real number, $t[x(s[m])]$ and $t[x(s[n])]$ have to be close for large m,n. This is *normally* achieved by making $t[p]$ 'continuous' as a function of rational numbers. That is, for rational numbers p and q that are close to each other, $t[p]$ and $t[q]$ are also close. It then means that $f(x)$ will be a continuous function of x, because when x and y are close, $x(s[m])$ and $y(s[m])$ must also be close, and then $t[x(s[m])]$ and $t[y(s[m])]$ will be close. This is not a rigorous proof, but it does show that the normal ways of defining a function always give continuous functions. It reflects the fact that we normally estimate the values of such a function by operations on the estimates of its argument.

Note that the step function

$$f(x) = \begin{cases} 0, & \text{for } x \in [0,1), \\ 1, & \text{for } x \in [1,2] \end{cases}$$

is a function on $[0,1) \cup [1,2]$, which is a subset of $[0,2]$ but not equal to $[0,2]$. For this function, the value $f(1)$ is not approximated by operating on an arbitrary sequence of rational numbers approaching to 1. Therefore, this is not a counter-example to the above informal argument. This also implies that it is very natural to consider partial functions defined on the *subsets* of intervals such as $[0,1) \cup [1,2]$. We will consider that in Lebesgue integration theory.

In applications, a function f may represent a distribution of some physical quantity, which must be discrete for finite and discrete phenomena above the Planck scale. For instance, f may represent the population on the Earth. Given a modulus of continuity ω, an instance of the continuity condition is

$$|t - t'| < \omega(n) \to |f(t) - f(t')| < 1/n. \tag{3.2}$$

Choosing an appropriate unit for population (e.g. billion), this can be literally true for the population on the Earth at any two moments t, t', as long as n is not too large. It reflects the fact that a discrete physical quantity can 'look continuous macroscopically'.

Then, recall that in strict finitism, according to the numerical interpretation of implication, when we use a universally quantified statement $\forall t \forall t' \forall n \varphi$ as a premise in deriving a conclusion, the proof really depends only on an instance of the universally quantified statement. It means that the idealized continuity condition is not strictly indispensable. For deriving our conclusion about the population, we need only instances like (3.2) above, which can be true of discrete population values. The proof will contain a construction of the required instances, namely, the required t, t', n. Then, by examining the elementary recursive function $\omega(n)$ more closely, we can check if the required instances like (3.2) are indeed literally true of those discrete population values. This will allow us to demonstrate that our conclusion about the population depends only on literally true premises about discrete population values, not on the idealized continuity condition.

3.3 Differentiation and Integration

For $I = [a,b]$ a compact interval, g is a derivative of f on I, if $g, f \in C(I, \mathbb{R})$, and there exists δ, such that $\delta(n) > 0$ for all $n > 0$, and

$$|f(y) - f(x) - g(x)(y-x)| \leq |y-x|/n$$

for all $x, y \in I$, $|x-y| \leq \delta(n)$. We will use the notations $g = f'$ and $g(x) = \mathrm{d}f(x)/\mathrm{d}x$. δ is called a *modulus of differentiability* for $g = f'$. f is *differentiable*, if there exists g such that $f' = g$. ([6], p. 44.)

Suppose that δ is a modulus of differentiability for $f' = g$ on $I = [a,b]$. Given $x \in I$, for n sufficiently large such that $\delta(2n) < (b-a)/4$, by deciding if $x < a + \frac{3}{4}(b-a)$ or $x > a + \frac{1}{4}(b-a)$, we can decide if $x + \delta(2n) \in I$ or $x - \delta(2n) \in I$. Let $t[x,a,b]$ be a term such that $t[x,a,b] = 1$ in the former case and $t[x,a,b] = -1$ in the latter case. Then, $x + t\delta(2n) \in I$ and

$$\left| \frac{f(x + t\delta(2n)) - f(x)}{t\delta(2n)} - g(x) \right| \leq 1/2n.$$

Therefore, we can approximate $g(x)$ up to $\pm\frac{1}{n}$ by approximating $\frac{f(x+\delta(2n)) - f(x)}{\delta(2n)}$ sufficiently. That is, we can construct f' as a term containing the putative modulus of differentiability. More specifically, define Df such that for n sufficiently large,

$$Df(x)(n) = \frac{f(x + t\delta(2n)) - f(x)}{t\delta(2n)}(2n).$$

Then, 'δ is a modulus of differentiability for $f' = g$' implies that for $x \in I$,

$$|Df(x)(n) - g(x)| \leq 1/n.$$

That is, $Df(x)$ is a real number and $g(x) = Df(x)$. Therefore, f is differentiable with δ as a modulus of differentiability, if and only if δ is a modulus of differentiability for $Df = f'$.

The basic properties of derivative are easily proved. For instance, on an appropriate interval, we have

$$(\lambda x.c)' = 0,\ (x^n)' = nx^{n-1}$$
$$(f+g)' = f' + g',\ \left(\sum_{i=0}^{n} f_i(x)\right)' = \sum_{i=0}^{n} f_i'(x),$$
$$(fg)' = f'g + fg',$$
$$(f \circ g)'(x) = f'(g(x))\, g'(x).$$

These can be directly verified from the definition.

Rolle's theorem and the mean value theorem must take an approximation format. ([6], pp. 47–48.)

Theorem 3.9. *If f is differentiable on $[a,b]$, $a < b$, and $f(a) = f(b)$, then for every $n > 0$, there exists $x \in [a,b]$ such that $|f'(x)| \le 1/n$; if f is differentiable on $[a,b]$, then for every $n > 0$, there exists $x \in [a,b]$ such that*

$$\left|f(b) - f(a) - f'(x)(b-a)\right| \le 1/n.$$

Proof. We prove the first half of the theorem. Let δ be a modulus of differentiability for f. Since f' is continuous, given n, we can divide $[a,b]$ into intervals $p_0 = a$, ..., $p_N = b$ such that $|f'(y) - f'(p_i)| \le 1/4n$ for $y \in [p_{i-1}, p_{i+1}]$, $i = 1$, ..., $N-1$, and such that $p_{i+1} - p_i < \delta(4n)$ for $i = 0, \ldots, N-1$. For each $f'(p_i)$, we can decide if $|f'(p_i)| < 1/n$ or $|f'(p_i)| > 1/2n$. Therefore, we can let

$$k \equiv \mu i \le N\left(\left|f'(p_i)\right| < 1/n\right),\ x \equiv p_k.$$

We will show that there exists $i \le N$ such that $|f'(p_i)| < 1/n$. It then follows that $|f'(x)| \le 1/n$. Since $|f'(p_i)| < 1/n$ or $|f'(p_i)| > 1/2n$ is decidable, we can show this by proving that $|f'(p_i)| > 1/2n$ for all $i \le N$ will lead to a contradiction. Therefore, suppose that $|f'(p_i)| > 1/2n$ for all $i \le N$. It means that we have m_i witnessing $f'(p_i) > 1/2n$ or $f'(p_i) < -1/2n$ for $i \le N$. Suppose that $f'(p_0) > 1/2n$. Note that 'm_i witnesses $f'(p_i) > 1/2n$' is quantifier-free. Therefore, we can use an induction to show that m_i witnesses $f'(p_i) > 1/2n$ for all $i \le N$, since $|f'(p_{i+1}) - f'(p_i)| \le 1/4n$ for $i = 0$, ..., $N-1$. Similarly, in case $f'(p_0) < -1/2n$, we have $f'(p_i) < -1/2n$ for all $i \le N$. Therefore, either $f'(p_i) > 1/2n$ for all $i \le N$, or $f'(p_i) < -1/2n$ for all $i \le N$. Both contradict the condition $f(a) = f(b)$. For instance, suppose that $f'(p_i) > 1/2n$ for all $i \le N$. Then, since $p_{i+1} - p_i < \delta(4n)$, we have

$$f(p_{i+1}) - f(p_i) \geq f'(p_i)(p_{i+1} - p_i) - \frac{1}{4n}(p_{i+1} - p_i)$$
$$\geq \frac{1}{4n}(p_{i+1} - p_i)$$

for $i < N$. Therefore, $f(b) \geq f(a) + (b-a)/4n$, which contradicts $f(a) = f(b)$.

The second half of the theorem follows from the first half in the same way as in the classical case. □

As a corollary, we have

Corollary 3.10. *Suppose that f is differentiable on an interval I. f is increasing if and only if $f'(x) \geq 0$ on the interval I. Moreover, if $f'(x) = 0$ on the interval I, then for some constant c, $f(x) = c$ for $x \in I$.*

For n a variable, g is the n-th derivative of f if

$$g \in C(I, \mathbb{R}) \wedge f \in C(I, \mathbb{R}) \wedge \exists F\left(F_0 = f \wedge \forall i < n\left(F_{i+1} = F_i'\right) \wedge g = F_n\right),$$

where F is intuitively the sequence $f = f^{(0)}, f^{(1)}, \ldots, f^{(n)}$. We also use the notation $g = f^{(n)}$. Note that F is a witness for $g = f^{(n)}$ and it contains all derivatives $f^{(m)}$ for $m < n$. $C^n(I, \mathbb{R})$ denotes the set of functions on I whose n-th derivative exists.

Note that $\forall n(f \in C^n(I, \mathbb{R}))$, or $f \in C^\infty(I, \mathbb{R})$, implies $\exists G \forall n\left(G_n = f^{(n)}\right)$. G is a witness for $f \in C^\infty(I, \mathbb{R})$. Then, the Taylor series $\sum_{n=0}^{\infty} \frac{f^{(n)}(a)}{n!}(x-a)^n$ can be constructed, using the witness G. Then, we can prove that for each n and $\varepsilon > 0$, there exists ζ, $\min(a, x) \leq \zeta \leq \max(a, x)$, such that

$$\left| f(x) - \sum_{i=0}^{n} \frac{f^{(i)}(a)}{i!}(x-a)^i - \frac{f^{(n+1)}(\zeta)}{n!}(x-\zeta)^n(x-a) \right| \leq \varepsilon.$$

The proof of this involves only straightforward calculations of derivatives and an application of Rolle's theorem above. See [6], p. 49. From this, it follows that

$$f(x) = \sum_{n=0}^{\infty} \frac{f^{(n)}(a)}{n!}(x-a)^n$$

on the interval $[a-r, a+r]$ if $\frac{r^n f^{(n+1)}(x)}{n!} \to 0$ on the interval.

Now, consider Riemann integration. A finite sequence of real numbers $P = (a_0, \ldots, a_n)$ is a partition of an interval $I = [a, b]$ if $a = a_0 \leq a_1 \leq \ldots \leq a_n = b$. Define the Riemann sum

$$S(f, P) \equiv_{df} \sum_{i=0}^{n-1} f(a_i)(a_{i+1} - a_i).$$

To define the integration, we choose a sequence of standard partitions (P_n),

$$P_n \equiv \left(a, \ldots, a + i\frac{b-a}{2^n}, \ldots, b\right).$$

It can be proved that if $f \in C(I,\mathbb{R})$ then $(S(f,P_n))_n$ is a Cauchy sequence. Therefore, we let

$$\int_a^b f(x)\,\mathrm{d}x \equiv_{df} \lim_{n\to\infty} S(f,P_n).$$

The construction of the limit on the right hand side actually depends on a modulus of Cauchyness for $(S(f,P_n))_n$, which in turn depends on a given modulus of continuity ω for f on $[a,b]$. So we should write the limit as $T[f,\omega,a,b]$. It can be shown that if ω' is also a modulus of continuity for f on $[a,b]$ then $T[f,\omega,a,b] =_{\mathbb{R}} T[f,\omega',a,b]$. So we can simply use the notation $\int_a^b f(x)\,\mathrm{d}x$. ([6], pp. 50–51.)

We use a special type of partitions, standard partitions, to define $\int_a^b f(x)\,\mathrm{d}x$. The following lemma shows that $\int_a^b f(x)\,\mathrm{d}x$ is actually independent of the choice.

Lemma 3.11. *Suppose that ω is a modulus of continuity for f, $n>0$, and $P = (a_0,\dots,a_m)$ is any partition such that* $\max_{i<m}(a_{i+1}-a_i) < \omega(2n)$. *Then, we have*

$$\left|S(f,P) - \int_a^b f(x)\,\mathrm{d}x\right| \le (b-a)/n.$$

Proof. We can choose k very large so that for each a_i, $i \le m$, there is a point $a'_{J(i)}$ in the standard partition $P_k \equiv (a'_0, a'_1, \dots)$ arbitrarily close to a_i. With that, we can make

$$\left| f(a_i)(a_{i+1}-a_i) - \sum_{j=J(i)}^{J(i+1)-1} f(a'_j)\left(a'_{j+1}-a'_j\right)\right| \le (a_{i+1}-a_i)/n + \varepsilon/m$$

for an arbitrarily small ε. Then, we have

$$|S(f,P) - S(f,P_k)| \le (b-a)/n + \varepsilon.$$

The result then follows. □

For arbitrary $a,b \in \mathbb{R}$, we define

$$\int_a^b f(x)\,\mathrm{d}x \equiv_{df} \int_c^b f(x)\,\mathrm{d}x - \int_c^a f(x)\,\mathrm{d}x,$$

where $c \equiv \min(a,b)$. When $a \le b$, this is equivalent to the above case.

The basic properties of integration can be easily verified, for instance,

$$\int_a^b cf(x)\,\mathrm{d}x = c\int_a^b f(x)\,\mathrm{d}x,$$

$$\int_a^b \sum_{i\le n} f_i(x)\,\mathrm{d}x = \sum_{i\le n}\int_a^b f_i(x)\,\mathrm{d}x,$$

$$(a \le b) \to \inf_{a\le x\le b}(f)(b-a) \le \int_a^b f(x)\,\mathrm{d}x \le \sup_{a\le x\le b}(f)(b-a),$$

$$\int_a^b |f(x)|\,\mathrm{d}x = 0 \to \forall x \in [a,b]\,(f(x) = 0)\,.$$

The first two directly follow from the definition, since the partial sum $S(f, P_n)$ is linear in f. The last two also follow from the definition directly. Moreover, we have

$$\int_a^b f(x)\,\mathrm{d}x = -\int_b^a f(x)\,\mathrm{d}x,$$
$$\int_a^c f(x)\,\mathrm{d}x = \int_a^b f(x)\,\mathrm{d}x + \int_b^c f(x)\,\mathrm{d}x$$

for any a, b, c.

The fundamental theorem of calculus holds. That is, suppose that f is continuous on some interval I, $a \in I$, and define $g(x) \equiv \int_a^x f(t)\,\mathrm{d}t$ for $x \in I$. Then, g is continuous on I and $g' = f$ on I. To see this, note that

$$g(y) - g(x) - f(x)(y - x) = \int_x^y f(t)\,\mathrm{d}t - f(x)(y - x)\,.$$

Then, by Lemma 3.11 above, we have

$$|g(y) - g(x) - f(x)(y - x)| \leq |y - x|/n$$

whenever $|y - x| \leq \omega(2n)$, where ω is a modulus of continuity for f.

Conversely, suppose that f is differentiable on I. Then,

$$\int_a^b f'(x)\,\mathrm{d}x = f(b) - f(a)\,.$$

To see this, note that $\left(\int_a^x f'(t)\,\mathrm{d}t\right)' = f'(x) = (f(x) - f(a))'$. Therefore, there exists a constant c such that $\int_a^x f'(t)\,\mathrm{d}t - f(x) - f(a) = c$ for $x \in I$. Let $x = a$. We see that $c = 0$.

From the fundamental theorem of calculus, it follows that if $\int_a^x f(t)\,\mathrm{d}t = 0$ for all $x \in I$, then $f(x) = 0$ on I.

Suppose that (f_n) is a sequence of continuous functions on $I \equiv [a,b]$ and suppose that it uniformly converges. Then, it is easy to see that

$$\lim_{n\to\infty} \int_a^b f_n(t)\,\mathrm{d}t = \int_a^b \lim_{n\to\infty} f_n(t)\,\mathrm{d}t\,.$$

Moreover, if f_n' exists for all n, and the sequences (f_n) and (f_n') uniformly converge to continuous functions on $I = [a,b]$, then

$$\left(\lim_{n\to\infty} f_n(x)\right)' = \lim_{n\to\infty} f_n'(x)\,.$$

To see this, note that for $x \in [a,b]$, by the fundamental theorem of calculus,

$$\int_a^x \left(\lim_{n\to\infty} f_n(t)\right)' \mathrm{d}t = \lim_{n\to\infty} f_n(x) - \lim_{n\to\infty} f_n(a).$$

On the other side, by the integration of limit and the fundamental theorem of calculus,

$$\int_a^x \lim_{n\to\infty} f_n'(t)\,\mathrm{d}t = \lim_{n\to\infty}\int_a^x f_n'(t)\,\mathrm{d}t = \lim_{n\to\infty}(f_n(x) - f(a)).$$

Therefore,

$$\int_a^x \left(\left(\lim_{n\to\infty} f_n(t)\right)' - \lim_{n\to\infty} f_n'(t)\right)\mathrm{d}t = 0$$

for all $x \in [a,b]$. The conclusion then follows.

$g = f'$ says that whenever $|x-y| \le \delta(n)$, we have

$$|f(y) - f(x) - g(x)(y-x)| \le |y-x|/n,$$

or

$$\left|\frac{f(y)-f(x)}{y-x} - g(x)\right| \le \frac{1}{n}$$

whenever $|x-y| \le \delta(n)$ and $x \neq y$. Again, this could be translated into literally true assertions about some finite and discrete physical quantities represented by f, g. For instance, suppose that f represents the population on the Earth, and suppose that $g(x)$ represents the population growth rate at the moment x, determined by measuring the population after a short period of time since the moment x. Although population literally grows in discrete jumps, as long as the short period of time for measuring the growth rate at a moment is not too short, growth rates are smooth at the macro-scale. Then, the inequality above can be translated into literally true assertions about the population if $|y-x|$ is not too small and n is not too large. Therefore, similar to the continuity condition, the instances of a differentiability condition that are actually required for deriving a conclusion in strict finitism can potentially be translated into literally true realistic assertions about discrete physical quantities. An instance of the differentiability condition, with x, y, n in the above inequality instantiated by concrete real numbers and a numeral, says only that the growth rate of $f(x)$ at x is approximately $g(x)$.

Also note that a Riemann sum $S(g, P_m)$ of the derivative g on an interval $[a,x]$ recovers f's total growth in that interval, with local growth rates evaluated at the partition points. This is seen from the definition of differentiation:

$$|f(a_{i+1}) - f(a_i) - g(a_i)(a_{i+1} - a_i)| \le |a_{i+1} - a_i|/n,$$

assuming that the partition is sufficiently fine. Adding them up, we have

$$|f(x) - f(a) - S(g, P_m)| \le (x-a)/n.$$

That is the fundamental theorem of calculus, $\int_a^x f'(y)\,\mathrm{d}y = f(x) - f(a)$. When translated into a realistic assertion about a discrete quantity, this means that the Riemann sum of discrete increments on small sub-intervals (obtained by the function representing growth rates) approximately recovers the total growth on the entire interval. That is the finitistic meaning of the fundamental theorem of calculus. It can be literally true for a discrete quantity as well, as long as that discrete quantity appears smooth at the macro-scale.

3.4 Certain Important Functions

Functions e^x, $\ln x$, a^x, $\sin x$, and $\cos x$ are defined as limits of series or integrations. Here we follow [6], pp. 55–58.

$$\mathrm{e}^x \equiv \sum_{n=0}^{\infty} \frac{x^n}{n!}, \ \ln(x) \equiv \int_1^x t^{-1} \mathrm{d}t, \ a^x \equiv \mathrm{e}^{x\ln(a)},$$

$$\cos x \equiv \sum_{n=0}^{\infty} (-1)^n \frac{x^{2n}}{(2n)!}, \ \sin x \equiv \sum_{n=0}^{\infty} (-1)^n \frac{x^{2n+1}}{(2n+1)!}.$$

The convergence of the power series for e^x, $\cos x$, and $\sin x$, and the existence of the integration for $\ln x$ are obvious.

The basic properties of these functions are easily proved. For instance, we have

$$\frac{\mathrm{d}}{\mathrm{d}x}\mathrm{e}^x = \mathrm{e}^x, \ \mathrm{e}^{x+y} = \mathrm{e}^x \mathrm{e}^y,$$

$$\frac{\mathrm{d}}{\mathrm{d}x}\ln x = x^{-1}, \ \ln xy = \ln x + \ln y,$$

$$\mathrm{e}^{\ln x} = \ln \mathrm{e}^x = x,$$

$$\frac{\mathrm{d}}{\mathrm{d}x}\cos x = -\sin x, \ \frac{\mathrm{d}}{\mathrm{d}x}\sin x = \cos x,$$

$$\sin(x+y) = \sin x \cos y + \cos x \sin y,$$

$$\cos(x+y) = \cos x \cos y - \sin x \sin y.$$

The first follows from taking derivative on a series. To see the second, note that $\frac{\mathrm{d}}{\mathrm{d}x}\left(\mathrm{e}^{x+y}/\mathrm{e}^x\right) = 0$ and $\mathrm{e}^{0+y}/\mathrm{e}^0 = \mathrm{e}^y$. The third follows from the fundamental theorem of calculus. To see the fifth, note that $\frac{\mathrm{d}}{\mathrm{d}x}\left(\mathrm{e}^{\ln x}/x\right) = 0$ and $\frac{\mathrm{d}}{\mathrm{d}x}\left(\ln \mathrm{e}^x\right) = 1$. The sixth and the seventh directly follow from taking derivative on a series. To see the last two equations, let

$$f(x) \equiv \sin(x+y) - (\sin x \cos y + \cos x \sin y).$$

Then, $f''(x) = -f(x)$ for all x. Therefore, $f \in C^\infty(I, \mathbb{R})$ for any interval I containing 0. Moreover, $f^{(n)}(0) = 0$ for any n. Using the Taylor series expansion for f, we see that $f(x) = 0$ on any interval I containing 0.

The zero of cos in the interval $[0,2]$ can be estimated as follows. First, by some simple calculations on the initial terms in $\cos x$ and $\sin x$, it can be estimated that $\cos 0 = 1$, $\cos 2 < -\frac{1}{3}$, and $\sin x > 0$ for $x \in (0,2)$. Since $\frac{\mathrm{d}}{\mathrm{d}x}\cos x = -\sin x$, $\cos x$ is strictly decreasing in $(0,2)$. Then, by the intermediate value theorem for monotonic continuous functions, the zero of cos in $[0,2]$ can be estimated. (Note that this simplifies the proof of this proposition on [6], pp. 59–60.) It is denoted by $\pi/2$. Then, it is trivial to show that sin is strictly increasing on $(-\pi/2, \pi/2)$, so arcsin can be defined on $(-\pi/2, \pi/2)$.

3.5 Functions of Several Variables

An n-tuple of real numbers is a finite sequence of real numbers (of the length n). $\mathbb{R}^n$ denotes the set of n-tuples of real numbers. For $\mathbf{x} = (x_1, ..., x_n) \in \mathbb{R}^n$, we define its norm

$$|\mathbf{x}| = \left(x_1^2 + ... + x_n^2\right)^{\frac{1}{2}}.$$

An open ball $S(\mathbf{x}, r)$ is a subset in the format

$$S(\mathbf{x}, r) = \{\mathbf{z} \in \mathbb{R}^n : |\mathbf{z} - \mathbf{x}| < r\}.$$

We can quantify over all open balls of $\mathbb{R}^n$ (although we cannot quantify over all subsets of $\mathbb{R}^n$), by which we mean a quantification over the centers and the radii of those open balls. Closed balls $Sc(\mathbf{x}, r)$ are defined similarly,

$$Sc(\mathbf{x}, r) = \{\mathbf{z} \in \mathbb{R}^n : |\mathbf{z} - \mathbf{x}| \leq r\}.$$

We will mostly treat balls in this section, but most claims can be generalized to subsets of $\mathbb{R}^n$ of other regular shapes, such as n-dimensional cubes, polyhedrons, ellipsoids and so on.

We can treat $\mathbb{R}^n$ as a vector space with an inner product. That is, for $\mathbf{x} = (x_1, ..., x_n), \mathbf{y} = (y_1, ..., y_n) \in \mathbb{R}^n$ and $a \in \mathbb{R}$, we define

$$\begin{aligned} \mathbf{x} + \mathbf{y} &= (x_1 + y_1, ..., x_n + y_n), \\ a\mathbf{x} &= (ax_1, ..., ax_n), \\ \mathbf{x} \cdot \mathbf{y} &= x_1 y_1 + ... x_n y_n. \end{aligned}$$

It is easy to construct these as terms in our language of strict finitism. Finite linear combinations $\sum_{j=1}^{k} a_j \mathbf{x}_j$ of vectors in $\mathbb{R}^n$ can be defined similarly. For $i = 1, ..., n$, let $\mathbf{e}_i^{(n)} \in \mathbb{R}^n$ be the vector with 1 as the i-th component and 0 as all other components. We will omit the superscript (n) when there is no ambiguity in the context. Apparently, $\mathbf{e}_1, ..., \mathbf{e}_n$ form a basis of $\mathbb{R}^n$. That is,

$$\mathbf{x} = (x_1, ..., x_n) = \sum_{i=1}^{n} x_i \mathbf{e}_i.$$

A unit vector $\mathbf{x}$ is one such that $|\mathbf{x}| = 1$. A linear transformation $\mathbf{A} : \mathbb{R}^n \to \mathbb{R}^m$ is a function such that for any finite sequence $\mathbf{x}_1, ..., \mathbf{x}_k$ of vectors in $\mathbb{R}^n$ and any sequence $r_1, ..., r_k$ of real numbers,

$$\mathbf{A}\left(\sum_{j=1}^{k} r_j \mathbf{x}_j\right) = \sum_{j=1}^{k} r_j \mathbf{A}\mathbf{x}_j.$$

$L(\mathbb{R}^n, \mathbb{R}^m)$ denotes the set of linear transformations from $\mathbb{R}^n$ to $\mathbb{R}^m$. Obviously, a linear transformation is uniquely determined by a matrix $A = (a_{ij})$, $a_{ij} = \mathbf{e}_i^{(m)} \cdot \mathbf{A}\mathbf{e}_j^{(n)}$, $i = 1,...,m$, $j = 1,...,n$. Then, $\mathbf{A}\mathbf{x} = \mathbf{y}$ with $y_i = \sum_{j=1}^{n} a_{ij}x_j$, for $i = 1,...,m$. We say that the matrix A represents $\mathbf{A}$. Moreover, if another linear transformation $\mathbf{B} : \mathbb{R}^m \to \mathbb{R}^l$ is such that $B = (b_{ki})$, then the composition $\mathbf{C} = \mathbf{B} \circ \mathbf{A}$ is represented by the product matrix $C = BA = (c_{kj})$, where $c_{kj} = \sum_{i=1}^{m} b_{ki}a_{ij}$. Most basic properties of vectors and linear transformations are easy to prove. The proofs involve only computations on sum and product of real numbers, and therefore they can be easily carried out within strict finitism.

A function $\mathbf{f} : \mathbb{R}^n \to \mathbb{R}^m$ is *(uniformly) continuous* on a closed ball U, if for any $k > 0$, there exists $l > 0$, such that $|\mathbf{f}(\mathbf{x}) - \mathbf{f}(\mathbf{y})| < 1/k$ whenever $\mathbf{x}, \mathbf{y} \in U$ and $|\mathbf{x} - \mathbf{y}| < 1/l$. $C(U, \mathbb{R}^m)$ denotes the set of continuous functions from U to $\mathbb{R}^m$. Most results for continuous functions of a single variable in the above sections hold for continuous functions on a closed ball as well. For instance, a continuous function $\mathbf{f}$ on a closed ball U must be bounded. For $f \in C(U, \mathbb{R})$, the approximate version of the intermediate value theorem holds as well. That is, if $f(\mathbf{x}) < r < f(\mathbf{y})$ for some $\mathbf{x}, \mathbf{y} \in U$, then for any $\varepsilon > 0$, there exists $t \in [0,1]$, such that $|f((1-t)\mathbf{x} + t\mathbf{y}) - r| < \varepsilon$.

Let U be a closed ball in $\mathbb{R}^n$ and $f \in C(U, \mathbb{R})$. The *partial derivatives* of f can be defined similarly as in the single variable case. Suppose that $g \in C(U, \mathbb{R})$, and suppose that for any $k > 0$, there exists $l > 0$, such that

$$|f(\mathbf{x} + \delta\mathbf{e}_i) - f(\mathbf{x}) - \delta g(\mathbf{x})| \le |\delta| / k$$

whenever $\mathbf{x}, \mathbf{x} + \delta\mathbf{e}_i \in U$ and $|\delta| < 1/l$. Then, we say that g is the partial derivative of f for the i-th argument on U, and we denote the fact as $g = \frac{\partial f}{\partial x_i} = \partial_i f$. Note that we always deal with partial derivatives on an entire closed ball (not a single point) and we consider only continuous partial derivatives. Arbitrary higher order partial derivatives $\partial_{i_k} ... \partial_{i_1} f$ can also be defined as in the single variable case. Note that as in the single variable case, a claim about the existence of $\partial_{i_k} ... \partial_{i_1} f$ requires $\partial_{i_h} ... \partial_{i_1} f$, $h = 1,...,k$, as the witnesses.

Let U be a closed ball in $\mathbb{R}^n$ and $\mathbf{f} \in C(U, \mathbb{R}^m)$. The *total differential* of $\mathbf{f}$ can be defined similarly as in the classical calculus. Suppose that $\mathbf{A} : U \to L(\mathbb{R}^n, \mathbb{R}^m)$ is a function such that for any $k > 0$, there exists $l > 0$, such that for any $\mathbf{x}, \mathbf{y} \in U$,

$$|\mathbf{f}(\mathbf{y}) - \mathbf{f}(\mathbf{x}) - \mathbf{A}(\mathbf{x})(\mathbf{y} - \mathbf{x})| \le |\mathbf{y} - \mathbf{x}| / k$$

whenever $|\mathbf{y} - \mathbf{x}| < 1/l$. Then, $\mathbf{A}$ is called a total differential of $\mathbf{f}$ on U, and it is also denoted as $\mathbf{f}'$.

Let $\mathbf{f} = (f_1, ..., f_m)$ and $\mathbf{y} = \mathbf{x} + \delta\mathbf{e}_j$ in the above inequality. Then,

$$\left|\mathbf{f}(\mathbf{x} + \delta\mathbf{e}_j) - \mathbf{f}(\mathbf{x}) - \delta\mathbf{f}'(\mathbf{x})\mathbf{e}_j\right| \le |\delta| / k.$$

Let $\mathbf{f}'(\mathbf{x})$ be represented by the matrix $(a_{ij}(\mathbf{x}))$ and consider the i-th component of the vector on the left hand side of the above inequality, we have

$$\left|f_i(\mathbf{x}+\delta\mathbf{e}_j)-f_i(\mathbf{x})-\delta a_{ij}(\mathbf{x})\right| \le \left|\mathbf{f}(\mathbf{x}+\delta\mathbf{e}_j)-\mathbf{f}(\mathbf{x})-\delta\mathbf{A}(\mathbf{x})\mathbf{e}_j\right| \le |\delta|/k.$$

Therefore, by the definition of partial derivatives, $a_{ij}(\mathbf{x}) = \partial_j f_i(\mathbf{x})$. That is, if a total differential $\mathbf{f}'$ exists on U, then it is unique and it must be represented by the matrix $(\partial_j f_i)$.

On the other side, suppose that all partial derivatives $\partial_j f_i$ exist on U. Let $\mathbf{A}(\mathbf{x})$ be the linear transformation represented by the matrix $A(\mathbf{x}) = (\partial_j f_i(\mathbf{x}))$ and let $\mathbf{y}-\mathbf{x} = \sum_{j=1}^{n}\delta_j\mathbf{e}_j$. Note that $\left|\delta_j\right| \le |\mathbf{y}-\mathbf{x}|$. Then,

$$\begin{aligned}
&\mathbf{f}(\mathbf{y})-\mathbf{f}(\mathbf{x})-\mathbf{A}(\mathbf{x})(\mathbf{y}-\mathbf{x})\\
&=\sum_{j=1}^{n}\left(\mathbf{f}\left(\mathbf{x}+\sum_{h=1}^{j-1}\delta_h\mathbf{e}_h+\delta_j\mathbf{e}_j\right)-\mathbf{f}\left(\mathbf{x}+\sum_{h=1}^{j-1}\delta_h\mathbf{e}_h\right)\right)-\sum_{j=1}^{n}\delta_j\mathbf{A}(\mathbf{x})\mathbf{e}_j\\
&=\sum_{j=1}^{n}\left(\mathbf{f}\left(\mathbf{x}+\sum_{h=1}^{j-1}\delta_h\mathbf{e}_h+\delta_j\mathbf{e}_j\right)-\mathbf{f}\left(\mathbf{x}+\sum_{h=1}^{j-1}\delta_h\mathbf{e}_h\right)-\delta_j\mathbf{A}(\mathbf{x})\mathbf{e}_j\right).
\end{aligned}$$

The i-th component of this vector is

$$\sum_{j=1}^{n}\left(f_i\left(\mathbf{x}+\sum_{h=1}^{j-1}\delta_h\mathbf{e}_h+\delta_j\mathbf{e}_j\right)-f_i\left(\mathbf{x}+\sum_{h=1}^{j-1}\delta_h\mathbf{e}_h\right)-\delta_j\partial_j f_i(\mathbf{x})\right).$$

Given $k>0$, for each j, since $\partial_j f_i$ exists,

$$\left|f_i\left(\mathbf{x}+\sum_{h=1}^{j-1}\delta_h\mathbf{e}_h+\delta_j\mathbf{e}_j\right)-f_i\left(\mathbf{x}+\sum_{h=1}^{j-1}\delta_h\mathbf{e}_h\right)-\delta_j\partial_j f_i\left(\mathbf{x}+\sum_{h=1}^{j-1}\delta_h\mathbf{e}_h\right)\right|<\frac{\left|\delta_j\right|}{2knm}$$

when δ_j is sufficiently small. Moreover, since $\partial_j f_i$ is uniformly continuous,

$$\begin{aligned}
&\left|\delta_j\partial_j f_i\left(\mathbf{x}+\sum_{h=1}^{j-1}\delta_h\mathbf{e}_h\right)-\delta_j\partial_j f_i(\mathbf{x})\right|\\
&=\left|\delta_j\right|\left|\partial_j f_i\left(\mathbf{x}+\sum_{h=1}^{j-1}\delta_h\mathbf{e}_h\right)-\partial_j f_i(\mathbf{x})\right|<\frac{\left|\delta_j\right|}{2knm}
\end{aligned}$$

when δ_h, $h=1,\ldots,j-1$, are sufficiently small. Then, it is easy to see that when $|\mathbf{y}-\mathbf{x}|$ is sufficiently small,

$$|\mathbf{f}(\mathbf{y})-\mathbf{f}(\mathbf{x})-\mathbf{A}(\mathbf{x})(\mathbf{y}-\mathbf{x})| \le |\mathbf{y}-\mathbf{x}|/k.$$

That is, $\mathbf{A}(\mathbf{x})$ is a total differential of $\mathbf{f}$. Note that this holds in strict finitism (unlike the classical case) because we consider only partial derivatives that are uniformly continuous.

When the total differential of $\mathbf{f}$ exists on U, we say that $\mathbf{f}$ is *differentiable* on U and we use $C^1(U,\mathbb{R}^m)$ to denote the set of differentiable functions from U to $\mathbb{R}^m$. Since $\mathbf{f}$ is differentiable if and only if all its first order partial derivatives exist, $C^1(U,\mathbb{R}^m)$ is also the set of functions from U to $\mathbb{R}^m$ with first order partial derivatives. Similarly, $C^\infty(U,\mathbb{R}^m)$ denotes the set of functions from U to $\mathbb{R}^m$ with arbitrary higher order partial derivatives.

The chain rule of differentiation also holds. Suppose that U,V are closed balls in $\mathbb{R}^n$ and $\mathbb{R}^m$ respectively and $\mathbf{f} \in C^1(U,\mathbb{R}^m)$, $\mathbf{g} \in C^1(V,\mathbb{R}^l)$, $\mathbf{f}(U) \subseteq V$. Let $\mathbf{h}(\mathbf{x}) = \mathbf{g}(\mathbf{f}(\mathbf{x}))$. We can show that $\mathbf{h}'(\mathbf{x}) = \mathbf{g}'(\mathbf{f}(\mathbf{x}))\mathbf{f}'(\mathbf{x})$, where the right hand side is the composition of linear operators. Since $\partial_j f_i$ and $\partial_i g_k$ are continuous and thus bounded on U and V respectively, it is easy to see that there exists M such that $|\mathbf{f}'(\mathbf{x})\mathbf{y}| \le M|\mathbf{y}|$ and $|\mathbf{g}'(\mathbf{f}(\mathbf{x}))\mathbf{z}| \le M|\mathbf{z}|$ for all $\mathbf{x} \in U$, $\mathbf{y} \in \mathbb{R}^n$, and $\mathbf{z} \in \mathbb{R}^m$. Therefore, given $k > 0$, when $|\mathbf{y}-\mathbf{x}|$ is sufficiently small,

$$|\mathbf{f}(\mathbf{y})-\mathbf{f}(\mathbf{x})| < |\mathbf{f}'(\mathbf{x})(\mathbf{y}-\mathbf{x})| + |\mathbf{y}-\mathbf{x}|/k \le \left(M+\frac{1}{k}\right)|\mathbf{y}-\mathbf{x}|$$

will also be arbitrarily small. Then, when $|\mathbf{y}-\mathbf{x}|$ is sufficiently small, we can have

$$\begin{aligned}
&\left|\mathbf{h}(\mathbf{y})-\mathbf{h}(\mathbf{x})-\mathbf{g}'(\mathbf{f}(\mathbf{x}))\mathbf{f}'(\mathbf{x})(\mathbf{y}-\mathbf{x})\right| \\
\le\ &|\mathbf{g}(\mathbf{f}(\mathbf{y}))-\mathbf{g}(\mathbf{f}(\mathbf{x}))-\mathbf{g}'(\mathbf{f}(\mathbf{x}))(\mathbf{f}(\mathbf{y})-\mathbf{f}(\mathbf{x}))| \\
&+\left|\mathbf{g}'(\mathbf{f}(\mathbf{x}))(\mathbf{f}(\mathbf{y})-\mathbf{f}(\mathbf{x})-\mathbf{f}'(\mathbf{x})(\mathbf{y}-\mathbf{x}))\right| \\
\le\ &|\mathbf{f}(\mathbf{y})-\mathbf{f}(\mathbf{x})|/k + M\left|\mathbf{f}(\mathbf{y})-\mathbf{f}(\mathbf{x})-\mathbf{f}'(\mathbf{x})(\mathbf{y}-\mathbf{x})\right| \\
\le\ &\left(M+\frac{1}{k}\right)\frac{|\mathbf{y}-\mathbf{x}|}{k} + M\frac{|\mathbf{y}-\mathbf{x}|}{k}.
\end{aligned}$$

Then, it follows that $\mathbf{h}'(\mathbf{x}) = \mathbf{g}'(\mathbf{f}(\mathbf{x}))\mathbf{f}'(\mathbf{x})$.

Compare the i-th component of the equality $\mathbf{h}'(\mathbf{x}) = \mathbf{g}'(\mathbf{f}(\mathbf{x}))\mathbf{f}'(\mathbf{x})$ we get the chain rule for partial derivatives: For $h(\mathbf{x}) = g(\mathbf{f}(\mathbf{x}))$,

$$\begin{aligned}
\partial_j h(\mathbf{x}) &= \sum_{k=1}^{m} \partial_k g(\mathbf{f}(\mathbf{x}))\,\partial_j f_k(\mathbf{x}), \text{ or} \\
\frac{\partial z}{\partial x_j} &= \sum_{k=1}^{m} \frac{\partial z}{\partial y_k}\frac{\partial y_k}{\partial x_j},
\end{aligned}$$

where $z = g(\mathbf{y})$, $\mathbf{y} = \mathbf{f}(\mathbf{x})$.

An application of partial derivative is about the differentiation of an integral. Suppose that $f(x,t)$ is continuous on $[a,b]\times[c,d]$ and $\frac{\partial f}{\partial t}(x,t)$ exists on $[a,b]\times[c,d]$. Then,

$$\frac{\mathrm{d}}{\mathrm{d}t}\int_a^b f(x,t)\,\mathrm{d}x = \int_a^b \frac{\partial f}{\partial t}(x,t)\,\mathrm{d}x.$$

To see this, note that given $k > 0$, for $s,t \in [c,d]$, when $|s-t|$ is sufficiently small,

$$\begin{aligned}
&\left|\int_a^b f(x,s)\,\mathrm{d}x - \int_a^b f(x,t)\,\mathrm{d}x - (s-t)\int_a^b \frac{\partial f}{\partial t}(x,t)\,\mathrm{d}x\right| \\
&= \left|\int_a^b \left(f(x,s) - f(x,t) - (s-t)\frac{\partial f}{\partial t}(x,t)\right)\mathrm{d}x\right| \\
&\leq \int_a^b \left|\left(f(x,s) - f(x,t) - (s-t)\frac{\partial f}{\partial t}(x,t)\right)\right|\mathrm{d}x \\
&\leq \int_a^b |s-t|/k\mathrm{d}x = |s-t|\frac{b-a}{k}.
\end{aligned}$$

Therefore, the above formula for the differentiation of an integral holds.

As another application, we derive the first-order *Taylor expansion* of a differentiable function of several variables, which will be used later in the book. Suppose that $U = Sc(\mathbf{z}, r)$ is a closed ball in $\mathbb{R}^n$ and $f \in C^\infty(U, \mathbb{R})$. We want to expand f in the format

$$f(\mathbf{x}) = f(\mathbf{z}) + \sum_{i=1}^{n} (x_i - z_i)\, g_i(\mathbf{x}),$$

where $g_i \in C^\infty(U, \mathbb{R})$. Without loss of generality, we may assume that $\mathbf{z} = \mathbf{0} = (0, ..., 0)$. For any $\mathbf{x} \in U$, $f(t\mathbf{x})$ is a differentiable function of $t \in [0,1]$. Applying the chain rule,

$$\frac{\mathrm{d}}{\mathrm{d}t} f(t\mathbf{x}) = \sum_{i=1}^{n} x_i \partial_i f(t\mathbf{x}).$$

By the fundamental theorem of calculus,

$$f(\mathbf{x}) - f(\mathbf{0}) = \int_0^1 \frac{\mathrm{d}}{\mathrm{d}t} f(t\mathbf{x})\,\mathrm{d}t = \sum_{i=1}^{n} x_i \int_0^1 \partial_i f(t\mathbf{x})\,\mathrm{d}t.$$

Therefore, we have the expansion above with $g_i(\mathbf{x}) = \int_0^1 \partial_i f(t\mathbf{x})\,\mathrm{d}t$. By taking partial derivatives on the integral, we see that $g_i \in C^\infty(U, \mathbb{R})$ since $f \in C^\infty(U, \mathbb{R})$.

3.6 Ordinary Differential Equations

We consider the initial value problem for first-order ordinary differential equations. Let $a, b > 0$, $(x_0, y_0) \in \mathbb{R}^2$,

$$D = [x_0 - a, x_0 + a] \times [y_0 - b, y_0 + b] \subset \mathbb{R}^2,$$

$F \in C(D, \mathbb{R})$. We want to find a function $y = f(x)$ such that f is differentiable on an interval $I = [x_0 - \alpha, x_0 + \alpha]$, $\alpha \leq a$, and $(x, f(x)) \in D$ for $x \in I$, and

$$f(x_0) = y_0, \;\; f'(x) = F(x, f(x)) \tag{3.3}$$

for $x \in I$.

We say that $F(x,y)$ satisfies the Lipschitz condition for y on D, if there exists K such that

$$|F(x,y_1) - F(x,y_2)| \leq K|y_1 - y_2|$$

for all $(x,y_1),(x,y_2) \in D$. If the partial derivative $\frac{\partial F}{\partial y}$ exists on D, then obviously F satisfies the Lipschitz condition for y on D, because we always assume that partial derivatives are continuous. In a *classical* proof of the existence and uniqueness of the solution for the equation (3.3) under Lipschitz condition (Coddington and Levinson [12]), we let

$$\begin{aligned} M &= \max\{F(x,y) : (x,y) \in D\}, \\ \alpha &= \min(a, b/M). \end{aligned}$$

For each $n > 0$, let N be such that $1/N < \min\left(\alpha^{-1}\omega_F(n), M^{-1}\alpha^{-1}\omega_F(n)\right)/2$, where ω_F is a modulus of continuity for F on D. Let $a_{n,i} = x_0 + \frac{i}{N}\alpha$ for $i = 0,...,N$. Then, we construct a sequence $b_{n,0},...,b_{n,N}$ of real numbers such that

$$\begin{aligned} b_{n,0} &= y_0, \\ b_{n,i+1} &= b_{n,i} + F(a_{n,i}, b_{n,i})\frac{\alpha}{N}, \end{aligned} \tag{3.4}$$

for $i = 0,...,N-1$. Note that $|b_{n,i+1} - b_{n,i}| \leq \frac{M\alpha}{N}$. Therefore, $|b_{n,i+1} - y_0| \leq M\alpha \leq b$. That is, $(a_{n,i}, b_{n,i}) \in D$ for all $i = 0,...,N$. Let $f_n(x)$ be the piecewise linear function defined on $[x_0, x_0 + \alpha]$ such that $f_n(a_{n,i}) = b_{n,i}$ for $i = 0,...,N$. Then we have

$$f_n(x) = f_n(a_{n,i}) + F(a_{n,i}, f_n(a_{n,i}))(x - a_{n,i})$$

on the small interval $[a_{n,i}, a_{n,i+1}]$. Therefore, $f_n'(x) = F(a_{n,i}, f_n(a_{n,i}))$ on the interval $[a_{n,i}, a_{n,i+1}]$, where $f_n'(a_{n,i})$ and $f_n'(a_{n,i+1})$ are understood as the right and left derivative respectively. For $x \in [a_{n,i}, a_{n,i+1}]$, $|x - a_{n,i}| \leq \alpha/N < \omega_F(n)/2$, and $|f_n(x) - f_n(a_{n,i})| \leq M\alpha/N < \omega_F(n)/2$, and therefore,

$$\left|f_n'(x) - F(x, f_n(x))\right| = |F(x, f_n(x)) - F(a_{n,i}, f_n(a_{n,i}))| < 1/n.$$

That is, f_n is a $1/n$ approximate solution of the equation (3.3) on $[a_{n,i}, a_{n,i+1}]$ for all i. Then one can prove that, under the Lipschitz condition, (f_n) uniformly converges to a solution of (3.3).

Because of the Lipschitz condition, the construction of $b_{n,i}$, $i = 1,...,N$, can be carried out within strict finitism. First, note that if a rational number q is an ε approximation to $b_{n,i}$, $|b_{n,i} - q| < \varepsilon$, then by the Lipschitz condition, $|F(a_{n,i}, b_{n,i}) - F(a_{n,i}, q)| < K\varepsilon$, and hence

$$\left|\left(b_{n,i} + F(a_{n,i}, b_{n,i})\frac{\alpha}{N}\right) - \left(q + F(a_{n,i}, q)\frac{\alpha}{N}\right)\right| < C\varepsilon$$

for some constant C. Therefore, if we compute a rational approximation to $q + F(a_{n,i}, q)\frac{\alpha}{N}$ up to the precision of $C\varepsilon$, we will get a rational approximation to $b_{n,i+1}$

up to the precision of $2C\varepsilon$. It means that if we start from a $(2C)^{-N}\varepsilon$ rational approximation to $b_{n,0}$, we will get a $(2C)^{-N+i}\varepsilon$ rational approximation to $b_{n,i}$ for each $i = 1, ..., N$. We can carry out the iteration in (3.4) within strict finitism since given an approximation q to $b_{n,i}$, we only need a $(2C)^{-N+i}\varepsilon$ rational approximation to $q + F(a_{n,i}, q)\frac{\alpha}{N}$. The numeral encoding this rational approximation can be bounded by a function within strict finitism. Therefore, real numbers satisfying (3.4) can be constructed.

Now we show that (f_n) uniformly converges to a solution of (3.3). Since f_n is a $1/n$ approximate solution on each $[a_{n,i}, a_{n,i+1}]$, that is, $|f_n'(x) - F(x, f_n(x))| < 1/n$ on each $[a_{n,i}, a_{n,i+1}]$, integrating from $a_{n,i}$ to $x \in [a_{n,i}, a_{n,i+1}]$, we get

$$\left| f_n(x) - f_n(a_{n,i}) - \int_{a_{n,i}}^{x} F(t, f_n(t))\, \mathrm{d}t \right| \leq (x - a_{n,i})/n.$$

In particular,

$$I_{n,i} = \left| f_n(a_{n,i+1}) - f_n(a_{n,i}) - \int_{a_{n,i}}^{a_{i+1}} F(t, f_n(t))\, \mathrm{d}t \right| \leq (a_{n,i+1} - a_{n,i})/n.$$

Therefore, for $x \in [a_{n,i}, a_{n,i+1}]$,

$$\begin{aligned}
&\left| f_n(x) - f_n(x_0) - \int_{x_0}^{x} F(t, f_n(t))\, \mathrm{d}t \right| \\
&\leq \sum_{j=0}^{i-1} I_{n,j} + \left| f_n(x) - f_n(a_{n,i}) - \int_{a_{n,i}}^{x} F(t, f_n(t))\, \mathrm{d}t \right| \\
&\leq (x - x_0)/n.
\end{aligned}$$

Note that $f_n(x_0) = y_0$. Since the left hand side of the above inequality is a continuous function of x on $[x_0, x_0 + \alpha]$, we have

$$\left| f_n(x) - y_0 - \int_{x_0}^{x} F(t, f_n(t))\, \mathrm{d}t \right| \leq \alpha/n \tag{3.5}$$

for all $x \in [x_0, x_0 + \alpha]$. Then, for $m, n > 0$ and $x \in [x_0, x_0 + \alpha]$, we have

$$\begin{aligned}
&|f_m(x) - f_n(x)| \\
&\leq \int_{x_0}^{x} |F(t, f_m(t)) - F(t, f_n(t))|\, \mathrm{d}t + (1/m + 1/n)\alpha \\
&\leq K \int_{x_0}^{x} |f_m(t) - f_n(t)|\, \mathrm{d}t + (1/m + 1/n)\alpha.
\end{aligned} \tag{3.6}$$

Denote $g(x) = \int_{x_0}^{x} |f_m(t) - f_n(t)|\, \mathrm{d}t$, $\varepsilon = (1/m + 1/n)$. By the fundamental theorem of calculus, the above inequality is

$$g'(x) \leq K g(x) + \varepsilon\alpha.$$

Multiply both sides by $e^{-K(x-x_0)}$ and integrate from x_0 to x, we get

$$g(x)\,e^{-K(x-x_0)} \leq \varepsilon\alpha \int_{x_0}^{x} e^{-K(x-x_0)}\,dt \leq \varepsilon\alpha^2.$$

Using the inequality (3.6) again, we see that, for $x \in [x_0, x_0+\alpha]$,

$$|f_m(x) - f_n(x)| \leq Kg(x) + \varepsilon\alpha \leq \varepsilon\alpha\left(1+\alpha K e^{K\alpha}\right). \tag{3.7}$$

Therefore, (f_n) uniformly converges on $[x_0, x_0+\alpha]$ to some continuous function f on $[x_0, x_0+\alpha]$. Let $n \to \infty$ in the inequality (3.5). Since F is (uniformly) continuous and (f_n) uniformly converges, we obtain

$$f(x) = y_0 + \int_{x_0}^{x} F(t, f(t))\,dt \tag{3.8}$$

for all $x \in [x_0, x_0+\alpha]$. The same construction can be carried out for the interval $[x_0-\alpha, x_0]$ and the resulted function $f(x)$ on the interval will satisfy this integration equation. Then it easily follows that $f(x_0) = y_0$ and, by the fundamental theorem of calculus, f satisfies the differential equation (3.3).

Uniqueness of the solution follows from a similar argument. Let f be a solution of (3.3). That is, f satisfies (3.8). Then, similar to (3.6), we have

$$|f(x) - f_n(x)| \leq K \int_{x_0}^{x} |f(t) - f_n(t)|\,dt + \alpha/n.$$

By the same argument we can derive

$$|f(x) - f_n(x)| \leq \frac{1}{n}\alpha\left(1+\alpha K e^{K\alpha}\right).$$

This means that (f_n) must uniformly converge to f. This also gives an estimate of the error of the approximation.

Moreover, we can show that the solution depends on its initial value y_0 uniformly continuously. Write (3.8) as

$$f(x, y_0) = y_0 + \int_{x_0}^{x} F(t, f(t, y_0))\,dt.$$

Then,

$$|f(x,y_0) - f(x,y_1)| \leq |y_0 - y_1| + K \int_{x_0}^{x} |f(t,y_0) - f(t,y_1)|\,dt.$$

This is again similar to (3.6). Therefore,

$$|f(x,y_0) - f(x,y_1)| \leq |y_0 - y_1|\left(1+\alpha K e^{K\alpha}\right),$$

which implies that $f(x, y_0)$ is uniformly continuous in its initial value y_0.

We summarize the conclusion as follows:

Theorem 3.12. *Suppose that $F(x,y)$ is continuous on the rectangle $D: |x-x_0| \leq a, |y-y_0| \leq b$, and suppose that $F(x,y)$ satisfies the Lipschitz condition for y on the rectangle. Let $M = \max\{F(x,y) : (x,y) \in D\}$ and $\alpha = \min(a, b/M)$. Then, there exists a unique function f differentiable on an interval $I = [x_0 - \alpha, x_0 + \alpha]$, such that $f(x_0) = y_0$ and $f'(x) = F(x, f(x))$ for $x \in I$. Moreover, f depends on its initial value y_0 uniformly continuously.*

Another classical approximation to the solution is by constructing functions f_n as follows: for $x \in [x_0, x_0 + \alpha]$, define

$$f_0(x) = y_0,$$
$$f_{n+1}(x) = y_0 + \int_{x_0}^{x} F(t, f_n(t))\,\mathrm{d}t.$$

This construction can be completed within strict finitism as well, although it will be a little more complex. Instead of constructing the sequence (f_n) directly, we will have to construct a sequence $(p_{n,i}, q_{n,i})$ of rational numbers such that, for each n, the sequence $(p_{n,i}, q_{n,i})$ approximates the graph of f_n. The Lipschitz condition also implies that we can approximate all f_n uniformly within strict finitism.

Now consider the initial value problem for a system of first-order differential equations. For convenience, we will redefine the norm on $\mathbb{R}^m$ as

$$|\mathbf{x}| = |x_1| + \ldots + |x_m|.$$

Suppose that $x_0 \in \mathbb{R}$, $\mathbf{y}_0 \in \mathbb{R}^m$, $D = [x_0 - a, x_0 + a] \times [\mathbf{y}_0 - \mathbf{b}, \mathbf{y}_0 + \mathbf{b}]$, and $\mathbf{F} : D \to \mathbb{R}^m$, where $\mathbf{F} = (F_1, \ldots, F_m)$, $F_i : D \to \mathbb{R}$, $i = 1, \ldots, m$. We want to find an interval $I = [x_0 - \alpha, x_0 + \alpha]$, $\alpha \leq a$, and a vector function $\mathbf{f} \in C^1(I, \mathbb{R}^m)$, such that $(x, \mathbf{f}(x)) \in D$ for $x \in I$, and

$$\mathbf{f}(x_0) = \mathbf{y}_0, \ \mathbf{f}'(x) = \mathbf{F}(x, \mathbf{f}(x)) \tag{3.9}$$

for $x \in I$.

We say that $\mathbf{F}(x, \mathbf{y})$ satisfies the Lipschitz condition for $\mathbf{y}$ on D, if there exists K such that

$$|\mathbf{F}(x, \mathbf{y}_1) - \mathbf{F}(x, \mathbf{y}_2)| \leq K |\mathbf{y}_1 - \mathbf{y}_2| \tag{3.10}$$

for all $(x, \mathbf{y}_1), (x, \mathbf{y}_2) \in D$. If all partial derivatives $\frac{\partial \mathbf{F}}{\partial y_k}$, $k = 1, \ldots, m$, exist on D, then $\mathbf{F}(x, \mathbf{y})$ satisfies the Lipschitz condition for $\mathbf{y}$ on D. To see this, suppose that $\mathbf{y}_j = (y_{j,1}, \ldots, y_{j,m})$, $j = 1, 2$. Let $\mathbf{z}_k = (y_{1,1}, \ldots, y_{1,k}, y_{2,k+1}, \ldots, y_{2,m})$, $k = 0, \ldots, m$. Then

$$|\mathbf{F}(x, \mathbf{y}_1) - \mathbf{F}(x, \mathbf{y}_2)| = \sum_{i=1}^{m} |F_i(x, \mathbf{y}_1) - F_i(x, \mathbf{y}_2)| \leq \sum_{i=1}^{m} \sum_{k=1}^{m} |F_i(x, \mathbf{z}_{k-1}) - F_i(x, \mathbf{z}_k)|,$$

$$\sum_{k=0}^{m-1} |\mathbf{z}_k - \mathbf{z}_{k+1}| = |\mathbf{y}_1 - \mathbf{y}_2|.$$

Since the partial derivative $\frac{\partial F_i}{\partial y_k}$ is continuous on D, there exists a constant $K_{i,k}$ such that

$$|F_i(x,\mathbf{z}_{k-1}) - F_i(x,\mathbf{z}_k)| \le K_{i,k} |\mathbf{z}_{k-1} - \mathbf{z}_k|.$$

Add up these inequalities, we see that there exists a constant K such that (3.10) holds.

When $\mathbf{F}(x,\mathbf{y})$ satisfies the Lipschitz condition for $\mathbf{y}$ on D, an approximation to the solution for (3.9) can be constructed almost exactly as before. For instance, (3.4) will be replaced by

$$\begin{aligned} \mathbf{b}_{n,0} &= \mathbf{y}_0, \\ \mathbf{b}_{n,i+1} &= \mathbf{b}_{n,i} + \mathbf{F}(a_{n,i}, \mathbf{b}_{n,i}) \frac{\alpha}{N}, \end{aligned} \tag{3.11}$$

and $f_n(x)$ will be replaced by vector functions

$$\mathbf{f}_n(x) = \mathbf{f}_n(a_{n,i}) + \mathbf{F}(a_{n,i}, \mathbf{f}_n(a_{n,i}))(x - a_{n,i}).$$

Other arguments can be straightforwardly translated over. Therefore, we also have the existence, uniqueness and uniform continuity theorem for the solution of the system of differential equations (3.9).

As in the classical case, the initial value problem of an n-th-order differential equation

$$\begin{aligned} f(x_0) &= y_0,\ f'(x_0) = y_1,\ \ldots,\ f^{(n-1)}(x_0) = y_{n-1}, \\ f^{(n)}(x) &= F\left(x, f(x), f'(x), \ldots, f^{(n-1)}(x)\right) \end{aligned}$$

can be translated into the initial value problem for a system of first-order differential equations

$$\begin{aligned} f_0(x_0) &= y_0,\ f_1(x_0) = y_1,\ \ldots,\ f_{n-1}(x_0) = y_{n-1}, \\ f_0'(x) &= f_1(x), \\ &\ldots \\ f_{n-2}'(x) &= f_{n-1}(x), \\ f_{n-1}'(x) &= F(x, f_0(x), f_1(x), \ldots, f_{n-1}(x)). \end{aligned}$$

Therefore, we also have the existence and uniqueness for the solution of the initial value problem of an n-th-order differential equation. The same holds for a system of n-th-order differential equations, since it can be similarly transformed into a system of first-order differential equations.

Most ordinary techniques for solving ordinary differential equations are available for strict finitism. In particular, if a technique is just for finding the analytic expression of a solution, it is usually available for strict finitism, since verifying a solution can usually be done by straightforward computations.

3.7 Case Study: A Population Growth Model

We will study the applicability of the logistic model for simulating population growth. We will show how the conclusion drawn in the application is a logical consequence of literally true realistic premises about finite real things, including discrete population values and the computational device for simulating them.

Recall that there are two realistic premises specifically about the population growth on the Earth. First, there is a premise about the population at some initial moment

There are N_0 people at the moment 0. (3.12)

Here, we assume a temporal unit and an origin point for determining temporal moments. We also assume a people-count unit, for instance, in billions. N_0 is a decimal numeral, expressing the people-count property in that unit. Then, there is a premise about the population growth

For each moment, if there are p people on the Earth at that moment, then there are $\alpha(N-p)p\delta_0$ more people on the Earth δ_0 (units) after the moment. (3.13)

Here, δ_0 (units) is an appropriately small temporal distance for measuring the population growth. It should not be too small so as to reveal discreteness in the population growth, but it should be small enough so that this estimate of the population growth is accurate. Moreover, N is a constant decimal numeral, intuitively, the largest population that the Earth can support.

This premise (3.13) corresponds to the logistic differential equation

$$\frac{\mathrm{d}p}{\mathrm{d}t} = \alpha(N-p)p, t \geq 0. \tag{3.14}$$

This is one of our mathematical premises. We solve this differential equation by a simple argument as follows. From (3.14), it follows that

$$\left(\frac{1}{p} + \frac{1}{N-p}\right)\frac{\mathrm{d}p}{\mathrm{d}t} = \alpha N. \tag{3.15}$$

If we consider $\ln \frac{p}{N-p} = \ln p - \ln(N-p)$ as a composite function of t, we see that

$$\frac{\mathrm{d}}{\mathrm{d}t}\ln\frac{p}{N-p} = \left(\frac{1}{p} + \frac{1}{N-p}\right)\frac{\mathrm{d}p}{\mathrm{d}t}. \tag{3.16}$$

Therefore, from (3.15) and (3.16), we have

$$\frac{\mathrm{d}}{\mathrm{d}t}\ln\frac{p}{N-p} = \alpha N. \tag{3.17}$$

This implies that for some constant C,

$$\ln \frac{p}{N-p} = \alpha Nt + C. \tag{3.18}$$

Therefore, for some constant C (different from the one above),

$$\frac{p}{N-p} = Ce^{\alpha Nt}. \tag{3.19}$$

That is, for some constant C,

$$p(t) = \frac{N}{1+Ce^{-\alpha Nt}}. \tag{3.20}$$

Then, from our initial condition

$$p(0) = N_0, \tag{3.21}$$

which is another mathematical premise, we have

$$p(t) = \frac{N}{1+Ce^{-\alpha Nt}},\, C = \frac{N-N_0}{N_0}. \tag{3.22}$$

Finally, we have

$$p(t_1) = \frac{N}{1+Ce^{-\alpha Nt_1}} \approx N_1, \tag{3.23}$$

where t_1 is some given concrete rational number representing a temporal moment after the initial moment, and N_1 is a decimal numeral with some finite precision closest to $\frac{N}{1+Ce^{-\alpha Nt_1}}$. (3.23) is our mathematical conclusion. From it we get our realistic conclusion

$$\text{There are } N_1 \text{ people at the moment } t_1. \tag{3.24}$$

Note that the proof steps from the mathematical premises (3.14) and (3.21) to the mathematical conclusion (3.23) are already within strict finitism. In particular, (3.16) comes from our definition of the function $\ln x$ and its basic properties in Sect. 3.4 above, together with the chain rule for taking derivative. (3.18) follows from the proposition that if $f'(x) = 0$ on an interval, then $f(x) = C$ for some constant on the interval. (3.19) comes from our definition of the function e^x and its basic properties in Sect. 3.4 above.

Now, the premises we use in this application are (3.12), (3.13), (3.14), and (3.21). (3.12) and (3.13) are literally true realistic premises about the initial population and the population growth rates. The mathematical conclusion (3.23) follows from (3.14) and (3.21), together with other axioms of strict finitism.

We assume that p is a term in strict finitism, considered as a concrete program, such that it represents the population on the Earth. That is, it makes the following bridging postulation literally true for decimal numerals t, n up to some finite precision:

$$\text{for each } t,\, n, \text{ there are } n\text{-people on the Earth} \\ \text{at the moment } t, \text{ if and only if } p(t) \approx n. \tag{3.25}$$

Note that t, n here are treated as constant terms in strict finitism representing rational numbers, and p is a term of the type $((o \to o) \to (o \to o))$, that is, the type of a real function, but we ambiguously apply it to rational numbers. Moreover, quantifications in (3.25) range over only finitely many decimal numerals with some limited precision.

Note that the premise (3.21) directly follows from this bridging hypothesis (3.25) and the literally true realistic premise (3.12). Similarly, the realistic conclusion (3.24) directly follows from (3.25) and the mathematical conclusion (3.23).

The premise (3.14) about p states that some term δ witnesses that the derivative of p is $\alpha(N-p)p$ on $[0,\infty)$, that is,

$$\exists\delta\forall x,y,n\left(\begin{array}{c} n>0\wedge x\geq 0\wedge y\geq 0\wedge|y-x|\leq\delta(n)\rightarrow \\ |p(y)-p(x)-\alpha(N-p(x))p(x)(y-x)|\leq|y-x|/n \end{array}\right). \tag{3.26}$$

Let $\delta(n)$ be δ_0 in the realistic premise (3.13) and consider dividing the temporal interval $[0,\infty)$ into sub-intervals of the length δ_0, with the end points 0, δ_0, $2\delta_0$, $3\delta_0$, ... We can see that the instances of (3.26) for that δ, for x,y among 0, δ_0, $2\delta_0$, $3\delta_0$, ..., and for sufficiently large n follow from the realistic premise (3.13) and the bridging hypothesis (3.25). That is,

$$\varphi\left[\delta_0,\overline{k}\delta_0,\overline{h}\delta_0,\overline{n}\right] \tag{3.27}$$

follows from the realistic premise (3.13) and the bridging hypothesis (3.25), where $\varphi[\delta(n),x,y,n]$ is the formula in the bracket in (3.26), and $\overline{k}$ and $\overline{h}$ are arbitrary numerals, and $\overline{n}$ is any sufficiently large numeral.

Recall that in strict finitism, when we derive a formula from a premise like $\exists\delta\forall x\varphi[\delta,x]$, we actually use only finitely many instances of the premise $\forall x\varphi[\delta,x]$ for an arbitrary δ. The general format of the instances of the premise is

$$\forall k\leq M[\delta]\,\varphi[\delta,Y(\delta,k)],$$

where the terms M,Y can be extracted from the proof. Applying this to the premise (3.14) used for deriving the conclusion (3.23) in strict finitism, we see that the proof needs only some instances of (3.26).

Here, we expect that the instances of (3.26) in (3.27) will be sufficient. To make sure about that, we will have to spell out the finitistic proof from the premises (3.14) and (3.21) to the conclusion (3.23), extract the terms like M,Y above, and extract the required instances of (3.26). In principle, we can start with the needed instance of (3.23) and trace back to get the required instances of (3.26). This will be tedious, but it is actually a mechanical procedure and can in principle be done by a computer. Here, we rely on our intuition to assure us that the instances in (3.27) will be sufficient.

In summary, this is our explanation of applicability in this example. The bridging hypothesis (3.25) says that some program p encodes the population at various discrete moments. This can always be realized by *some* program p, because the programs in strict finitism are rich enough. The mathematical premise (3.21) about the program p follows from the bridging hypothesis (3.25) and the realistic premise (3.12). The realistic premise (3.13) and the bridging hypothesis (3.25) imply the instances (3.27) of the differential equation (3.14) about the program p. These instances, together with (3.21) and the axioms of strict finitism, are sufficient for deriving the mathematical conclusion (3.23) about the program p. The axioms of strict finitism are literally true assertions about programs in general. Finally, the realistic conclusion (3.24) directly follows from the bridging hypothesis (3.25) and the mathematical conclusion (3.23) about p. Therefore, the realistic conclusion (3.24) actually follows from the literally true realistic premises (3.12), (3.13), (3.25) and other axioms of strict finitism, all of which are literally true assertions about concrete real things.

Chapter 4
Metric Space

Abstraction and idealization are two major techniques to allow building simplified mathematical models in the sciences. Using a continuous function to represent discrete population values is an instance of idealization. Idealization helps to ignore and smooth over insignificant details, and therefore it simplifies the models, but it produces models that do not exactly represent real things, especially when idealization to infinity and continuity is used. Therefore, the applicability of idealization is not logically transparent, because one cannot straightforwardly translate the mathematical premises about an idealized model into literally true realistic assertions about finite real things without modifying the logical structures of those mathematical premises. Abstraction does not have this problem. Abstraction means presenting concepts, thoughts and proofs in some highly schematic and abstract format, and then they can be instantiated into more and more concrete concepts, thoughts and proofs in a few stages, with their logical structures preserved. The resulted thoughts and proofs can become much more complex than the original ones. That is how abstraction helps to simplify the presentation of a theory.

The theory of metric space in strict finitism is a typical case of using this technique of abstraction in strict finitism. It is presented as statements about an arbitrary set and an arbitrary function on the set (i.e., the metric function). Recall that a set is a pair of formulas and a function is a term. Therefore, the theory actually consists of schematic statements with an arbitrary pair of formulas and an arbitrary term of some formats. Applying this general theory of metric spaces to a more concrete metric space, for instance, the metric space of real numbers or the metric space of continuous functions, means instantiating the definitions, theorems and proofs in the general theory with more concrete formulas and terms. This chapter shows that some abstract mathematics can also be developed within strict finitism.

We mostly follow the ideas in Chap. 4 of Bishop and Bridges [6], with necessary modifications to fit into our framework here. We will not again give references to the pages in that book for every notion defined and every theorem proved here.

F. Ye, *Strict Finitism and the Logic of Mathematical Applications*, Synthese Library 355,
DOI 10.1007/978-94-007-1347-5_4,

4.1 Basic Definitions

Suppose that X is a set. ρ is a metric on X, or (X,ρ) is a metric space, if $\rho : X \times X \to \mathbb{R}^{+0}$, and for all $x,y,z \in X$, (i) $\rho(x,y) = 0 \leftrightarrow x =_X y$, and (ii) $\rho(x,y) = \rho(y,x)$, and (iii) $\rho(x,y) \leq \rho(x,z) + \rho(z,y)$. Note that in strict finitism, '(X,ρ) is a metric space' is a schematic assertion about a term ρ and two arbitrary formulas defining a set X. Therefore, we do not quantify over metric spaces in strict finitism.

If ρ is a metric on X, and Y is a subset of X, then ρ is also a metric on Y, called the induced metric on Y. The following notions are defined in the same format as in classical mathematics:

(X,ρ) is bounded, if and only if there exists M such that $\rho(x,y) \leq M$ for all $x,y \in X$.

c is a bound of (X,ρ), if and only if $\rho(x,y) \leq c$ for all $x,y \in X$.

Y is a bounded subset of X, if and only if Y is a subset of X and there exists M such that $\rho(x,y) \leq M$ for all $x,y \in Y$.

A sequence (x_n) of elements of X converges to an element y of X, if and only if for any $k > 0$, there exists N, such that $\rho(x_n,y) \leq 1/k$ for $n > N$.

f is a uniformly continuous function from the metric space (X,ρ) to the metric space (X',ρ'), if and only if for any $k > 0$, there exists $N > 0$, such that $\rho'(f(x),f(y)) \leq 1/k$, whenever $x,y \in X$ and $\rho(x,y) \leq 1/N$.

A sequence (f_n) of functions from a set S to the metric space (X,ρ) converges uniformly to a function $f : S \to X$, if and only if for any $k > 0$, there exists $N > 0$, such that whenever $n > N$, $\rho(f_n(x),f(x)) \leq 1/k$ for all $x \in S$.

Two metrics ρ_1,ρ_2 on X are equivalent, if and only if the identity function on X is uniformly continuous as a function from (X,ρ_1) to (X,ρ_2) and as a function from (X,ρ_2) to (X,ρ_1).

Note that some of these notions require witnesses. For instance, the bound M is a witness for the property '(X,ρ) is bounded', and convergence and uniform continuity each needs a modulus of convergence or a modulus of continuity.

The product of metric spaces (X_1,ρ_1), ..., (X_n,ρ_n) is $(\prod_{i=1}^{n} X_i,\rho)$ with

$$\rho(x,y) = \sum_{i=1}^{n} \rho_i(x_i,y_i).$$

The product of a sequence of metric spaces (X_n,ρ_n) bounded by 1 is $(\prod_{n=0}^{\infty} X_n,\rho)$ with

$$\rho(x,y) = \sum_{n=0}^{\infty} 2^{-n-1}\rho_n(x_n,y_n).$$

It is easy to verify that these do define metrics on relevant sets.

Instances of metric spaces include $(\mathbb{R}^n,d)$ with $d(x,y) = \left(\sum_{i=1}^{n}(x_i - y_i)^2\right)^{\frac{1}{2}}$ and $(C([a,b],\mathbb{R}),d)$ with $d(f,g) = \left(\int_a^b |f(x) - g(x)|^2 dx\right)^{\frac{1}{2}}$. To show that they satisfy the conditions for a metric, we need Cauchy-Schwarz inequality and Minkowski inequality: if (a_i) and (b_i) are sequences of real numbers, then

$$\left|\sum_{i=0}^{n} a_i b_i\right| \le \left(\sum_{i=0}^{n} a_i^2\right)^{1/2} \left(\sum_{i=0}^{n} b_i^2\right)^{1/2},$$

$$\left(\sum_{i=0}^{n} (a_i + b_i)^2\right)^{1/2} \le \left(\sum_{i=0}^{n} a_i^2\right)^{1/2} + \left(\sum_{i=0}^{n} b_i^2\right)^{1/2}.$$

These can be proved as in the classical case, using the properties for finite sum of real numbers in Sect. 3.2. We will omit the details (see [6], pp. 83–84). Then, any constructions and verifications done on an arbitrary metric space will apply to these two instances of metric space.

The open ball $S(x,r)$ and closed ball $Sc(x,r)$ are defined as (parameterized) subsets of X:

$$(y \in S(x,r)) \equiv_{df} \rho(y,x) < r,$$
$$(y \in Sc(x,r)) \equiv_{df} \rho(y,x) \le r.$$

Y is an open subset of X, if Y is a subset and for each $x \in Y$, there exists $r > 0$ such that $S(x,r) \subseteq Y$. Y is a closed subset of X, if $Y \subseteq X$, and for each $x \in X$,

$$\forall r \exists y (y \in S(x,r) \cap Y) \rightarrow x \in Y.$$

Then, it is easy to see that $S(x,r)$ is open and $Sc(x,r)$ is closed. It is also trivial to show that if a sequence of points in a closed subset converges, the limit point also belongs to the closed subset. Similarly, suppose that a subset Y is such that for any sequence (x_n), if $x_n \in Y$ for all n, and (x_n) has a limit x, then $x \in Y$. Then, Y is a closed subset. We can use the axiom of choice to get a sequence converging to x from the condition $\forall r \exists y (y \in S(x,r) \cap Y)$.

Y is a dense subset of X, if Y is a subset, and for each $x \in X$ and $r > 0$, there exists $y \in Y$ such that $y \in S(x,r)$. Given a subset Y, the interior of Y is the set $int(Y)$:

$$(x \in int(Y)) \equiv_{df} \exists r > 0 (S(x,r) \subseteq Y),$$

and the closure $\overline{Y}$ of Y is:

$$\left(x \in \overline{Y}\right) \equiv_{df} \forall r > 0 \exists y (y \in Y \cap S(x,r)).$$

It is easy to see that $int(Y) \subseteq Y \subseteq \overline{Y}$, and Y is open iff $int(Y) = Y$, and Y is closed iff $Y = \overline{Y}$. Notice that here we quantify over the centers and radiuses of open balls, instead of arbitrary subsets. The latter is not available in strict finitism.

The subset A of X is *located* in X ([6], p. 88), if A is not empty and

$$\forall x \in X \exists r \in \mathbb{R} (r = \inf\{\rho(x,y) : y \in A\}).$$

We will use the notation $\rho(x,A)$ for the term such that

$$\rho(x,A) = \inf\{\rho(x,y) : y \in A\}$$

if it is assumed or proved that $\exists r \in \mathbb{R}\,(r = \inf\{\rho(x,y) : y \in A\})$. The metric complement $X - A$ of a located subset A is a subset of X defined by

$$(x \in X - A) \equiv_{df} x \in X \wedge \rho(x,A) > 0,$$

where the right hand side uses the term $\rho(x,A)$ whose construction depends on the witness for A to be located.

4.2 Completeness

Let (X,ρ) be a metric space. A sequence (x_n) in X is a Cauchy sequence if for any $k > 0$, there exists N, such that $\rho(x_m,x_n) \leq 1/k$ for all $m,n \geq N$. X is complete if every Cauchy sequence in X converges. From the definition it is easy to see that a closed subset of a complete metric space is complete and a complete subset of a metric space is closed. It is also straightforward to prove that the product of a sequence of complete metric spaces is still complete.

Given (X,ρ), the completion of it, $\left(\widetilde{X},\widetilde{\rho}\right)$, can be constructed: $\widetilde{X}$ is the set of sequences (x_n) such that

$$\forall n,m > 0\left(x_n \in X \wedge \rho(x_n,x_m) \leq n^{-1} + m^{-1}\right),$$

and the equality on $\widetilde{X}$ is defined by

$$\left(x =_{\widetilde{X}} y\right) \equiv_{df} \forall n > 0\left(\rho(x_n,y_n) \leq 2n^{-1}\right).$$

For $x,y \in \widetilde{X}$ it can be easily verified that $(\rho(x_n,y_n))$ is a Cauchy sequence of real numbers:

$$\forall k > 0 \forall n,m \geq 4k\ \left(|\rho(x_n,y_n) - \rho(x_m,y_m)| \leq 1/k\right).$$

So we can construct the term $\lim_{n\to\infty} \rho(x_n,y_n)$ (with $N \equiv \lambda k.4k$ as the witness of Cauchyness for $(\rho(x_n,y_n))$) and define

$$\widetilde{\rho} \equiv_{df} \lambda xy. \lim_{n\to\infty} \rho(x_n,y_n).$$

Then, it is easy to prove that $\widetilde{\rho}$ is a metric on $\widetilde{X}$.

To see that $\left(\widetilde{X},\widetilde{\rho}\right)$ is complete, let (z^n) be a Cauchy sequence in $\left(\widetilde{X},\widetilde{\rho}\right)$. From the above we see that for $n \geq 4k$, we have $|\rho(x_n,y_n) - \widetilde{\rho}(x,y)| \leq 1/k$. Let $N(k)$ be a modulus of Cauchyness for (z^n). That is, for any $k > 0$ and $m,n \geq N(k)$, we have $\widetilde{\rho}(z^m,z^n) \leq 1/k$. Then, we can let $(x_k) \equiv \left(\left(z^{N(2k)}\right)_{8k}\right)$. Then, for any $n \leq m$, it is easy to see that

$$\begin{aligned}\rho(x_n,x_m) &= \rho\left(\left(z^{N(2n)}\right)_{8n},\left(z^{N(2m)}\right)_{8m}\right)\\ &\le \rho\left(\left(z^{N(2n)}\right)_{8n},\left(z^{N(2n)}\right)_{8m}\right)+\rho\left(\left(z^{N(2n)}\right)_{8m},\left(z^{N(2m)}\right)_{8m}\right)\\ &\le \frac{1}{8n}+\frac{1}{8m}+\widetilde{\rho}\left(z^{N(2n)},z^{N(2m)}\right)+\frac{1}{2m}\\ &\le \frac{1}{n}+\frac{1}{m}.\end{aligned}$$

Therefore, $(x_k) \in \widetilde{X}$. It is obvious that (z^n) converges to (x_k).

Furthermore, the inclusion map $i : X \to \widetilde{X}$ can be defined and it is easy to see that when X is complete, i is an isomorphism.

Lemma 4.1. *Suppose that f is a uniformly continuous function from a dense subset Y of a metric space (X,ρ) to a complete metric space (X',ρ'). Then, f can be extended into a uniformly continuous function f' from the whole space (X,ρ) to (X',ρ').*

The natural extension in classical or constructive mathematics is already within strict finitism. We omit the details.

When A is complete and located, and therefore $\rho(x,A)$ exists, we may not be able to find $a \in A$ such that $\rho(x,a) = \rho(x,A)$. However, we have a weaker result, whose proof in [6], p. 92, needs some improvements.

Lemma 4.2. *If A is a complete, non-void, located subset of X, and x a point of X, then there exists a point $a \in A$ such that if $\rho(x,a) > 0$ then $\rho(x,A) > 0$.*

Proof. To prove this, we construct a Cauchy sequence (a_n) of elements of A and take the limit as a. First, the assumption of locatedness implies that for $n > 0$, if $\rho(x,A) < 1/n$, there exists $y \in A$ such that $\rho(x,y) < 1/n$. It follows that there exists δ such that for $m,n > 0$, if $\rho(x,A)(m) < 1/n - 1/m$ (that is, if m witnesses $\rho(x,A) < 1/n$), then $\delta(m,n) \in A \wedge \rho(x,\delta(m,n)) < 1/n$.

Since A is non-void, there exists $a_0 \in A$. By a bounded minimalization, we can construct a term $h[m]$ such that for $m > 0$, if for all $n = 1,...,m$ we have $\rho(x,A)(4n) < \frac{3}{4n}$, then $h[m] = m$, and otherwise $h[m]$ is the least n such that $\rho(x,A)(4(n+1)) \ge \frac{3}{4(n+1)}$. Notice that $\rho(x,A)(4n) < \frac{3}{4n}$ implies that $4n$ witnesses $\rho(x,A) < 1/n$, while $\rho(x,A)(4(n+1)) \ge \frac{3}{4(n+1)}$ implies that $\rho(x,A) \ge \frac{2}{4(n+1)}$.

Then, we let $a_m = \delta(4h[m],h[m])$. Intuitively, this is what we were doing. If for all $n = 1,...,m$, $4n$ witnesses $\rho(x,A) < 1/n$, then we let a_m be the point $\delta(4m,m) \in A$, which is such that $\rho(x,a_m) < 1/m$; otherwise, we have found an $n < m$ such that $\rho(x,A) \ge \frac{2}{4(n+1)}$, and either $n = 0$ or $4n$ witnesses $\rho(x,A) < 1/n$, and we let $a_m = a_n$ for the minimum such n, which means that the sequence (a_m) stops at this a_n.

Given m, m', $m > m' > 0$, there are three decidable cases (1) $h[m] = m$, $h[m'] = m'$; (2) $h[m'] = m'$, $m' \le h[m] < m$; (3) $h[m] = h[m'] < m'$. In case (1) we have $\rho(x,a_m) < 1/m$ and $\rho(x,a_{m'}) < 1/m'$, which implies $\rho(a_m,a_{m'}) \le 2/m'$. For the other two cases, we also have $\rho(a_m,a_{m'}) \le 2/m'$. Therefore, (a_n) is a Cauchy sequence. By the completeness of A, (a_n) converges to a limit a.

Clearly, if $\rho(x,a) > 0$, then $\rho(x,a_m) > 1/m$ for some m. Then, there must be $n \leq m$ such that $\rho(x,A)(4n) \geq 3/4n$, and hence $\rho(x,A) > 1/4n$, for otherwise $h[m] = m$, $\rho(x,A)(4m) < 3/4m$, and by the construction of δ, $a_m = \delta(4h[m], h[m]) = \delta(4m,m)$ is such that $\rho(x,a_m) < 1/m$. □

4.3 Total Boundedness and Compactness

A metric space (X,ρ) is separable, if there exists a sequence (c_n) of elements of X such that for any $x \in X$ and $m > 0$, there exists n such that $\rho(x,c_n) < 1/m$. (X,ρ) is totally bounded, if for any $l > 0$, there exists a finite sequence x of elements in X, such that for any $y \in X$, there exists $i < lh(x)$ such that $\rho((x)_i, y) \leq 1/l$. Total boundedness implies separability. Because, total boundedness implies that we have a term $t[l]$ that gives a sequence of $lh(t[l])$ points that constitute a $1/l$ approximation. Then, we can construct a term $T[k]$ that enumerates all the points in $t[l]$ for $l > 0$, by letting $T[k] = (t[l])_i$, where $l > 0$ and $i \leq lh(t[l])$ are appropriately calculated from k by an elementary recursive function so that l and i can range over all $l > 0$ and $i \leq lh(t[l])$.

Lemma 4.3. *The following propositions about separability and total boundedness hold:*

(1) The product of a sequence of separable or totally bounded metric spaces is still separable or totally bounded.

(2) The image of a totally bounded space under a uniformly continuous function is still totally bounded.

(3) If $f : X \to \mathbb{R}$ is uniformly continuous and X is totally bounded, then $\sup f$ *and* $\inf f$ *exists.*

(4) Any totally bounded subset of a metric space is located.

(5) Any located subset of a totally bounded metric space is totally bounded.

(6) When X is totally bounded, the diameter of X exists:

$$diam(X) \equiv \sup\{\rho(x,y) : x, y \in X\}.$$

Proof. To see that (1) holds, let (X_n, ρ_n) be a sequence of metric spaces bounded by 1. First, note that to get a 2^{-N} approximation to $\prod_{n=0}^{\infty} X_n$, we can ignore X_n for $n > N$. Now, suppose that each (X_n, ρ_n) is totally bounded. Given any $l > 0$, let $N > 0$ be such that $2^{-N} < 1/2l$. Then, for each $n \leq N$, we can construct a finite sequence δ_n of finitely many points in X_n that constitute a $1/2l$ approximation to X_n. For each $n > N$, arbitrarily fix a point $x_n \in X_n$. Then, a $1/l$ approximation to $\prod_{n=0}^{\infty} X_n$ can consist of all the sequences $\left((\delta_0)_{i_0}, \ldots, (\delta_N)_{i_N}, x_{N+1}, x_{N+2}, \ldots\right)$, where $0 \leq i_0 < lh(\delta_0)$, ..., $0 \leq i_N < lh(\delta_N)$.

Next, suppose that each (X_n, ρ_n) is separable. That is, for each n, there exists a sequence δ_n such that $(\delta_n)_k$, $k = 0, 1, 2, \ldots$, constitute a dense subset of X_n. Note that for each N, all

$$x^N_{i_0,\dots,i_N} \equiv \left((\delta_0)_{i_0}, \dots, (\delta_N)_{i_N}, (\delta_{N+1})_0, (\delta_{N+2})_0, \dots \right)$$

with $i_0, \dots, i_N$ ranging over $0, 1, 2, \dots$ constitute a 2^{-N} approximation to $\prod_{n=0}^{\infty} X_n$. We can arrange all $x^N_{i_0,\dots,i_N}$ for all N and all $i_0, \dots, i_N$ ranging over $0, 1, 2, \dots$ into a sequence. This will be a dense sequence in $\prod_{n=0}^{\infty} X_n$. To see this, for any $x \in \prod_{n=0}^{\infty} X_n$ and $l > 0$, first choose N such that $2^{-N} < 1/2l$. Then, for each $n \le N$, we can (uniformly) choose i_n such that $\rho\left((\delta_n)_{i_n}, x_n\right) \le 1/2l$. It is easy to see that $\rho\left(x^N_{i_0,\dots,i_N}, x\right) \le 1/l$. Note that this construction does not require any recursion.

(2) follows from the definitions directly. (3) follows from Corollary 3.4 in Chap. 3. (4) follows from (3).

To see that (5) holds, let A be a located subset of a totally bounded metric space (X, ρ). A is non-empty. Therefore, we can choose $a \in A$. Given $l > 0$, let δ be a finite sequence of points of X that constitute a $1/4l$ approximation to X. Construct a finite sequence δ' of points in A as follows: $lh(\delta') = lh(\delta)$. For each $i < lh(\delta)$, $\rho((\delta)_i, A)$ exists. We can decide if $\rho((\delta)_i, A) > 1/4l$ or $\rho((\delta)_i, A) < 1/2l$. In the former case, let $(\delta')_i = a$, and in the latter case, there exists $a_i \in A$ such that $\rho((\delta)_i, a_i) < 1/2l$ and we let $(\delta')_i = a_i$. Then, for any $y \in A$, choose $i < lh(\delta)$ such that $\rho((\delta)_i, y) < 1/4l$. We must have $\rho((\delta)_i, A) < 1/2l$. Therefore, $\rho((\delta)_i, a_i) < 1/2l$. Then, $\rho(a_i, y) < 1/l$. Therefore, δ' is a $1/l$ approximation to A.

Finally, (6) is similar to (3) and (4). □

A metric space is compact if it is complete and totally bounded. Obviously, the product of compact spaces is compact.

Recall that we cannot quantify over all subsets of a set, because a subset is defined as an arbitrary pair of formulas. However, we can quantify over all compact subsets of a metric space. For a subset K of a metric space X, the witness for K's being totally bounded will give a sequence (x_n) of elements of K approximating all elements of K. Since K is closed, K is equal to the closure of the subset $\{x_n : n \in \mathbb{N}\}$. Therefore, instead of quantifying over all compact subsets of X, we can quantify over all sequences of elements of X.

The constructive proof in [6], pp. 96–98, of the following lemma requires some modifications.

Lemma 4.4. *Given a compact space X, for any $l > 0$, there exist finitely many compact subsets $X_1, \dots, X_n$ of X, such that each X_i has a diameter at most $1/l$ and the union of them all is X.*

Proof. X_is are constructed as a single family of subsets $X_{l,i}$, indexed by (l, i), $l > 0$, $i < n = N(l)$. Therefore, this does not really quantify over subsets. (We will assume that l is given and will suppress the subscript l below.) From the assumption, we have a function δ such that for $i \ge 0$, $\delta(i)$ is a finite sequence of elements of X that constitute a $3^{-i-2}l^{-1}$ approximation to X. Let $\delta(0) = \langle x_1, \dots, x_n \rangle$. The idea is to construct a sequence of finite sets $X_j^0, X_j^1, \ldots$, for each $j = 1, \ldots, n$. Let $X_j^0 = \{x_j\}$. Given X_j^i, X_j^{i+1} is obtained as follows: for each y in the sequence $\delta(i+1)$, decide if $\rho\left(y, X_j^i\right) < 3^{-i-1}l^{-1}$ or $\rho\left(y, X_j^i\right) > \frac{1}{2}3^{-i-1}l^{-1}$, and add y to X_j^{i+1} in the former case. Notice that these two cases can be distinguished by considering if

$$\rho\left(y, X_j^i\right)\left(8 \cdot 3^{i+1} l\right) < 3^{-i-1}l^{-1} - 4^{-1}3^{-i-1}l^{-1}$$

or not. Then X_j is the closure of the union $\cup_{i=0}^{\infty} X_j^i$. ([6], pp. 96–98)

However, this construction requires recursive constructions of sets on apparent. Since the elements of X_j^i are selected from the sequence $\delta(i)$, we can replace X_j^i by the indices of its members in the sequence $\delta(i)$. So, instead of constructing a sequence $X_j^0, X_j^1, \ldots$ of finite subsets of X, we construct a sequence of finite sequences of natural numbers, Z_j^i, $i = 0, 1, \ldots$, such that $Z_j^0 = \langle j \rangle$, and Z_j^{i+1} is the sequence of k, $k < lh(\delta(i+1))$, such that for some $h < lh\left(Z_j^i\right)$ we have

$$\rho\left((\delta(i))_{\left(Z_j^i\right)_h}, (\delta(i+1))_k\right)\left(8 \cdot 3^{i+1} l\right) < 3^{-i-1}l^{-1} - 4^{-1}3^{-i-1}l^{-1}.$$

So, Z_j^i is to be the sequence of indices of elements of X_j^i in the sequence $\delta(i)$. Given the term $\delta(i)$, Z_j^i can be constructed by bounded primitive recursion. We then define X_j^i as a single family of subsets:

$$x \in X_j^i \equiv \exists h < lh\left(Z_j^i\right)\left(x = (\delta(i))_{\left(Z_j^i\right)_h}\right).$$

Let $Y_j \equiv \cup_{i=0}^{\infty} X_j^i$ and let X_j be the closure of Y_j.

From the construction, we have $\rho\left(x, X_j^i\right) < 3^{-i-1}l^{-1}$ for each $x \in X_j^{i+1}$. It follows that for $x \in X_j^{i'}$, $i' > i$,

$$\rho\left(x, X_j^i\right) < \sum_{h=i}^{i'-1} 3^{-h-1}l^{-1} < 3^{-i}l^{-1}.$$

Therefore, $\cup_{h=0}^{i} X_j^h$ is a $3^{-i}l^{-1}$ approximation to Y_j and Y_j is totally bounded. Therefore, X_j is totally bounded.

For each $y \in X$, there exists j such that $\rho\left(y, (\delta(0))_j\right) < 3^{-2}l^{-1}$. By an induction on m we can show that $\rho\left(y, X_j^m\right) < 3^{-m-2}l^{-1}$. The inductive step from m to $m+1$ is the following. Since $\delta(m+1)$ is a $3^{-(m+1)-2}l^{-1}$ approximation to X, there exists index k such that $\rho\left(y, (\delta(m+1))_k\right) < 3^{-(m+1)-2}l^{-1}$. Therefore,

$$\rho\left((\delta(m+1))_k, X_j^m\right) < 3^{-(m+1)-2}l^{-1} + 3^{-m-2}l^{-1} < \frac{1}{2}3^{-m-1}l^{-1}.$$

It follows that k must be in the sequence Z_j^{m+1}. Therefore, $\rho\left(y, X_j^{m+1}\right) < 3^{-(m+1)-2}l^{-1}$. Note that $\rho\left(y, X_j^{m}\right) < 3^{-m-2}l^{-1}$ is a Σ_1^0 formula. However, in the inductive step, the witness for $\rho\left(y, X_j^{m+1}\right) < 3^{-(m+1)-2}l^{-1}$ comes from the witness for $\rho\left(y, (\delta(m+1))_k\right) < 3^{-(m+1)-2}l^{-1}$ and does not depend on the witness for the inductive assumption $\rho\left(y, X_j^{m}\right) < 3^{-m-2}l^{-1}$. Therefore, the construction of witnesses is iteratively bounded and can be obtained by a bounded primitive recursion. Then it follows that $y \in X_j$ and $X = \cup_{j=1}^{n} X_j$. □

The following corollary will be useful. Its proof is from [6], p. 98.

Corollary 4.5. *If $f : X \to \mathbb{R}$ is a continuous function on a compact space X, then there exists a sequence of real numbers (α_n) such that if β is a real number, $\beta >$ $\inf\{f(x) : x \in X\}$, and $\beta \neq \alpha_n$ for all n, then $\{x \in X : f(x) \leq \beta\}$ is a compact set. In particular, for any $x_0 \in X$, there exists a sequence of real numbers (α_n) such that if β is a real number, $\beta > 0$, and $\beta \neq \alpha_n$ for all n, then the closed ball $Sc(x_0, \beta)$ is compact.*

Proof. For $l > 0$, let $X_{l,1}, \ldots, X_{l,N(l)}$ be compact subsets of X such that $diam\left(X_{l,i}\right) <$ $1/l$ and $X = \cup_i X_{l,i}$. Let $c_{l,i} = \inf\left\{f(x) : x \in X_{l,i}\right\}$. Let (α_n) be any arrangement of $c_{l,i}$ into a sequence. Suppose that β satisfies the condition. We only need to show that $\{x \in X : f(x) \leq \beta\}$ is totally bounded. Let $l > 0$. For each i, we have either $\beta < c_{l,i}$ or $\beta > c_{l,i}$. In the latter case, we can choose $x_{l,i} \in X_{l,i}$ such that $c_{l,i} \leq f\left(x_{l,i}\right) < \beta$. Let Y_l be the set of such $x_{l,i}$. Note that Y_l is defined uniformly for l. Then, let $x \in X$ be such that $f(x) \leq \beta$. We have $x \in X_{l,i}$ for some i. Therefore, $c_{l,i} \leq f(x) \leq \beta$. We must have $\beta > c_{l,i}$ and $x_{l,i} \in Y_l$. Now, $\rho\left(x, x_{l,i}\right) \leq diam\left(X_{l,i}\right) < 1/l$. So, Y_l is a $1/l$ approximation to $\{x \in X : f(x) \leq \beta\}$. □

Recall that by Cantor's theorem (i.e., Theorem 3.2), there is always a β in any interval such that $\beta \neq \alpha_n$ for all n.

4.4 The Stone-Weierstrass Theorem

We will verify that the constructive proof of this theorem in [6], pp. 104–108, is available to strict finitism.

For a compact space X and any metric space Y, the set $C(X,Y)$ of (uniformly) continuous functions from X to Y is a metric space with the metric

$$\rho(f,g) \equiv_{df} \sup\{\rho(f(x), g(x)) : x \in X\}.$$

This exists because X is totally bounded and f, g are uniformly continuous. $C(X,\mathbb{R})$ will be denoted as $C(X)$. The norm $\|f\|$ for $f \in C(X)$ is

$$\|f\| \equiv_{df} \sup\{|f(x)| : x \in X\},$$

and therefore $\rho(f,g) = \|f - g\|$.

A polynomial p of degree m in n variables can be represented by a sequence of $(m+1)^n+2$ real numbers, where the first two numbers are m and n and the others are the coefficients. For a sequence x of n real numbers, the value of p at x can be represented by a term $p(x)$. Similarly, for a sequence $h=\langle h_1,...,h_n\rangle$ of n functions in $C(X)$, the composition $p\circ h$ can be constructed and is also a function in $C(X)$. Then, for G a subset of $C(X)$, the algebra $\mathscr{A}(G)$ of functions generated from G is the subset of $C(X)$ such that $f\in\mathscr{A}(G)$ if and only if for some polynomial p and some sequence h of functions from G, $f=p\circ h$.

A subset G of $C(X)$ is separating if (1) for any $\varepsilon>0$ there exists $\delta>0$ such that for any $x,y\in X$ with $\rho(x,y)\geq\varepsilon$, there exists $g\in G$ such that $|g(z)|\leq\varepsilon$ for $z\in Sc(x,\delta)$, and $|g(z)-1|\leq\varepsilon$ for $z\in Sc(y,\delta)$; and (2) for any $\varepsilon>0$ there exists $\delta>0$ such that for any $x\in X$, there exists $g\in G$ such that $|g(z)-1|\leq\varepsilon$ for $z\in Sc(x,\delta)$.

First, it can be proved that the function $f(x)=|x|$ can be approximated on $[-1,1]$ by polynomials $p(x)$ such that $p(0)=0$. The proof in [6] is based on the power series expansion of $(1-t)^{\frac{1}{2}}$ and Taylor's theorem. It is already finitistic. Then, it can be directly verified that if G is a subset of $C(X)$ and f,g belong to the closure of $\mathscr{A}(G)$, then $|f|$, $\max\{f,g\}$, and $\min\{f,g\}$ also belong to the closure of $\mathscr{A}(G)$, and if $\inf\{|f(x)|:x\in X\}>0$, then f^{-1} also belongs to the closure $\mathscr{A}(G)$ ([6], Lemma 5.11, 5.12, pp. 105–106).

Then, we have Stone-Weierstrass Theorem:

Theorem 4.6. *For a separating subset G of $C(X)$, $\mathscr{A}(G)$ is dense in $C(X)$.*

Its proof in [6], pp. 106–108, is already within strict finitism. Note that the proof there does not use any recursive construction, except for using the finite sum and finite product of functions in $C(X)$. For the latter, it implicitly assumes the following proposition.

Lemma 4.7. *If H is the closure of $\mathscr{A}(G)$ in $C(X)$, and for all $i\leq n$, $f_i\in H$, then $\sum_{i=0}^n f_i\in H$ and $\prod_{i=0}^n f_i\in H$.*

Proof. We check the second conclusion. Since for each $i\leq n$ there exists M such that $\|f_i\|\leq M$, it follows that there exists $M>1$ such that for all $i\leq n$, $\|f_i\|\leq M$. Since $f_i\in H$ for all $i\leq n$, there exists g such that for all $i\leq n$ and all $k\geq 0$, $g(i)_k\in\mathscr{A}(G)$, and for all $i\leq n$, $\|g(i)_k-f_i\|\to 0$ as $k\to\infty$. Then, $\prod_{i=0}^n g(i)_k\in\mathscr{A}(G)$ and we can show that $\|\prod_{i=0}^n g(i)_k-\prod_{i=0}^n f_i\|\to 0$ as $k\to\infty$. For this, we need the following assertion: if for all $i\leq n$, $a_i\in\mathbb{R}$, $b_i\in\mathbb{R}$, $|a_i|\leq M$, $|b_i|\leq M$, and $|a_i-b_i|\leq r$, then

$$\left|\prod_{i=0}^n a_i-\prod_{i=0}^n b_i\right|\leq(n+1)M^n r.$$

This can be easily proved just as Lemma 3.5 and 3.6. □

As corollaries, it follows that if G is the subset of $C(X)$ of functions $f(x)=\rho(x,x_0)$ for some $x_0\in X$, then $\mathscr{A}(G)$ is dense in $C(X)$, and it also follows that the set of polynomial functions on a compact subset $X\subseteq\mathbb{R}^n$ is dense in $C(X)$ ([6], pp. 108–109).

Applying the Stone-Weierstrass theorem in such a general format will consist of a few steps. For instance, consider approximating continuous functions on an interval $[a,b]$ by polynomials. First, the construction in the proof of the theorem is instantiated by letting $X \equiv [a,b]$ and $G \equiv$ the set of polynomial functions. We can construct a witness for the claim that G is separating. Then, for a concrete function $f \in C([a,b])$ represented by a term, together with its modulus of continuity, the construction in the proof gives a $g[n]$ such that for $n > 0$, $\|g[n] - f\| < 1/n$. $g[n]$ is a polynomial, which means that it gives a degree and a sequence of coefficients, each depending on n. Now, when we apply these to real things, the physics quantity represented by f will not be literally continuous, but as before the proof gives a concrete term that relates a degree of approximation of f by $g[n]$ to a 'degree of continuity' of f. Therefore, in a concrete application, the resulted approximation of a discrete, finite quantity by a polynomial could still be literally true.

Chapter 5
Complex Analysis

We will show that some basic notions and results of complex analysis can be developed in strict finitism, including the notions of integration, differentiation, and analytic functions, and including results such as Cauchy's integral theorem, Cauchy's integral formula, some properties regarding maximum values and zeros, and the Fundamental Theorem of Algebra. We will again follow the major ideas in Chap. 5 of Bishop and Bridges [6]. Some of the proofs have to be revised to fit into our more restrictive framework here.

5.1 Basic Notions

A complex number $x+y\mathrm{i}$ will be represented as a pair $\langle x,y\rangle$ of real numbers. The set $\mathbb{C}$ of complex numbers is defined as

$$(z\in\mathbb{C})\equiv_{df} Seq(z)\wedge lh(z)=2\wedge(z)_0\in\mathbb{R}\wedge(z)_1\in\mathbb{R},$$
$$(z=_{\mathbb{C}}z')\equiv_{df}(z)_0=_{\mathbb{R}}(z')_0\wedge(z)_1=_{\mathbb{R}}(z')_1.$$

Therefore, a complex number is represented by a term of the type $(o\to o)$. We will write $z=\langle x,y\rangle$ as $x+y\mathrm{i}$, and write $(z)_0$ as $\mathrm{Re}(z)$, and write $(z)_1$ as $\mathrm{Im}(z)$. We will ignore the type difference and treat a real or rational number x as a complex number $x+0\mathrm{i}$. Similarly, for rational numbers a,b, we also call $a+b\mathrm{i}\equiv\langle a,b\rangle$ a complex number, by which we mean $\langle\lambda m.a,\ \lambda m.b\rangle$. We will frequently use complex numbers with rational real and imaginary parts to approximate an arbitrary complex number. We will call them *rational complex numbers*. Note that a rational complex number actually has the base type o.

Common functions on $\mathbb{C}$ or $\mathbb{C}\times\mathbb{C}$, such as norm, conjugation, sum, product, finite sum, finite product and so on, are easily defined, and the basic properties of them are also simple. The standard inequality on $\mathbb{C}$ is

$$w\neq z\equiv_{df}|w-z|>0.$$

F. Ye, *Strict Finitism and the Logic of Mathematical Applications*, Synthese Library 355, DOI 10.1007/978-94-007-1347-5_5, © Springer Science+Business Media B.V. 2011

For convenience we will make this convention: When in a context we treat $z = \langle x, y \rangle$ as a complex number, we let $z(m) \equiv \langle x(2m), y(2m) \rangle$, instead of $\langle x(m), y(m) \rangle$. So, $z(m)$ is a complex number and $|z - z(m)| < 1/m$. That is, $z(m)$ is a $1/m$ approximation to z in the norm of complex numbers.

We treat $\mathbb{C}$ as a metric space with the metric $\rho(z, w) = |z - w|$. Therefore, notions such as uniform continuity, uniform convergence, completeness, compactness and so on are available. It is obvious that $\mathbb{C}$ is complete and separable, and that finite discs, rectangles and many other regular subsets of $\mathbb{C}$ are totally bounded. Recall that a compact subset of a metric space is a complete and totally bounded subset, and thus a compact set is located. On the complex plane, obviously, all closed discs and rectangles are compact subsets. We usually treat only quite regular compact subsets.

For K a compact subset of $\mathbb{C}$, U any subset, and $r > 0$, we denote

$$K_r \equiv_{df} \{z \in \mathbb{C} : \rho(z, K) \leq r\},$$
$$(K \Subset U) \equiv_{df} \exists r > 0 (K_r \subseteq U).$$

In case $K \Subset U$, we say that K is well-contained in U. If f is continuous on K, we define

$$\|f\|_K \equiv_{df} \sup\{|f(z)| : z \in K\}.$$

Recall that we can quantify over compact subsets, by quantifying over sequences of points generating the compact subset. When we refer to an arbitrary compact subset of $\mathbb{C}$ in a claim, it can be either understood as a schematic claim involving an arbitrary formula defining the subset, together with the condition that it is compact, or understood as a universal quantification over sequences of complex numbers, together with the condition that they generate compact sets. On the other side, all references to open subsets are understood as schematic claims.

By quantifying over compact subsets, we can define continuity on open sets:

Definition 5.1. Let U be an open subset of $\mathbb{C}$. $f : U \to \mathbb{C}$ is continuous on U, if it is continuous on any compact subset K of $\mathbb{C}$ such that $K \Subset U$.

We will also consider functions from $\mathbb{R}$ or intervals of real numbers to $\mathbb{C}$. If $\gamma : [a, b] \to \mathbb{C}$ is such a function, then there are $\alpha, \beta : [a, b] \to \mathbb{R}$ such that $\gamma(t) = \alpha(t) + \beta(t)\mathrm{i}$ for $t \in [a, b]$. The notions of continuity, convergence and so on apply to such functions as functions from a metric space to another. Moreover, when γ is continuous, for a another continuous $\eta : [a, b] \to \mathbb{C}$, we define $\gamma' = \eta$ as: there exists δ, a modulus of differentiability, such that for any $n > 0$, $t', t \in [a, b]$, if $|t' - t| \leq \delta(n)$, then

$$|\gamma(t') - \gamma(t) - \eta(z)(t' - t)| \leq |t' - t| / n.$$

It is easy to see that if γ' exists, then we must have

$$\gamma'(t) = \alpha'(t) + \beta'(t)\mathrm{i}.$$

Similarly, the Riemann integration on γ is defined as

$$\int_a^b \gamma(t)\,\mathrm{d}t \equiv_{df} \int_a^b \alpha(t)\,\mathrm{d}t + \mathrm{i}\int_a^b \beta(t)\,\mathrm{d}t.$$

Note that from the basic inequalities of real numbers, we have

$$\left|\int_a^b \gamma(t)\,\mathrm{d}t\right| \le \int_a^b |\gamma(t)|\,\mathrm{d}t. \tag{5.1}$$

Then, we can define the derivative of a complex function. It has the same format as the derivative of a real function.

Definition 5.2. Let K be a compact subset of $\mathbb{C}$, $f, g : K \to \mathbb{C}$ be continuous functions. g is a derivative of f on K, if there exists δ, a modulus of differentiability, such that for any $n > 0$, $z, w \in K$, if $|w - z| \le \delta(n)$, then

$$|f(w) - f(z) - g(z)(w - z)| \le |w - z|/n.$$

We denote this fact as $f' = g$, or $g(z) = \mathrm{d}f(z)/\mathrm{d}z$. Let U be an open subset of $\mathbb{C}$, $f, g : U \to \mathbb{C}$ be continuous functions. g is a derivative of f on U, if for any compact subset K of $\mathbb{C}$ such that $K \Subset U$, $f' = g$ on K.

The basic properties of derivative can be easily verified. They include the chain rule in two formats. First, if K_1, K_2 are compact subsets, and $f : K_1 \to K_2$ and $g : K_2 \to \mathbb{C}$ are continuous, and f', g' exist, then $(g \circ f)'$ exists, and for $z \in K_1$,

$$(g \circ f)'(z) = g'(f(z))\,f'(z).$$

This can be verified by the definition directly. If U_1, U_2 are open subsets, $f : U_1 \to U_2$ and $g : U_2 \to \mathbb{C}$ are continuous, and for any compact subset $K \Subset U_1$ we have $f(K) \Subset U_2$, and f', g' exist, then $(g \circ f)'$ exists and the above chain rule holds. Next, if $\gamma : [a, b] \to K$ and $f : K \to \mathbb{C}$ are differentiable, then we also have

$$(f \circ \gamma)'(t) = f'(\gamma(t))\,\gamma'(t).$$

Partial derivatives are defined similarly:

Definition 5.3. Let K be a compact subset of $\mathbb{C}$, and let $f, g : K \to \mathbb{C}$ be continuous functions. g is a partial derivative of f with respect to x on K, denoted as $g = f'_x$, if there exists δ, a modulus of partial differentiability with respect to x, such that for any $n > 0$, $x + y\mathrm{i}, x' + y\mathrm{i} \in K$, if $|x' - x| \le \delta(n)$, then

$$\left|f(x' + y\mathrm{i}) - f(x + y\mathrm{i}) - g(x + y\mathrm{i})(x' - x)\right| \le \left|x' - x\right|/n.$$

$g = f'_y$ on K, and partial derivatives on open subsets are defined similarly.

From the definitions it directly follows that if f is differentiable, then f has partial derivatives and $f_x = f'$, $f_y = \mathrm{i}f'$. Reversely, by the definitions, it is easy to verify that if f is continuous on an open set U and f_x, f_y exists on U such that $f_y = \mathrm{i}f_x$ on U, then f' exists on U. We omit the details here.

Now, consider integration. A path γ on the parameter interval $[a,b]$, $a \le b$, is a continuous function $\gamma\colon [a,b] \to \mathbb{C}$ such that there exists a finite sequence $\langle t_0,...,t_n\rangle$ of real numbers, $n > 0$, such that $t_0 \equiv a \le t_1 \le ... \le t_n \equiv b$ and γ is differentiable on each $[t_i,t_{i+1}]$, $i = 0,...,n-1$. Intervals $[t_i,t_{i+1}]$ are the intervals of differentiability of the path. γ is a path in a subset A, if $\gamma\colon [a,b] \to A$. Sometimes we ignore the parameter interval and simply say 'γ is a path'. The path is a closed path or a loop if $\gamma(a) = \gamma(b)$. Note that we do not require strict inequalities $a < b$ or $t_i < t_{i+1}$ in the definition. This allows us to claim that if γ is a path on $[a,b]$, and $a \le a' \le b' \le b$, then γ is a path on $[a',b']$.

If $\delta : [c,d] \to \mathbb{C}$ is another path, then γ and δ are equivalent, if $a = b$ and $c = d$ and $\delta(c) = \gamma(a)$, or $a < b$ and $c < d$ and for $t \in [c,d]$

$$\delta(t) = \gamma\left(a + \frac{t-c}{d-c}(b-a)\right).$$

Note that given $[c,d]$, $c < d$, this path δ can always be defined. We also consider γ and δ equivalent, if we can divide $[a,b]$ and $[c,d]$ into an equal number of sub-intervals so that γ and δ are equivalent on each sub-interval in the above sense.

If $\delta : [c,d] \to \mathbb{C}$ is another path such that $\gamma(b) = \delta(c)$, then we can construct the concatenation $\gamma + \delta : [a, b + (d-c)] \to \mathbb{C}$, such that

$$(\gamma+\delta)(t) = \gamma(t), \text{ for } t \in [a,b],$$
$$(\gamma+\delta)(t) = \delta(c+t-b), \text{ for } t \in [b, b+(d-c)].$$

It is also called the sum of γ and δ. The reverse $-\gamma$ of the path γ is also a path $-\gamma\colon [a,b] \to \mathbb{C}$,

$$-\gamma(t) \equiv_{df} \gamma(a+b-t).$$

We write $\gamma + (-\delta)$ as $\gamma - \delta$.

For $a < b$, a path γ is a linear path on $[a,b]$ from z_1 to z_2, if

$$\gamma(t) = z_1 + (z_2 - z_1)\frac{t-a}{b-a}.$$

In particular, $\gamma(t) = z_1 + (z_2 - z_1)t$ is the linear path on $[0,1]$ from z_1 to z_2.

The length $|\gamma|$ of γ is defined as

$$|\gamma| \equiv_{df} \int_a^b |\gamma'(t)|\,\mathrm{d}t.$$

Here, the integration on the right hand side is understood as the sum of Riemann integrations, $\sum_{i=0}^{n-1} \int_{t_i}^{t_{i+1}} |\gamma'(t)|\,\mathrm{d}t$, on the intervals of differentiability. Note that $\gamma'(t)$ is continuous on $[t_i,t_{i+1}]$. In particular, for a linear path γ on $[a,b]$ from z_1 to z_2, we have $|\gamma| = |z_2 - z_1|$. Moreover, we have $|\gamma + \delta| = |\gamma| + |\delta|$.

We use $car(\gamma)$ to denote the closure of the range $\gamma([a,b])$ of γ. Since γ is uniformly continuous, $car(\gamma)$ is a compact subset and $\rho(z, car(\gamma))$ always exists.

Then, we can define integration.

Definition 5.4. If γ is a path on $[a,b]$, and f is a continuous function on $car(\gamma)$, then the integration of f along the path γ is defined as

$$\int_{\gamma} f(z)\,\mathrm{d}z \equiv_{df} \int_{a}^{b} f(\gamma(t))\,\gamma'(t)\,\mathrm{d}t,$$

where the integration on the right hand side is the sum of Riemann integrations on the intervals of differentiability for the path γ.

Note that the term $\int_{\gamma} f(z)\,\mathrm{d}z$ actually contains the witnesses for f's being a continuous function on $car(\gamma)$ and γ's being a path with the parameter interval $[a,b]$, although we do not indicate these in the notation. If $z_1 \neq z_2$ and γ is the linear path on $[0,1]$ from z_1 to z_2, then we write

$$\int_{z_1}^{z_2} f(z)\,\mathrm{d}z \equiv_{df} \int_{\gamma} f(z)\,\mathrm{d}z$$

By the definition, we have

Lemma 5.5. *If f is differentiable on $car(\gamma)$, then*

$$\int_{\gamma} f'\mathrm{d}z = f(\gamma(b)) - f(\gamma(a)).$$

In particular, if γ is a loop, then

$$\int_{\gamma} f'\mathrm{d}z = 0. \tag{5.2}$$

Proof. By the chain rule of differentiation,

$$\int_{\gamma} f'\mathrm{d}z = \int_{a}^{b} f'(\gamma(t))\,\gamma'(t)\,\mathrm{d}t = \int_{a}^{b} (f\circ\gamma)'(t)\,\mathrm{d}t = f(\gamma(b)) - f(\gamma(a)).$$

□

In particular, for any polynomial function $p(z)$, there is a polynomial function f such that $f' = p$. Therefore, when γ is a loop, for any polynomial $p(z)$, $\int_{\gamma} p\mathrm{d}z = 0$.

If γ and δ are equivalent, then obviously $\int_{\gamma} f(z)\,\mathrm{d}z = \int_{\delta} f(z)\,\mathrm{d}z$. Similarly, by the definitions,

$$\int_{\gamma+\delta} f(z)\,\mathrm{d}z = \int_{\gamma} f(z)\,\mathrm{d}z + \int_{\delta} f(z)\,\mathrm{d}z, \tag{5.3}$$

$$\int_{-\gamma} f(z)\,\mathrm{d}z = -\int_{\gamma} f(z)\,\mathrm{d}z.$$

Moreover, if (γ_n) is a sequence of paths such that γ_n is on $[a_n, a_{n+1}]$ and $\gamma_n(a_{n+1}) = \gamma_{n+1}(a_{n+1})$. Then, for each N, we can construct the path $\sum_{n=0}^{N-1} \gamma_n : [a_0, a_N] \to \mathbb{C}$ and show that

$$\int_{\sum_{n=0}^{N-1}\gamma_n} f(z)\,\mathrm{d}z = \sum_{n=0}^{N-1}\int_{\gamma_n} f(z)\,\mathrm{d}z. \tag{5.4}$$

The path $\sum_{n=0}^{N-1}\gamma_n$ is constructed directly. For any $t \in [a_0, a_N]$, $\left(\sum_{n=0}^{N-1}\gamma_n\right)(t)$ is approximated by comparing the approximation $t(m)$ with each $a_n(m)$. If it has been decided that $a_n < t < a_{n+1}$ for some n, then we use the approximation to $\gamma_n(t)$. Otherwise, use the least n such that it is still not decided that $a_n < t$. Note that $\sum_{n=0}^{N}\gamma_n$ is not defined recursively from $\sum_{n=0}^{N-1}\gamma_n$, but it follows that

$$\sum_{n=0}^{N}\gamma_n = \sum_{n=0}^{N-1}\gamma_n + \gamma_N. \tag{5.5}$$

Then, the equality (5.4) above follows from (5.5), (5.3) and Lemma 3.5.

From the definition of $|\gamma|$ and (5.1), we have

$$\left|\int_\gamma f\mathrm{d}z\right| \le \|f\|_\gamma |\gamma|,$$

where $\|f\|_\gamma \equiv \sup\{|f(\gamma(t))| : t \in [a,b]\}$.

5.2 Differentiable and Analytic Functions

The exponential function is defined by

$$\mathrm{e}^z \equiv \mathrm{e}^x(\cos y + \mathrm{i}\sin y), \text{ for } z = x + y\mathrm{i},$$

where $\mathrm{e}^x, \cos y, \sin y$ are real functions defined in Chap. 3. Then, it is easy to verify that partial derivatives $(\mathrm{e}^z)_x$ and $(\mathrm{e}^z)_y$ exist and $(\mathrm{e}^z)_y = \mathrm{i}(\mathrm{e}^z)_x$. Therefore,

$$\mathrm{d}\mathrm{e}^z/\mathrm{d}z = (\mathrm{e}^z)_x = \mathrm{e}^z.$$

Using relevant equations for the real functions $\mathrm{e}^x, \cos y, \sin y$, it is easy to verify that

$$\mathrm{e}^{z_1+z_2} = \mathrm{e}^{z_1}\mathrm{e}^{z_2}.$$

Using the exponential function, we can calculate a simple integration:

Lemma 5.6. *Let γ be a closed path in $\mathbb{C}$ and $\rho(z_0, car(\gamma)) > 0$. Then for some n,*

$$\int_\gamma \frac{\mathrm{d}z}{(z-z_0)} = 2n\pi\mathrm{i}.$$

Proof. We may assume that $z_0 = 0$ and γ is on $[0,1]$. Let

$$w \equiv \int_\gamma z^{-1} \mathrm{d}z = \int_0^1 \gamma^{-1}(t)\,\gamma'(t)\,\mathrm{d}t.$$

First, we want to show that $\mathrm{e}^w = 1$. Some direct calculation using the chain rule and the fundamental theorem of calculus for real functions shows that

$$\frac{\mathrm{d}}{\mathrm{d}u}\left(\gamma(u)\,\mathrm{e}^{-\int_0^u \gamma^{-1}(t)\gamma'(t)\mathrm{d}t}\right) = 0.$$

Therefore, $\gamma(u)\,\mathrm{e}^{-\int_0^u \gamma^{-1}(t)\gamma'(t)\mathrm{d}t}$ is a constant. Considering its value at $u = 0$, we have $\mathrm{e}^{\int_0^u \gamma^{-1}(t)\gamma'(t)\mathrm{d}t} = \gamma(u)/\gamma(0)$. Since γ is a closed path, letting $u = 1$, we see that $\mathrm{e}^w = 1$ holds. That is,

$$\mathrm{e}^{\mathrm{Re}\,w}\left(\cos(\mathrm{Im}\,w) + \mathrm{i}\sin(\mathrm{Im}\,w)\right) = 1.$$

Therefore, $\mathrm{Im}\,w = 2n\pi$ and $\mathrm{Re}\,w = 0$. □

n in the lemma is called the winding number of γ around z_0 and is denoted as $j(\gamma, z_0)$. This winding number $j(\gamma, z_0)$ does not change when z_0 moves without crossing the path γ:

Lemma 5.7. *Suppose that γ is a closed path on the interval $[0,1]$, $\rho(z_0, car(\gamma)) > 0$, and $\rho(z_1, car(\gamma)) > 0$, and suppose that there is a continuous function $\eta : [0,1] \to \mathbb{C}$ such that $\eta(0) = z_0$, $\eta(1) = z_1$, and for some $\delta > 0$, $|\eta(s) - \gamma(t)| \geq \delta$ for all $s, t \in [0,1]$. Then, $j(\gamma, z_0) = j(\gamma, z_1)$.*

Proof. From the assumption, it is easy to see that the function $h : [0,1] \to \mathbb{C}$,

$$h(s) \equiv \int_\gamma \frac{\mathrm{d}z}{(z - \eta(s))} = \int_a^b \frac{\gamma'(t)\,\mathrm{d}t}{(\gamma(t) - \eta(s))},$$

is continuous. Since it takes only discrete values $2n\pi\mathrm{i}$ by the lemma above, it must be constant. □

Lemma 5.8. *Let γ be a closed path, and let f be a continuous function on $car(\gamma)$. Suppose that the function g on $-car(\gamma)$ is defined by*

$$g(z) \equiv \frac{1}{2\pi\mathrm{i}} \int_\gamma \frac{f(\zeta)}{(\zeta - z)^n} \mathrm{d}\zeta,$$

where $n > 0$. Then, g is differentiable and its derivative is given by

$$g'(z) \equiv \frac{n}{2\pi\mathrm{i}} \int_\gamma \frac{f(\zeta)}{(\zeta - z)^{n+1}} \mathrm{d}\zeta.$$

Proof. Suppose that K is a compact set and $K_r \subseteq -car(\gamma)$ for some $r > 0$. We have $\rho(z, car(\gamma)) \geq r$ for any $z \in K$. For $w, z \in K$,

$$
\begin{aligned}
& g(w)-g(z)-g'(z)(w-z) \\
&= \frac{1}{2\pi i}\int_\gamma f(\zeta)\left(\frac{1}{(\zeta-w)^n}-\frac{1}{(\zeta-z)^n}-\frac{n(w-z)}{(\zeta-z)^{n+1}}\right)d\zeta \\
&= \frac{(w-z)}{2\pi i}\int_\gamma f(\zeta)\left(\frac{\sum_{k=0}^{n-1}(\zeta-z)^{n-k-1}(\zeta-w)^k}{(\zeta-w)^n(\zeta-z)^n}-\frac{n}{(\zeta-z)^{n+1}}\right)d\zeta \\
&= \frac{(w-z)}{2\pi i}\int_\gamma \frac{f(\zeta)}{(\zeta-z)^n}\left(\frac{\sum_{k=0}^{n-1}(\zeta-w)^k\left((\zeta-z)^{n-k}-(\zeta-w)^{n-k}\right)}{(\zeta-w)^n(\zeta-z)}\right)d\zeta.
\end{aligned}
$$

Note that $|\zeta-w|,|\zeta-z|\geq r$ and $|\zeta-w|$, $|\zeta-z|$, $|f(\zeta)|$ are bounded above, while $\left|(\zeta-z)^{n-k}-(\zeta-w)^{n-k}\right|$ can be made arbitrarily small when $|w-z|$ is small. It is easy to see that the condition for derivative holds. □

Since $g(z)\equiv\int_\gamma\frac{d\zeta}{(\zeta-z)}$ is constant for z in a small disc in $-car(\gamma)$ by Lemma 5.7, $g'(z)=0$. Therefore, we have

Corollary 5.9. *Let γ be a closed path in $\mathbb{C}$, and suppose that $\rho(z_0,car(\gamma))>0$. Then, for $n>1$,*

$$\int_\gamma\frac{dz}{(z-z_0)^n}=0.$$

Lemma 5.10. *Suppose that U is an open set and f is differentiable on each disc $Sc(z,r)\Subset U$ for $r>0$. Then, f is differentiable on U.*

Proof. For each $z\in U$, there exists $t=t(z)$ such that $Sc(z,t)\subseteq U$. Therefore, f is differentiable on $Sc(z,t/2)$. Let $g_z=f'$ on $Sc(z,t/2)$. Define the function g on U by $g(z)\equiv_{df}g_z(z)$. We want to show that $f'=g$ on U.

Let K be a compact set such that $K_r\Subset U$ for some $r>0$. There exists a finite $r/2$ approximation $\{z_1,...,z_n\}$ to K. For each i, $Sc(z_i,r)\subseteq K_r$. Therefore, f is differentiable on $Sc(z_i,r)$. Let f_i' be the derivative of f on $Sc(z_i,r)$. Since the derivative is unique, for each $z\in Sc(z_i,r/2)$ we have $f_i'=g_z$ on $Sc(z,\min(t(z)/2,r/2))$. In particular, $f_i'(z)=g_z(z)=g(z)$. Now, let δ be the common modulus of differentiability for all f_i', $i=1,...,n$. For any $N>0$, $w,z\in K$ such that $|w-z|\leq\min(r/2,\delta(N))$, there exists i such that $z\in Sc(z_i,r/2)$ and therefore $w\in Sc(z_i,r)$. Since f_i' is the derivative of f on $Sc(z_i,r)$ with the modulus of differentiability δ,

$$\left|f(w)-f(z)-f_i'(z)(w-z)\right|\leq|w-z|/N.$$

Since $f_i'(z)=g(z)$,

$$|f(w)-f(z)-g(z)(w-z)|\leq|w-z|/N.$$

This means that $f'=g$ on K. So, $f'=g$ on U. □

Now consider analytic functions. A polygonal path $\gamma\equiv poly(z_0,...,z_{n-1},z_0)$ is a closed path with the parameter interval $[0,n]$ such that for some sequence

$\langle z_0, ..., z_{n-1} \rangle$ of complex numbers, γ restricted to $[i, i+1]$ is a linear path on $[i, i+1]$ from $\gamma(i) = z_i$ to $\gamma(i+1) = z_{i+1}$, for $i = 0, ..., n-1$, where $z_n = z_0$. We can also take *poly* as a function from the set of finite sequences of complex numbers to the set of polygonal paths. Note that its length

$$|poly(z_0, ..., z_{n-1}, z_0)| = \sum_{i=0}^{n-1} |z_{i+1} - z_i|, \text{ where } z_n = z_0.$$

Let S be a subset of $\mathbb{C}$. S is convex, if for any finite sequence $\langle z_1, ..., z_n \rangle$ of complex numbers in S and any finite sequence $\langle a_1, ..., a_n \rangle$ of real numbers such that $a_i \in [0,1]$ for each i and $\sum_{i=1}^{n} a_i = 1$, we have $\sum_{i=1}^{n} a_i z_i \in S$. If S is convex, the closure of S is also convex.

Given a finite sequence $\langle z_1, ..., z_n \rangle$ of complex numbers, the set $span(\langle z_1, ..., z_n \rangle)$ spanned by $\langle z_1, ..., z_n \rangle$ is the closure of

$$\left\{ \sum_{i=1}^{n} a_i z_i : \langle a_1, ..., a_n \rangle \in [0,1]^{<\infty} \wedge \sum_{i=1}^{n} a_i = 1 \right\}. \tag{5.6}$$

It is easy to verify that the set (5.6) is convex, and therefore $span(\langle z_1, ..., z_n \rangle)$ is convex. Moreover, it is easy to see that the set (5.6) is totally bounded. Therefore, $span(\langle z_1, ..., z_n \rangle)$ is compact.

If $\gamma \equiv poly(z_0, ..., z_{n-1}, z_0)$, then we write $span(\langle z_0, ..., z_{n-1} \rangle)$ as $span(\gamma)$. Then, we can define analytic functions.

Definition 5.11. A continuous function f on an open set U is analytic on U, if $\int_\gamma f \mathrm{d}z = 0$ for any triangular path $\gamma = poly(z_1, z_2, z_3, z_1)$ such that $span(\gamma) \Subset U$.

Lemma 5.12. *Any differentiable function f on U is analytic on U.*

Proof. We follow the proof in [6], but we need some new constructions. First, note that by the (uniform) continuity of relevant functions, if $\gamma_1 \equiv poly(z_1, z_2, z_3, z_1)$ and $\eta_1 \equiv poly(w_1, w_2, w_3, w_1)$ are two paths such that $span(\gamma) \Subset U$ and $span(\eta) \Subset U$ and $\max\{|w_i - z_i| : i = 1,2,3\}$ is sufficiently small, then $\left| \int_\gamma f \mathrm{d}z - \int_\eta f \mathrm{d}z \right|$ will also be arbitrarily small. Therefore, it suffices to prove that for any triangular path $\gamma_1 \equiv poly(z_1, z_2, z_3, z_1)$ determined by *rational* complex numbers z_1, z_2, z_3 whose span is well-contained in U, we have $\int_{\gamma_1} f(z) \mathrm{d}z = 0$.

Then, given such a triangular path γ_1, given any $l > 0$, we will construct a sequence (γ_n) of paths by constructing a sequence $(\langle z_{n,1}, z_{n,2}, z_{n,3} \rangle)$ of triples of *rational* complex numbers and letting $\gamma_n \equiv poly(z_{n,1}, z_{n,2}, z_{n,3}, z_{n,1})$. The sequence (γ_n) of paths will satisfy the following:

(a) $span(\gamma_n) \subseteq span(\gamma_1) \Subset U$;
(b) $|\gamma_{n+1}| = \frac{1}{2} |\gamma_n|$;
(c) $\left| \int_{\gamma_{n+1}} f(z) \mathrm{d}z \right| > \frac{1}{4} \left| \int_{\gamma_n} f(z) \mathrm{d}z \right| - 2^{-3n} l^{-1}$.

The sequence is constructed as follows. Suppose that $\langle z_{n,1}, z_{n,2}, z_{n,3} \rangle$ and $\gamma_n \equiv poly(z_{n,1}, z_{n,2}, z_{n,3}, z_{n,1})$ have been constructed. Write

$$\gamma_{n,1} \equiv poly\left(z_{n,1}, \frac{1}{2}(z_{n,1}+z_{n,2}), \frac{1}{2}(z_{n,1}+z_{n,3}), z_{n,1}\right),$$
$$\gamma_{n,2} \equiv poly\left(z_{n,2}, \frac{1}{2}(z_{n,2}+z_{n,3}), \frac{1}{2}(z_{n,1}+z_{n,2}), z_{n,2}\right),$$
$$\gamma_{n,3} \equiv poly\left(z_{n,3}, \frac{1}{2}(z_{n,1}+z_{n,3}), \frac{1}{2}(z_{n,2}+z_{n,3}), z_{n,3}\right),$$
$$r_{n,4} \equiv poly\left(\frac{1}{2}(z_{n,1}+z_{n,2}), \frac{1}{2}(z_{n,2}+z_{n,3}), \frac{1}{2}(z_{n,1}+z_{n,3}), \frac{1}{2}(z_{n,1}+z_{n,2})\right).$$

It is easy to see that $\left|\gamma_{n,i}\right| = \frac{1}{2}\left|\gamma_n\right|$ for each i, and

$$\int_{\gamma_n} f(z)\,\mathrm{d}z = \sum_{i=1}^{4} \int_{\gamma_{n},i} f(z)\,\mathrm{d}z.$$

Computing $\left|\int_{\gamma_{n,i}} f(z)\,\mathrm{d}z\right| \left(K_0 2^{3n} l\right)$ for some constant K_0, $i = 1,2,3,4$, we can choose i such that

$$\begin{aligned}\left|\int_{\gamma_{n,i}} f(z)\,\mathrm{d}z\right| &> \frac{1}{4}\left(\sum_{i=1}^{4}\left|\int_{\gamma_{n,i}} f(z)\,\mathrm{d}z\right|\right) - 2^{-3n}l^{-1}\\ &\geq \frac{1}{4}\left|\int_{\gamma_n} f(z)\,\mathrm{d}z\right| - 2^{-3n}l^{-1}.\end{aligned}$$

Then let $\gamma_{n+1} = \gamma_{n,i}$. Obviously, the construction can be done by a bounded primitive recursion and the conditions (a), (b), (c) above are satisfied.

From these conditions, it follows that $\left|\gamma_{n+1}\right| = 2^{-n}\left|\gamma_1\right|$ and

$$\left|\int_{\gamma_{n+1}} f(z)\,\mathrm{d}z\right| > 2^{-2n}\left(\left|\int_{\gamma_1} f(z)\,\mathrm{d}z\right| - l^{-1}\sum_{i=1}^{n} 2^{-i}\right).$$

(Note that we resort to Lemma 3.6 here.) From the last inequality, it follows that

$$\left|\int_{\gamma_1} f(z)\,\mathrm{d}z\right| < 2^{2n}\left|\int_{\gamma_{n+1}} f(z)\,\mathrm{d}z\right| + l^{-1}.$$

Note that we have $|w-z| \leq \left|\gamma_{n+1}\right| = 2^{-n}\left|\gamma_1\right|$ for $w,z \in span\left(\gamma_{n+1}\right)$. Since f is differentiable on $span\left(\gamma_1\right)$, we can choose n sufficiently large so that for $z, z_0 \in span\left(\gamma_{n+1}\right)$,

$$\left|f(z) - f(z_0) - f'(z_0)(z-z_0)\right| \leq |z-z_0|/l.$$

Fix an $x_0 \in span\left(\gamma_{n+1}\right)$. Note that $f(z_0) - f'(z_0)(z-z_0)$ is a polynomial in z. Therefore,

$$\int_{\gamma_{n+1}} \left(f(z_0) + f'(z_0)(z-z_0)\right)\mathrm{d}z = 0.$$

Then,

$$\begin{aligned}
\left|\int_{\gamma_1} f(z)\,\mathrm{d}z\right| &< 2^{2n}\left|\int_{\gamma_{n+1}} f(z)\,\mathrm{d}z\right| + l^{-1} \\
&= 2^{2n}\left|\int_{\gamma_{n+1}} \left(f(z) - f(z_0) - f'(z_0)(z - z_0)\right)\mathrm{d}z\right| + l^{-1} \\
&\leq 2^{2n}\left\|f(z) - f(z_0) - f'(z_0)(z - z_0)\right\|_{\gamma_{n+1}} \left|\gamma_{n+1}\right| + l^{-1} \\
&= 2^{2n} l^{-1} \left\|z - z_0\right\|_{\gamma_{n+1}} \left|\gamma_{n+1}\right| + l^{-1} \\
&\leq 2^{2n} l^{-1} \left|\gamma_{n+1}\right|^2 + l^{-1} = l^{-1}\left(\left|\gamma_1\right|^2 + 1\right).
\end{aligned}$$

Since l is arbitrary, we see that $\int_{\gamma_1} f(z)\,\mathrm{d}z = 0$. Therefore, f is analytic on U. □

Lemma 5.13. *If U is a convex open set, and f is analytic on U, and the function g on U is defined by $g(z) \equiv \int_{z_0}^{z} f(\zeta)\,\mathrm{d}\zeta$ for some $z_0 \in U$, then $g' = f$ on U.*

Proof. Let K be a compact set and $K_r \Subset U$ for some $r > 0$. Given any $w, z \in K$, since U is convex and open, $span(z_0, w, z) \Subset U$. Since f is analytic,

$$\int_{z_0}^{w} f(\zeta)\,\mathrm{d}\zeta + \int_{w}^{z} f(\zeta)\,\mathrm{d}\zeta + \int_{z}^{z_0} f(\zeta)\,\mathrm{d}\zeta = 0.$$

Therefore,

$$\begin{aligned}
|g(w) - g(z) - f(z)(w - z)| &= \left|\int_{z}^{w} f(\zeta)\,\mathrm{d}\zeta - \int_{z}^{w} f(z)\,\mathrm{d}\zeta\right| \\
&\leq \|f(\zeta) - f(z)\|_{span(z,w)} |w - z|.
\end{aligned}$$

When $|w - z| \leq r$, we have $span(z, w) \subseteq K_r$. By the assumption, f is continuous on K_r. Let $\delta(n)$ be a modulus of continuity of f on K_r. Then, given $n > 0$, when $|w - z| \leq \min(r, \delta(n))$, we have $|\zeta - z| \leq \min(r, \delta(n))$ for $\zeta \in span(z, w)$. Therefore, $\|f(\zeta) - f(z)\|_{span(z,w)} \leq 1/n$, that is,

$$|g(w) - g(z) - f(z)(w - z)| \leq |w - z|/n.$$

So, $g' = f$ on K. □

From this lemma and Lemma 5.5, it easily follows that

Corollary 5.14. *If U is a convex open set, and f is analytic on U, then $\int_\gamma f\mathrm{d}z = 0$ for every closed path γ in U.*

Definition 5.15. Let $K \subseteq \mathbb{C}$ be a compact set, and let $\gamma_0, \gamma_1 : [0,1] \to K$ be two closed paths on $[0,1]$ in K. γ_0 and γ_1 are homotopic in K, if there exists a continuous function $\sigma : [0,1] \times [0,1] \to K$, such that $\gamma_0(t) = \sigma(0,t)$, and $\gamma_1(t) = \sigma(1,t)$ for $t \in [0,1]$, and for each $s \in [0,1]$, $\sigma_s : [0,1] \to K$, $\sigma_s(t) \equiv \sigma(s,t)$, is a closed path. γ_0 and γ_1 are homotopic in an open set U, if they are homotopic in a compact set

$K \Subset U$. σ is called a homotopy of γ_0 and γ_1. When γ_1 is a constant path (a point), we say that γ_0 is null-homotopic.

Since every path is equivalent to a unique path on $[0,1]$, the definition can be generalized to arbitrary paths.

We have Cauchy integral theorem.

Theorem 5.16. *Suppose that f is analytic on an open set U, and γ_0, γ_1 are homotopic closed paths in U. Then,*

$$\int_{\gamma_0} f(z)\,\mathrm{d}z = \int_{\gamma_1} f(z)\,\mathrm{d}z.$$

In particular, if γ_0 is null-homotopic then $\int_{\gamma_0} f(z)\,\mathrm{d}z = 0$.

Proof. We may assume that γ_0 and γ_1 are paths on $[0,1]$. Let K be a compact subset such that $K_r \subseteq U$ for some $r > 0$ and γ_0 and γ_1 are paths in K. Let σ be a homotopy of γ_0 and γ_1. Since σ is uniformly continuous on $[0,1] \times [0,1]$, there exists natural number $N > 0$, such that $|\sigma(s',t') - \sigma(s,t)| \leq 1/2r$ whenever $(s',t'),(s,t) \in [0,1] \times [0,1]$ and $|s'-s| \leq 1/N$ and $|t'-t| \leq 1/N$. For each $i = 0,\ldots,N$, $j = 0,\ldots,N-1$, let $\gamma_{i,j}$ be the path that is the restriction of $\sigma_{i/N}$ to $\left[\frac{j}{N}, \frac{j+1}{N}\right]$. For each $j = 0,\ldots,N$, $i = 0,\ldots,N-1$, let $\eta_{i,j}$ be the linear path from $\sigma\left(\frac{i}{N}, \frac{j}{N}\right)$ to $\sigma\left(\frac{i+1}{N}, \frac{j}{N}\right)$. Then,

$$\delta_{i,j} \equiv \gamma_{i,j} + \eta_{i,j+1} - \gamma_{i+1,j} - \eta_{i,j}$$

is a closed path, and it is in K_r since $\sigma\left(\frac{i}{N}, \frac{j}{N}\right) \in K$ and $\left|\sigma(s',t') - \sigma\left(\frac{i}{N}, \frac{j}{N}\right)\right| \leq 1/2r$ for any point $\sigma(s',t')$ on the path. Moreover, it is easy to see that the convex set $span(\delta_{i,j}) \Subset U$. Therefore, by the Corollary above,

$$\int_{\gamma_{i,j}} f(z)\,\mathrm{d}z + \int_{\eta_{i,j+1}} f(z)\,\mathrm{d}z - \int_{\gamma_{i+1,j}} f(z)\,\mathrm{d}z - \int_{\eta_{i,j}} f(z)\,\mathrm{d}z = 0.$$

Note that $\eta_{i,0} = \eta_{i,N}$. Adding up these equations for $i,j = 0,\ldots,N-1$, we have

$$\int_{\gamma_0} f(z)\,\mathrm{d}z - \int_{\gamma_1} f(z)\,\mathrm{d}z = 0.$$

□

An open set U is connected if any two points in U are connected by a path in U. U is simply connected, if further any closed path in U is null-homotopic. The basic properties of simply connected open sets follow from the above lemmas easily. For instance, suppose that f is analytic on a simply connected open set U. Then, for any closed path γ in U, $\int_\gamma f(z)\,\mathrm{d}z = 0$, and if γ_0, γ_1 are two paths from the same starting point to the same end point, then $\int_{\gamma_0} f(z)\,\mathrm{d}z = \int_{\gamma_1} f(z)\,\mathrm{d}z$. Moreover, fix a point $z_0 \in U$. For any $z \in U$, there exists a path γ from z_0 to z, and let $g(z) \equiv \int_\gamma f(z)\,\mathrm{d}z$. Then, $g' = f$ on U.

Lemma 5.17. *Suppose that f is differentiable on an open set U and $z_0 \in U$. Then, there exists an analytic function g on U such that for $z \in U$, $z \neq z_0$,*

$$g(z) = \frac{f(z) - f(z_0)}{z - z_0}.$$

Proof. Recall that $z(m)$, $m \geq 0$, are rational complex numbers approximating z. Given any $z \in U$, by modifying each $z(m)$ slightly if necessary, we can construct a sequence (q_m) of rational complex numbers approximating z such that $q_m \neq z_0(m)$ for all m. Let

$$g_m(z) = \frac{f(q_m) - f(z_0(m))}{q_m - z_0(m)}.$$

$(g_m(z))_m$ is a Cauchy sequence. To see this, let $Sc(z_0, r) \Subset U$ and let δ be a modulus of differentiability for f on $Sc(z_0, r)$. We may assume that $\delta(N) < r$ for any N. For any $N > 0$, we can first decide if $|z - z_0| > \delta(2N)/2$ or $|z - z_0| < \delta(2N)$. In the former case, $|q_m - z_0(m)| > \delta(2N)/2$ for all sufficiently large m. Then, since f is continuous, we see that $|g_m(z) - g_n(z)| < 1/N$ for all sufficiently large m, n. In the latter case, $|q_m - z_0(m)| < \delta(2N)$ and $q_m, z_0(m) \in Sc(z_0, r)$ for all sufficiently large m. Then, since δ is a modulus of differentiability for f on $Sc(z_0, r)$,

$$\left| \frac{f(q_m) - f(z_0(m))}{q_m - z_0(m)} - f'(z_0) \right| \leq 1/2N$$

for all sufficiently large m. Therefore, we have again $|g_m(z) - g_n(z)| < 1/N$ for all sufficiently large m, n. Therefore, $(g_m(z))_m$ is a Cauchy sequence.

Let $g(z) \equiv \lim_{m \to \infty} g_m(z)$. Obviously, $g(z) = \frac{f(z) - f(z_0)}{z - z_0}$ for $z \neq z_0$. Moreover, $|g(z) - f'(z_0)| \leq 1/2N$ when $|z - z_0| \leq \delta(2N)$. $g(z)$ is continuous on U. To see this, let $K \subseteq U$ be a compact set. For any $w, z \in K$, if $|w - z| \leq \delta(2N)/4$, by deciding if $|z - z_0| \leq 3\delta(2N)/4$ or $|z - z_0| \geq \delta(2N)/2$, we have either $|z - z_0| \leq \delta(2N)$ and $|w - z_0| \leq \delta(2N)$, or $|z - z_0| \geq \delta(2N)/4$ and $|w - z_0| \geq \delta(2N)/4$. In the former case, we already have $|g(w) - g(z)| \leq 1/N$. In the latter case, $g(z) = \frac{f(z) - f(z_0)}{z - z_0}$ and $g(w) = \frac{f(w) - f(z_0)}{w - z_0}$. Then, using the modulus of continuity for f on K, it is easy to see that we will have $|g(w) - g(z)| \leq 1/N$ when $|w - z|$ is sufficiently small.

To see that g is analytic on U, let $\gamma = poly(z_1, z_2, z_3, z_1)$ be any triangular path such that $K = span(\gamma) \Subset U$. Let $r > 0$ be such that $K_r \Subset U$. We have $\rho(z_0, K) > r/2$ or $\rho(z_0, K) < r$. In the former case, $K_{r/2} \Subset U - \{z_0\}$. $g(z) = \frac{f(z) - f(z_0)}{z - z_0}$ is differentiable on $K_{r/2}$. By Lemma 5.12, it is analytic on $K_{r/2}$. Therefore, $\int_\gamma g(z)\,\mathrm{d}z = 0$. In the latter case, $z_0 \in K_r$, and

$$\gamma = poly(z_0, z_1, z_2, z_0) + poly(z_0, z_2, z_3, z_0) + poly(z_0, z_3, z_1, z_0),$$

and the spans of these three polygons are well-contained in U. So, it suffices to prove $\int_\gamma g(z)\,\mathrm{d}z = 0$ for a path γ like $poly(z_0, z_1, z_2, z_0)$. For that, given any $\varepsilon > 0$, we first approximate $z_1 - z_0$, $z_2 - z_0$. If both $|z_1 - z_0|$ and $|z_2 - z_0|$ are sufficiently small, then $|\gamma|$ can be sufficiently small so that $\left|\int_\gamma g(z)\,\mathrm{d}z\right| < \varepsilon$. If $|z_1 - z_0| > 0$, then choose a point w on the line from z_0 to z_1 but sufficiently close to z_0. We will have $\int_\gamma g(z)\,\mathrm{d}z$ sufficiently close to $\int_{poly(w,z_1,z_2,w)} g(z)\,\mathrm{d}z$. But the latter equals 0 because $z_0 \in \mathbb{C} - span(w,z_1,z_2,w)$. Therefore, we will still have $\left|\int_\gamma g(z)\,\mathrm{d}z\right| < \varepsilon$. So, $\int_\gamma g(z)\,\mathrm{d}z = 0$. □

This can be generalized to multiple points.

Lemma 5.18. *Suppose that f is differentiable on an open set U, and $z_1,\ldots,z_n \in U$ is a finite sequence of points in U such that $z_i \neq z_j$ for $i \neq j$, $i,j = 1,\ldots,n$, and $f(z_i) = 0$ for $i = 1,\ldots,n$. Then, there exists an analytic function g on U such that for $z \in U$, $z \neq z_0$, ..., z_n,*

$$g(z) = \frac{f(z)}{(z-z_1)\ldots(z-z_n)}.$$

Proof. The construction in the last lemma can be done for $z_1,\ldots,z_n$ simultaneously. For instance, we can choose q_m such that $q_m \neq z_1(m)$, ..., $q_m \neq z_n(m)$. Similarly, we can assume that $\delta(N) < |z_i - z_j|/4$ for all N, for all $i \neq j$, and then in proving Cauchyness and continuity and so on, we can decide if $|z - z_i| > \delta(2N)/2$ or $|z - z_i| < \delta(2N)$ for each i. □

Then, we have Cauchy's integral formula.

Theorem 5.19. *Suppose that f is differentiable on the open set U, and $z \in U$, and γ is a closed path in $U - \{z\}$ and is null-homotopic in U. Then,*

$$j(\gamma,z)f(z) = (2\pi\mathrm{i})^{-1}\int_\gamma \frac{f(\zeta)}{(\zeta - z)}\mathrm{d}\zeta.$$

Proof. By the lemma above, $\frac{f(\zeta)-f(z)}{(\zeta-z)}$ can be extended into an analytic function on U. Since γ is null-homotopic in U, by the Cauchy integral theorem above,

$$\int_\gamma \frac{f(\zeta)-f(z)}{(\zeta - z)}\mathrm{d}\zeta = 0.$$

Then, by Lemma 5.6,

$$\int_\gamma \frac{f(\zeta)}{(\zeta - z)}\mathrm{d}\zeta = \int_\gamma \frac{f(z)}{(\zeta - z)}\mathrm{d}\zeta = (2\pi\mathrm{i})\,j(\gamma,z)f(z).$$

□

We also have Cauchy's integral formula for any order of derivative of a differentiable function f.

Theorem 5.20. *Suppose that f is differentiable on the open set U and $z \in U$. Then, $f^{(n)}(z)$ exists for all n. Moreover, if γ is a closed path in U and is null-homotopic in U, then for any $z \in U$ such that $z \in U - car(\gamma)$, we have, for $n \geq 0$,*

$$j(\gamma,z)f^{(n)}(z) = (2\pi\mathrm{i})^{-1}n!\int_\gamma \frac{f(\zeta)}{(\zeta - z)^{n+1}}\mathrm{d}\zeta.$$

Proof. Recall that 'f has any order of derivatives on U' means that for any $n > 0$, there exists a sequence $\langle f_0, ..., f_n \rangle$ of functions on U such that $f_0 = f$ and $f_{i+1} = f_i'$ on U for $i = 0, ..., n-1$. Now, suppose that f is differentiable on U. For any $z \in U$, there exists $r > 0$ such that $Sc(z, r) \Subset U$. Define a function f_n on U by

$$f_n(z) \equiv (2\pi i)^{-1} n! \int_{\gamma_0} \frac{f(\zeta)}{(\zeta - z)^{n+1}} d\xi,$$

where γ_0 is the circular path of radius $r > 0$ about z. f_n is a function on U, for the right hand side is independent of the choice of r. We have $f_0 = f$ by the Cauchy integral formula above.

We want to show that $f_n' = f_{n+1}$ on U. By Lemma 5.10, it suffices to show that $f_n' = f_{n+1}$ on any $Sc(z_0, r) \Subset U$. So, suppose that $Sc(z_0, r) \Subset U$ and choose $r' > r$ such that $Sc(z_0, r') \Subset U$. Let γ' be the circular path of radius r' about z_0. For each $z \in Sc(z_0, r)$, choose s sufficiently small such that $Sc(z, s) \subseteq Sc(z_0, r')$; then γ' is homotopic in U with the circular path γ_z of radius s about z. Therefore, by the Cauchy integral theorem,

$$f_n(z) = (2\pi i)^{-1} n! \int_{\gamma_z} \frac{f(\zeta)}{(\zeta - z)^{n+1}} d\xi = (2\pi i)^{-1} n! \int_{\gamma'} \frac{f(\zeta)}{(\zeta - z)^{n+1}} d\xi.$$

Similarly,

$$f_{n+1}(z) = (2\pi i)^{-1} (n+1)! \int_{\gamma'} \frac{f(\zeta)}{(\zeta - z)^{n+2}} d\xi$$

on $Sc(z_0, r)$. Then, by Lemma 5.8, $f_n' = f_{n+1}$ on $Sc(z_0, r)$. This completes the proof that f has derivatives of any order and $f_n = f^{(n)}$.

Now, for each $z_0 \in U - car(\gamma)$, $j(\gamma, z)$ is an integer and is a constant k_0 on a disk $Sc(z_0, r) \Subset U - car(\gamma)$. When $k_0 > 0$, for $z \in Sc(z_0, r)$ let

$$f_n(z) \equiv k_0^{-1} (2\pi i)^{-1} n! \int_{\gamma} \frac{f(\zeta)}{(\zeta - z)^{n+1}} d\xi.$$

Then, it similarly follows that on $Sc(z_0, r)$ we have $f_0 = f$ and $f_n' = f_{n+1}$ for $n \geq 0$, and therefore $f_n = f^{(n)}$. Henceforth, Cauchy's integral formula holds for z on $Sc(z_0, r)$. Similarly, when $k_0 = 0$, let

$$f_n(z) \equiv (2\pi i)^{-1} n! \int_{\gamma} f(\xi)(\xi - z)^{-n-1} d\xi.$$

Then, $f_0 = 0$ and again $f_n' = f_{n+1} = 0$ for $n \geq 0$ on $Sc(z_0, r)$. Cauchy's integral formula holds again. □

Corollary 5.21. *If f is analytic on an open set U, then it is differentiable on U.*

Proof. For each disc $Sc(z, r) \Subset U$, f is analytic on $Sc(z, r)$. By Lemma 5.13, $f = g'$ for some g on $Sc(z, r)$. Then, by the theorem above, f is differentiable on $Sc(z, r)$. Therefore, by Lemma 5.10, f is differentiable on U. □

Take a circular path around z as γ in the theorem, we have Cauchy's inequality.

Corollary 5.22. *If f is differentiable on $Sc(z,r)$, then*

$$\left|f^{(n)}(z)\right| \leq n!r^{-n}\sup\{|f(w)| : |w-z| = r\}.$$

Taylor's theorem is also available.

Theorem 5.23. *If f is differentiable on $S(z_0,r)$, $r>0$, then on $S(z_0,r)$,*

$$f(z) = \sum_{n=0}^{\infty} \frac{f^{(n)}(z_0)}{n!}(z-z_0)^n.$$

Proof. Let r' be any real number such that $0<r'<r$. It suffices to show that $\sum_{n=0}^{\infty}\frac{f^{(n)}(z_0)}{n!}(z-z_0)^n$ uniformly converges to f on $Sc(z_0,r')$. Let γ be the circular path around z_0 of the radius $(r'+r)/2$. By the theorems above

$$f(z) = (2\pi\mathrm{i})^{-1}\int_\gamma \frac{f(\zeta)}{(\zeta-z)}\mathrm{d}\zeta, \text{ for } z \in Sc(z_0,r'),$$
$$f^{(n)}(z_0) = (2\pi\mathrm{i})^{-1}n!\int_\gamma \frac{f(\zeta)}{(\zeta-z_0)^{n+1}}\mathrm{d}\zeta.$$

Therefore, for $z\in Sc(z_0,r')$,

$$\begin{aligned}
&f(z) - \sum_{n=0}^{N}\frac{f^{(n)}(z_0)}{n!}(z-z_0)^n \\
&= (2\pi\mathrm{i})^{-1}\int_\gamma f(\zeta)\left(\frac{1}{(\zeta-z)} - \sum_{n=0}^{N}\frac{(z-z_0)^n}{(\zeta-z_0)^{n+1}}\right)\mathrm{d}\zeta \\
&= (2\pi\mathrm{i})^{-1}\int_\gamma f(\zeta)\left(\frac{1}{(\zeta-z)} - \frac{1-\left(\frac{z-z_0}{\zeta-z_0}\right)^{N+1}}{(\zeta-z)}\right)\mathrm{d}\zeta \\
&= \frac{(z-z_0)^{N+1}}{2\pi\mathrm{i}}\int_\gamma \frac{f(\zeta)}{(\zeta-z)(\zeta-z_0)^{N+1}}\mathrm{d}\zeta.
\end{aligned}$$

Note that for $z\in Sc(z_0,r')$ and ζ on the path γ, $|z-z_0|\leq r'$, $|\zeta-z_0| = (r'+r)/2$, $|\zeta-z|\geq(r-r')/2$, and since f is continuous on $S(z_0,r)$, $|f(\zeta)|\leq C$ for some constant C. Therefore,

$$\left|f(z) - \sum_{n=0}^{N}\frac{f^{(n)}(z_0)}{n!}(z-z_0)^n\right| \leq \frac{(r'+r)C}{(r-r')}\left(\frac{r'}{(r'+r)/2}\right)^{N+1}.$$

So, $\sum_{n=0}^{\infty}\frac{f^{(n)}(z)}{n!}(z-z_0)^n$ converges to f uniformly on $Sc(z_0,r')$. □

5.3 Maximum Value and Zero

We define the boundary of $Sc(z,r)$ as

$$\Gamma(z,r) \equiv_{df} \{w : |w-z| = r\}.$$

We will also use $\Gamma(z,r)$ to denote the circular path around z of the radius r. Recall that for a compact set K, $\|f\|_K$ is the supreme of $|f|$ on K. We define

$$m(f,K) \equiv_{df} \inf\{f(z) : z \in K\}.$$

We will consider only discs in this section, but the constructions can be generalized to sets of other regular shapes. The following lemma shows that a differentiable function takes its maximum value at the boundary of a disc.

Lemma 5.24. *Suppose that f is differentiable on $Sc(z_0,r)$, then*

$$\|f\|_{Sc(z_0,r)} = \|f\|_{\Gamma(z_0,r)}.$$

Proof. For any $\varepsilon > 0$, since f is continuous on $Sc(z_0,r)$, we can find $\delta > 0$, $0 < \delta < r$, such that for z, $r-\delta \le |z-z_0| \le r$, we have $|f(z)| \le \|f\|_{\Gamma(z_0,r)} + \varepsilon$. We may assume that $\delta < 1$. Find a $w \in Sc(z_0,r)$ such that

$$|f(w)| \ge \|f\|_{Sc(z_0,r)} - \delta\varepsilon. \tag{5.7}$$

Let $s \equiv r - |w-z_0|$. If $s < \delta$, then $|f(w)| \le \|f\|_{\Gamma(z_0,r)} + \varepsilon$ and thus

$$\|f\|_{Sc(z_0,r)} \le \|f\|_{\Gamma(z_0,r)} + \varepsilon(1+\delta) < \|f\|_{\Gamma(z_0,r)} + 2\varepsilon.$$

Then, assume that $s > 0$. The circle $\Gamma(w,s)$ meets the circle $\Gamma(z_0,r)$ at a point a. Let γ_1 be the arc of $\Gamma(w,s)$ centered at a with the length $|\gamma_1| = \delta$. Then, γ_1 lies inside $Sc(z_0,r) - Sc(z_0,r-\delta)$ and $|f(z)| \le \|f\|_{\Gamma(z_0,r)} + \varepsilon$ for $z \in \gamma_1$. Let $\gamma_2 \equiv \Gamma(w,s) - \gamma_1$. By Cauchy's integral formula,

$$\begin{aligned}
|f(w)| &= (2\pi)^{-1} \left| \int_{\Gamma(w,s)} \frac{f(\zeta)}{(\zeta - w)} \mathrm{d}\zeta \right| \\
&\le (2\pi)^{-1} \left(\left| \int_{\gamma_1} \frac{f(\zeta)}{(\zeta - w)} \mathrm{d}\zeta \right| + \left| \int_{\gamma_2} \frac{f(\zeta)}{(\zeta - w)} \mathrm{d}\zeta \right| \right) \\
&\le (2\pi)^{-1} \left(s^{-1} \left(\|f\|_{\Gamma(z_0,r)} + \varepsilon \right) |\gamma_1| + s^{-1} \|f\|_{Sc(z_0,r)} |\gamma_2| \right) \\
&= (2\pi s)^{-1} \left(\left(\|f\|_{\Gamma(z_0,r)} + \varepsilon \right) \delta + \|f\|_{Sc(z_0,r)} (2\pi s - \delta) \right).
\end{aligned}$$

From this and (5.7), we get $\|f\|_{Sc(z_0,r)} \le \|f\|_{\Gamma(z_0,r)} + (1+2\pi s)\varepsilon$. Since ε is arbitrary, we have $\|f\|_{Sc(z_0,r)} = \|f\|_{\Gamma(z_0,r)}$. □

The following lemmas give a condition about when the function $f(z)$ has a zero in $Sc(z_0,r)$.

Lemma 5.25. *Suppose that f is differentiable on $Sc(z_0,r)$ and $m(f,Sc(z_0,r)) < m(f,\Gamma(z_0,r))$. Then, $m(f,Sc(z_0,r)) = 0$.*

Proof. Given any $n>0$, we can decide if $m(f,Sc(z_0,r)) > 1/2n$ or $m(f,Sc(z_0,r)) < 1/n$. Suppose that $m(f,Sc(z_0,r)) > 1/2n$. Then, $f^{-1}(z)$ is differentiable on $\|f\|_{Sc(z_0,r)}$. However, by the assumption,

$$\|f^{-1}\|_{Sc(z_0,r)} = m(f,Sc(z_0,r))^{-1} > m(f,\Gamma(z_0,r))^{-1} = \|f^{-1}\|_{\Gamma(z_0,r)}.$$

This contradicts the last lemma. Therefore, $m(f,Sc(z_0,r)) < 1/n$ for any $n>0$. That is, $m(f,Sc(z_0,r)) = 0$. □

Note that this lemma does not conclude that there exists z such that $f(z)=0$. The following lemmas will give that, under a stronger condition.

Definition 5.26. A function f on an open set U is non-zero on U, if there exists $z \in U$ such that $f(z) \neq 0$. f is strongly non-zero on U, if for each $z \in U$ and each $\varepsilon > 0$ such that $Sc(z,\varepsilon) \Subset U$, we have $\|f\|_{Sc(z,\varepsilon)} > 0$.

Lemma 5.27. *Suppose that $f(z) \equiv \sum_{i=0}^{n} a_i z^i$ is a polynomial function. Then, the following are equivalent:*
(1) f is strongly non-zero on $\mathbb{C}$.
(2) f is non-zero on $\mathbb{C}$.
(3) There exists some m such that $a_m \neq 0$.

Proof. It is trivial that (1) implies (2) and (2) implies (3).

Suppose that $|a_m| > 0$. We first show that given any $\varepsilon > 0$, there exists $\delta \leq \varepsilon$ such that $f(\delta) \neq 0$. Let $A \equiv 1 + \sum_{i=0}^{n} |a_i|$. Let $\delta \equiv \min\left(\varepsilon, \frac{|a_m|}{4A}, 1\right)$. Then, we have

$$\left| \sum_{i=m+1}^{n} a_i \delta^i \right| \leq \delta^{m+1} \sum_{i=m+1}^{n} |a_i| \leq \frac{|a_m|\,\delta^m}{4}.$$

For each $i = 0,\ldots,m-1$, we decide if $|a_i| < \frac{|a_m|\delta^{m-i}}{4n}$ or $|a_i| > \frac{|a_m|\delta^{m-i}}{8n}$. If the former holds for all $i = 0,\ldots,m-1$, we have

$$\left| \sum_{i=0}^{m-1} a_i \delta^i \right| \leq \sum_{i=0}^{m-1} |a_i|\,\delta^i \leq \frac{|a_m|\,\delta^m}{4}.$$

Therefore,

$$|f(\delta)| \geq |a_m|\,\delta^m - \left| \sum_{i=0}^{m-1} a_i \delta^i \right| - \left| \sum_{i=m+1}^{n} a_i \delta^i \right| \geq \frac{|a_m|\,\delta^m}{2} > 0.$$

Otherwise, let m' be the least $i < m$ such that $|a_i| > \frac{|a_m|\delta^{m-i}}{8n} > 0$. We can repeat the process for a_i. The construction can be performed by a bounded primitive recursion. Therefore, we can construct δ such that $f(\delta) \neq 0$.

Now, for any $w \in \mathbb{C}$, we can write $f(z)$ as $\sum_{i=0}^{n} b_i (z-w)^i$. The above argument shows that $\sum_{i=0}^{n} b_i (\delta - w)^i \neq 0$ for some δ. Therefore, there exists $b_m \neq 0$. Then, the same argument shows that for some $\delta < \varepsilon$, $\sum_{i=0}^{n} b_i \delta^i \neq 0$. That is, $f(w+\delta) \neq 0$. So, $\|f\|_{Sc(w,\varepsilon)} > 0$. □

The following three lemmas are preparations for locating zeros.

Lemma 5.28. *Suppose that h is differentiable on $Sc(w,r)$, and $r > 2\delta > 0$, $0 \leq t \leq r - 2\delta$. Let $z_1, \ldots, z_n \in Sc(w,\delta)$, and let*

$$f(z) \equiv (z - z_1)\ldots(z - z_n)\, h(z)$$

for $z \in Sc(w,r)$. Then,

$$\|f\|_{Sc(w,t)} \leq \left(\frac{t+\delta}{r-\delta}\right)^n \|f\|_{Sc(w,r)}.$$

Proof. Note that for $z \in Sc(w,t)$, we have $|z - z_n| \leq t + \delta$, and for $z \in \Gamma(w,r)$, we have $|z - z_n| \geq r - \delta$. Therefore, applying Lemma 5.24, we have

$$\|f\|_{Sc(w,t)} \leq (t+\delta)^n \|h\|_{Sc(w,t)} = (t+\delta)^n \|h\|_{\Gamma(w,t)},$$
$$\|f\|_{Sc(w,r)} \geq (r-\delta)^n \|h\|_{Sc(w,r)} \geq (r-\delta)^n \|h\|_{Sc(w,t)} = (r-\delta)^n \|h\|_{\Gamma(w,t)}.$$

Therefore, the inequality holds. □

Lemma 5.29. *Suppose that f is differentiable on $Sc(w,r)$, and $r > 2\delta > 0$, $0 \leq t \leq r - 2\delta$, $d > 0$, $\alpha > 0$, and suppose that $z_1, \ldots, z_n \in Sc(w,\delta)$ are such that $|z_k - z_j| \geq d$ for $k, j = 1, \ldots, n$ and $k \neq j$, and $|f(z_k)| \leq \alpha$ for $k = 1, \ldots, n$. Then,*

$$\|f\|_{Sc(w,t)} \leq \left(\frac{t+\delta}{r-\delta}\right)^n \|f\|_{Sc(w,r)} + 2\alpha n \left(\frac{r+\delta}{d}\right)^{n-1}.$$

Proof. Let g be defined by

$$g(z) \equiv \sum_{k=1}^{n} f(z_k) \prod_{j \neq k} \frac{z - z_j}{z_k - z_j}.$$

Then, $g(z_k) = f(z_k)$. For $z \in Sc(w,r)$, we have $|z - z_j| \leq r + \delta$. Therefore,

$$|g(z)| \leq n\alpha \left(\frac{r+\delta}{d}\right)^{n-1}. \tag{5.8}$$

Let $h(z) \equiv f(z) - g(z)$. Then, $h(z_k) = 0$ for $k = 1, \ldots n$. By Lemma 5.18, there exists a differentiable function h' such that

$$h(z) = (z - z_1)\ldots(z - z_n)\, h'(z).$$

Then, by the lemma above,

$$\begin{aligned}\|h\|_{Sc(w,t)} &\le \left(\frac{t+\delta}{r-\delta}\right)^n \|h\|_{Sc(w,r)} \\ &\le \left(\frac{t+\delta}{r-\delta}\right)^n \left(\|f\|_{Sc(w,r)} + \|g\|_{Sc(w,r)}\right) \\ &\le \left(\frac{t+\delta}{r-\delta}\right)^n \|f\|_{Sc(w,r)} + \|g\|_{Sc(w,r)} \,.\end{aligned}$$

So, by (5.8),

$$\begin{aligned}\|f\|_{Sc(w,t)} &\le \|h\|_{Sc(w,t)} + \|g\|_{Sc(w,t)} \\ &\le \left(\frac{t+\delta}{r-\delta}\right)^n \|f\|_{Sc(w,r)} + 2\alpha n \left(\frac{r+\delta}{d}\right)^{n-1} .\end{aligned}$$

□

For the following lemma, note that we have here the condition that f is *strongly* non-zero.

Lemma 5.30. *Suppose that f is a strongly non-zero differentiable function on an open set U, and r, δ, w are such that $r > 3\delta$ and $Sc(w,r) \Subset U$. Then, there exists s such that $\delta/2 < s < \delta$ and $m(f, \Gamma(w,s)) > 0$.*

Proof. Let $t \equiv r - 3\delta > 0$. Since f is strongly non-zero, $\|f\|_{Sc(w,t)} > 0$. Choose $N > 1$ such that

$$N > (r/\delta - 2)\left(\frac{2\|f\|_{Sc(w,r)}}{\|f\|_{Sc(w,t)}} - 1\right).$$

Then, we have

$$\frac{2\|f\|_{Sc(w,r)}}{\|f\|_{Sc(w,t)}} < 1 + \frac{\delta}{r-2\delta} N < \left(1 + \frac{\delta}{r-2\delta}\right)^N .$$

Therefore,

$$\|f\|_{Sc(w,r)} \left(\frac{t+\delta}{r-\delta}\right)^N < \frac{1}{2}\|f\|_{Sc(w,t)} \,.$$

(Note that the construction of N depends on $\|f\|_{Sc(w,t)}$, which depends on r, δ, and the dependency could be exponential. That is, the construction may not be iteratively bounded.) Let $d \equiv \frac{\delta}{2(N+1)}$. Choose α such that

$$2\alpha N \left(\frac{r+\delta}{d}\right)^{N-1} < \frac{1}{2}\|f\|_{Sc(w,t)} \,.$$

(Again, note that the construction of α from r, δ may not be iteratively bounded.) For each $k = 1, \ldots, N$, let $r_k = \frac{\delta}{2} + kd$. Then, $\frac{\delta}{2} < r_1 < \ldots < r_N < \delta$ and $\left|r_k - r_j\right| \ge d$ for $k \ne j$, $k, j = 1, \ldots, N$. For each $k = 1, \ldots, N$, we decide if $m(f, \Gamma(w, r_k)) < \alpha$ or $m(f, \Gamma(w, r_k)) > \alpha/2$. Suppose that $m(f, \Gamma(w, r_k)) < \alpha$ for all $k = 1, \ldots, N$. Then, for each k, we can find $z_k \in \Gamma(w, r_k)$ such that $|f(z_k)| < \alpha$. We have $|z_k - z_j| \geq d$, for $k \neq j$, $k, j = 1, ..., N$. Therefore, by Lemma 5.29,

$$\|f\|_{Sc(w,t)} \leq \left(\frac{t+\delta}{r-\delta}\right)^N \|f\|_{Sc(w,r)} + 2\alpha N \left(\frac{r+\delta}{d}\right)^{N-1} < \|f\|_{Sc(w,t)},$$

a contradiction. So, there must be some k, such that $m(f, \Gamma(w, r_k)) > \alpha/2$. Then, we can let $s \equiv r_k = \frac{\delta}{2} + kd$. □

Note that this proof contains some constructions that are potentially not iteratively bounded. However, we will see that these constructions will not be iterated in finding the zeros of a function, mostly because of the assumption that the function is strongly non-zero.

Theorem 5.31. *Suppose that f is a strongly non-zero differentiable function on an open set U, and $Sc(w,r) \Subset U$, $m(f, \Gamma(w,r)) > m(f, Sc(w,r)) = 0$. Then, there exists $z \in Sc(w,r)$ such that $f(z) = 0$.*

Proof. By the continuity of f, there exists $\varepsilon > 0$ such that

$$m(f, \Gamma(w, r-\varepsilon)) > m(f, Sc(w, r-\varepsilon)) = 0.$$

Moreover, we may assume that $r > 2\varepsilon$ and $Sc(w, r+2\varepsilon) \subseteq U$. Let (w_k) be a sequence of points in $Sc(w,r)$ such that $w_0 = w$ and for each $N > 0$, $w_0, ..., w_{k(N)}$ constitute an $1/N$ approximation to $Sc(w,r)$. We want to construct sequences (j_n), (z_n) such that $z_n = w_{j_n}$ with $z_0 = w_0 = w$, and for $n \geq 0$,

(1) $m(f, Sc(z_n, r_n)) = 0$, where $r_0 = r - \varepsilon$ and $r_n = 3^{-n}\varepsilon$ for $n > 0$, and
(2) $|z_{n+1} - z_n| < r_n + 3^{-(n+1)}\varepsilon$, and
(3) $|z_n - z_0| < r - 2 \cdot 3^{-n}\varepsilon$.

Then, it is easy to see that (z_n) converges to some $z \in Sc(w,r)$ and $f(z) = 0$.

Suppose that j_n and z_n have been constructed and (1) and (3) hold. Suppose that $w_0, ..., w_N$ constitute a $2^{-1}3^{-(n+1)}\varepsilon$ approximation to $Sc(w,r)$. For each $k = 0, ..., N$, $w_k \in Sc(w,r)$. Therefore, $Sc(w_k, 2 \cdot 3^{-n}\varepsilon) \subseteq Sc(w, r+2\varepsilon) \subseteq U$. Now, $2 \cdot 3^{-n}\varepsilon > 3 \cdot 3^{-(n+1)}\varepsilon$. By the lemma above, there exists s_k such that $2^{-1}3^{-(n+1)}\varepsilon < s_k < 3^{-(n+1)}\varepsilon$ and $c_k \equiv m(f, \Gamma(w_k, s_k)) > 0$. Since $m(f, Sc(z_n, r_n)) = 0$, we can find $\zeta \in Sc(z_n, r_n)$ such that $|f(\zeta)| < c_k$ for all $k = 0, ..., N$. Note that $Sc(z_0, r_0) \subseteq Sc(w,r)$. Then, since $|z_n - z_0| \leq r - 2 \cdot 3^{-n}\varepsilon$, for $n > 0$, we also have

$$Sc(z_n, r_n) \subseteq Sc\left(w, r - 2 \cdot 3^{-n}\varepsilon + r_n\right) = Sc\left(w, r - 3^{-n}\varepsilon\right) \subseteq Sc(w,r).$$

Since $w_0, ..., w_N$ constitute a $2^{-1}3^{-(n+1)}\varepsilon$ approximation to $Sc(w,r)$, there exists k such that

$$\zeta \in Sc\left(w_k, 2^{-1}3^{-(n+1)}\varepsilon\right) \subseteq Sc(w_k, s_k).$$

Then, since $|f(\zeta)| < c_k = m(f, \Gamma(w_k, s_k))$, by Lemma 5.25, $m(f, Sc(w_k, s_k)) = 0$. We let $j_{n+1} = k$ and $z_{n+1} = w_k$. Note that $Sc(w_k, s_k) \subseteq Sc\left(w_k, 3^{-(n+1)}\varepsilon\right)$ and

$r_{n+1} = 3^{-(n+1)}\varepsilon$. Therefore, $m(f, Sc(z_{n+1}, r_{n+1})) = 0$. Note that $\zeta \in Sc(z_n, r_n)$ and $\zeta \in Sc\left(z_{n+1}, 2^{-1}3^{-(n+1)}\varepsilon\right)$. Therefore,

$$|z_{n+1} - z_n| \leq r_n + 2^{-1}3^{-(n+1)}\varepsilon < r_n + 3^{-(n+1)}\varepsilon.$$

In particular, $|z_1 - z_0| \leq r - 2 \cdot 3^{-1}\varepsilon$. For $n > 0$, from the above and $|z_n - z_0| \leq r - 2 \cdot 3^{-n}\varepsilon$, it follows that $|z_{n+1} - z_0| \leq r - 2 \cdot 3^{-(n+1)}\varepsilon$.

This completes the construction. Note that the construction of s_k (actually depending on n) for $k = 0, ..., N$ is not iterated. It is uniformly constructed from (w_k) and $3^{-(n+1)}\varepsilon$. The same is true for all c_k. Then, the construction of ζ requires estimating $m(f, Sc(z_n, r_n))$ up to some precision depending on all c_k. This can be done by first uniformly estimating $f(w_k)$ up to some sufficiently large k and up to some precision, depending on all c_k and a modulus of continuity for f. Then, to find ζ, $z_n = w_{j_n}$ only serves to pick out those $w_k \in Sc(z_n, r_n)$. Similarly, the derivation of $m(f, Sc(z_{n+1}, r_{n+1})) = 0$ from $|f(\zeta)| < c_k$ does not depend on any results in the previous step of constructing j_n, except for using $z_n = w_{j_n}$ to pick out those $w_k \in Sc(z_n, r_n)$. Note that for $w_0, ..., w_N$ to constitute a $2^{-1}3^{-(n+1)}\varepsilon$ approximation to $Sc(w, r)$, N depends on n uniformly, not on the previously constructed j_n and j_n is always bounded by N. Therefore, all constructions are available with bounded primitive recursion. □

Note that without the assumption that f is strongly non-zero, we will have to make sure that $m(f, \Gamma(z_n, r_n)) > 0$ at each step, and then the construction of s_k above will depend on the value of $m(f, \Gamma(z_n, r_n))$ and will not be uniform. This may require iterating some not iteratively bounded operation. (cf. Ye [40])

We say that a polynomial $\sum_{j=0}^{n} a_j z^j$ has at least k degree, if $a_j \neq 0$ for some $j \geq k$. Then, we have the Fundamental Theorem of Algebra.

Theorem 5.32. *If a polynomial $p(z) \equiv \sum_{j=0}^{n} a_j z^j$ has at least $k > 0$ degree, then there exist $z_1, ..., z_k$ and a polynomial $q(z)$ such that*

$$p(z) = (z - z_1) \dots (z - z_k) q(z).$$

Proof. Suppose that $a_{j_0} \neq 0$ and $j_0 \geq k$. Consider the polynomial

$$p_1(w) \equiv a_n + a_{n-1}w + \dots + a_0 w^n + a_0 w^n.$$

By the same proof as in Lemma 5.27, we can find $\delta > 0$ such that for some $j \geq j_0$,

$$|a_j| \delta^{n-j} > |a_0| \delta^n + \sum_{l=0}^{j-1} |a_l| \delta^{n-l} + \sum_{l=j+1}^{n} |a_l| \delta^{n-l}.$$

Let $r = \delta^{-1}$. Then,

$$|a_j| r^j > |a_0| + \sum_{l=0}^{j-1} |a_l| r^l + \sum_{l=j+1}^{n} |a_l| r^l.$$

Therefore,

$$|p(r)| \geq |a_j| r^j - \sum_{l=0}^{j-1} |a_l| r^l - \sum_{l=j+1}^{n} |a_l| r^l > |p(0)|.$$

That is, $m(p,\Gamma(0,r)) > |p(0)|$. By Lemma 5.25, we have

$$m(p,\Gamma(0,r)) > m(p,Sc(0,r)) = 0.$$

Note that estimating r requires estimating each $|a_l|$ up to the precision of $\left(\frac{|a_{j_0}|}{A}\right)^n$, where $A = 1 + \sum_{l=0}^{n} |a_l|$, and the size of r is bounded by the scale $\left(\frac{A}{|a_{j_0}|}\right)^n$.

Then, by the theorem above, we can construct $z_1 \in Sc(0,r)$ such that $p(z_1) = 0$. Write $p(z)$ as $p(z) = (z - z_1) q_1(z) + c$, where $q_1(z) \equiv \sum_{j=0}^{n-1} b_j z^j$. We have $c = 0$ since $p(z_1) = 0$. Then, we have $a_n = b_{n-1}$, and $a_j = b_{j-1} - z_1 b_j$ for $j = 1, ..., n-1$, and $a_0 = z_1 b_0$. From $a_{j_0} \neq 0$, we see that $b_{j_0-1} \neq 0$ or $b_{j_0} \neq 0$. Therefore, $q_1(z)$ is at least $k-1$ degree. Moreover, in the former case, the estimate of $|b_{j_0-1}|$ is of the same scale as $|a_{j_0}|$, and in the latter case, the estimate of $|b_{j_0}|$ is of the same scale as $|a_{j_0}|/r = \frac{|a_{j_0}|^{n+1}}{A^n}$. Furthermore, each b_j is a sum of terms like $a_l z_1^{l'}$, where $|z_1|$ is bounded by r above. Therefore, to approximate b_j up to $b_j(m)$ requires approximations $a_l(Km)$ and $z_1(Km)$ where K is of the same scale as $r^n A = \frac{A^{n^2+1}}{|a_{j_0}|^{n^2}}$. These operations are all iteratable. Therefore, we can proceed recursively for k times and get $p(z) = (z - z_1) \ldots (z - z_k) q(z)$ for some $q(z)$. □

Chapter 6
Integration

We will generalize Riemann integration into a more general notion of integration, namely, Lebesgue integration. We will define Lebesgue integrable functions and measurable functions and prove some of their common properties. In particular, measurable functions include the characteristic functions of intervals and step functions. This extends the scope of functions treated in the previous chapters, which so far are limited to continuous functions. We will follow some ideas in Chap. 6 of Bishop and Bridges [6], but we have to make many changes in order to fit into strict finitism. We will consider only functions of real numbers and will simplify some of the notions. This allows us to see the finitistic content of Lebesgue integration more clearly. Moreover, we will consider only functions of a single variable. Extension to multiple variable cases is straightforward.

6.1 Lebesgue Integration

Let $f \in C(\mathbb{R},\mathbb{R})$ be a continuous function from $\mathbb{R}$ to $\mathbb{R}$, and let $I = [a,b]$ be a compact interval. We say that I is a compact support for f if $f(x) = 0$ for $x < a$ or $x > b$. Let $C(\mathbb{R}) \subseteq C(\mathbb{R},\mathbb{R})$ be the set of functions in $C(\mathbb{R},\mathbb{R})$ with compact supports. $C(\mathbb{R})$ is closed under finite linear combinations and finite products, and $f \in C(\mathbb{R})$ implies $|f| \in C(\mathbb{R})$.

Note that each $f \in C(\mathbb{R})$ has a compact support I for f as a witness. Therefore, we can define a function from $C(\mathbb{R})$ to $\mathbb{R}$ by

$$f \mapsto \int f \mathrm{d}\mu \equiv_{df} \int_a^b f(x)\,\mathrm{d}x, \text{ for } f \in C(\mathbb{R}),$$

where $I = [a,b]$ is the compact interval witnessing $f \in C(\mathbb{R})$. It is easy to verify that this does define a function on $C(\mathbb{R})$. $\int f \mathrm{d}\mu$ is also called the Riemann integration of f. It is easy to see that $\int f \mathrm{d}\mu$ is linear for f and $|\int f \mathrm{d}\mu| \le \int |f| \mathrm{d}\mu$. We want to gen-

F. Ye, *Strict Finitism and the Logic of Mathematical Applications*, Synthese Library 355, DOI 10.1007/978-94-007-1347-5_6, © Springer Science+Business Media B.V. 2011

eralize this notion into an integration of functions beyond $C(\mathbb{R})$, namely, Lebesgue integration on $\mathbb{R}$.

We need a lemma, which is a simplified version of Theorem 1.10 on [6], p. 220.

Lemma 6.1. *Suppose that* $f,\ f_n \in C(\mathbb{R}),\ f(x) \geq 0,\ f_n(x) \geq 0$ *for all* n *and all* $x \in \mathbb{R}$, *and suppose that* $\sum_{n=0}^{\infty} \int f_n \mathrm{d}\mu < \int f \mathrm{d}\mu$. *Then, there exists* $x \in \mathbb{R}$ *such that* $\sum_{n=0}^{\infty} f_n(x) < f(x)$.

Proof. We first show that the assumption implies that there exists $x \in \mathbb{R}$ such that $\sum_{n=0}^{N} f_n(x) \leq f(x)$ for all N. Let $I = [a,b]$ be a compact support for f. Then, $\sum_{n=0}^{\infty} \int_a^b f_n(x)\,\mathrm{d}x < \int_a^b f(x)\,\mathrm{d}x$. Let $c = (a+b)/2$. Then, we have

$$\sum_{n=0}^{\infty} \int_a^c f_n(x)\,\mathrm{d}x + \sum_{n=0}^{\infty} \int_c^b f_n(x)\,\mathrm{d}x < \int_a^c f(x)\,\mathrm{d}x + \int_c^b f(x)\,\mathrm{d}x.$$

Note that in general, if m witnesses $a+b < c+d$, that is, $(a+b)(m) < (c+d)(m) - \frac{2}{m}$, then $2m$ will either witness $a < c$ or witness $b < d$. Therefore, repeatedly dividing the interval into halves, we will get a sequence of intervals I_k such that the length of I_{k+1} is a half of that of I_k and $\sum_{n=0}^{\infty} \int_{I_k} f_n(x)\,\mathrm{d}x < \int_{I_k} f(x)\,\mathrm{d}x$. The witness for the $(k+1)$-th inequality is twice of the witness for the kth inequality. Therefore, the construction is a bounded primitive recursion. For any k, we have $\int_{I_k} \sum_{n=0}^{k} f_n(x)\,\mathrm{d}x < \int_{I_k} f(x)\,\mathrm{d}x$. From the properties of Riemann integration, we have $x_k \in I_k$ such that $\sum_{n=0}^{k} f_n(x_k) < f(x_k)$. The sequence (x_k) converges to some $x \in I$. Note that for $k > N$, $\sum_{n=0}^{N} f_n(x_k) < f(x_k)$. Therefore, $\sum_{n=0}^{N} f_n(x) \leq f(x)$ for all N.

Note that this does not guarantee that $\sum_{n=0}^{\infty} f_n(x)$ converges. To assure the convergence, we first choose $\delta > 0$ such that $\sum_{n=0}^{\infty} \int f_n \mathrm{d}\mu + \delta < \int f \mathrm{d}\mu$. Since $\sum_{n=0}^{\infty} \int f_n \mathrm{d}\mu$ converges, we have a sequence (N_k) such that $\sum_{n=N_k}^{\infty} \int f_n \mathrm{d}\mu < \delta/2^{2k}$. Then, consider the sequence

$$(g_n) = \left(f_0, 2\sum_{n=N_1}^{N_2} f_n, f_1, 2^2 \sum_{n=N_2}^{N_3} f_n, f_2, 2^3 \sum_{n=N_3}^{N_4} f_n, \ldots \right).$$

We have $\sum_{n=0}^{\infty} \int g_n \mathrm{d}\mu \leq \sum_{n=0}^{\infty} \int f_n \mathrm{d}\mu + \delta < \int f \mathrm{d}\mu$. Therefore, the argument above gives a $x \in I$ such that for all k,

$$\sum_{n=0}^{k} f_n(x) + 2^k \sum_{n=N_k}^{N_{k+1}} f_n(x) \leq f(x).$$

This implies that $\sum_{n=N_k}^{N_{k+1}} f_n(x) \leq f(x)/2^k$. Therefore, $\sum_{n=0}^{\infty} f_n(x)$ converges, and the inequality above again implies that $\sum_{n=0}^{\infty} f_n(x) \leq f(x)$.

To get the strict inequality $\sum_{n=0}^{\infty} f_n(x) < f(x)$, we can choose a small $\varepsilon > 0$ and a function $h \in C(\mathbb{R})$, such that $h(x) \geq 0$ for all x, and $h(x) = \varepsilon$ for x in the compact support of f, but ε is so small that we still have

$$\int h\mathrm{d}\mu + \sum_{n=0}^{\infty} \int f_n \mathrm{d}\mu < \int f \mathrm{d}\mu.$$

Then, the same argument above will give an x such that $h(x) + \sum_{n=0}^{\infty} f_n(x) \leq f(x)$. Note that x belongs to the compact support of f. Therefore, $h(x) = \varepsilon$, which implies the strict inequality $\sum_{n=0}^{\infty} f_n(x) < f(x)$. □

Lebesgue integrable functions will be partial functions on $\mathbb{R}$. To construct Lebesgue integration, first define an index set Γ for the family of the domains of Lebesgue integrable functions:

$$(f_n) \in \Gamma \equiv_{df} \forall n (f_n \in C(\mathbb{R})) \wedge \left(\sum_{n=0}^{\infty} \int |f_n| \mathrm{d}\mu \text{ converges} \right).$$

Then, define the family

$$\mathscr{D} \equiv_{df} \left\{ D_{(f_n)} : (f_n) \in \Gamma \right\}$$

of subsets of $\mathbb{R}$ indexed by Γ:

$$x \in D_{(f_n)} \equiv_{df} x \in \mathbb{R} \wedge \sum_{n=0}^{\infty} |f_n(x)| \text{ converges}.$$

$\mathscr{D}$ will be the family of the domains of Lebesgue integrable functions.

Suppose that the sequences (f_n^1), ..., (f_n^m) are in the index set Γ. Let f_n be defined by $f_n(x) = \sum_{i=1}^{m} |f_n^i(x)|$. Then, the sequence $(f_n) \in \Gamma$ and

$$D_{(f_n)} \subseteq D_{(f_n^1)} \cap \ldots \cap D_{(f_n^m)}.$$

That is, the family $\mathscr{D}$ is closed under finite intersection.

Recall that $\mathscr{F}(\mathscr{D}, \mathbb{R})$ denotes the set of all partial functions with domains in the family $\mathscr{D}$ and with $\mathbb{R}$ as the range. Finally, we can define the set of Lebesgue integrable functions on $\mathbb{R}$.

Definition 6.2. The set of Lebesgue integrable functions $L_1 = L_1(\mathbb{R})$ is defined by

$$(((f_n), f) \in L_1) \equiv_{df} ((f_n), f) \in \mathscr{F}(\mathscr{D}, \mathbb{R}) \wedge$$
$$\exists (g_n) \in \Gamma \left(D_{(g_n)} \subseteq D_{(f_n)} \wedge \forall x \in D_{(g_n)} \left(f(x) = \sum_{n=0}^{\infty} g_n(x) \right) \right).$$

Therefore, $L_1 \subseteq \mathscr{F}(\mathscr{D}, \mathbb{R})$. We will simply call f an integrable function and denote this fact as $f \in L_1$, and we will call $D_{(f_n)}$ the domain of f and denote it as $dmn(f)$. Among the witnesses for $f \in L_1$ is a sequence $(g_n) \in \Gamma$ satisfying the condition in the definition. This will be called a representation sequence of f.

Also notice that any $(g_n) \in \Gamma$ is a representation of some $g \in L_1$. g has the domain $D_{(g_n)}$ and $g(x) = \sum_{n=0}^{\infty} g_n(x)$ for $x \in D_{(g_n)}$. In the definition above, we require only $D_{(g_n)} \subseteq D_{(f_n)}$. This aims to make integrable functions more general, for we also want to consider f an integrable function in the cases where f extends that g defined by

$g(x) = \sum_{n=0}^{\infty} g_n(x)$. This is to reflect the classical idea that for an integrable function, its values on a set of measure zero are irrelevant.

Suppose that $f \in L_1$, and (g_n) is a representation of f, and $f' \in L_1$, and (g'_n) is a representation of f', and suppose that $f =_{L_1} f'$, that is, $D_{(f_n)} = D_{(f'_n)}$ and $f(x) = f'(x)$ for $x \in D_{(f_n)}$. The definition implies that $\sum_{n=0}^{\infty} \int |g_n| \, d\mu$ and $\sum_{n=0}^{\infty} \int |g'_n| \, d\mu$ converge. Therefore, $\sum_{n=0}^{\infty} \int g_n d\mu$ and $\sum_{n=0}^{\infty} \int g'_n d\mu$ converge. Now, suppose that $\sum_{n=0}^{\infty} \int g_n d\mu \neq \sum_{n=0}^{\infty} \int g'_n d\mu$. Choose $\varepsilon > 0$ such that

$$\left| \sum_{n=0}^{\infty} \int g_n d\mu - \sum_{n=0}^{\infty} \int g'_n d\mu \right| > \varepsilon.$$

Since $\sum_{n=0}^{\infty} \int |g_n| \, d\mu$ and $\sum_{n=0}^{\infty} \int |g'_n| \, d\mu$ converge, there exists N such that

$$\left| \sum_{n=N}^{\infty} \int g_n d\mu \right| + \left| \sum_{n=N}^{\infty} \int g'_n d\mu \right| \leq \sum_{n=N}^{\infty} \int |g_n| \, d\mu + \sum_{n=N}^{\infty} \int |g'_n| \, d\mu < \varepsilon/2.$$

Then,

$$\begin{aligned} \int \left| \sum_{n=0}^{N} g_n - \sum_{n=0}^{N} g'_n \right| d\mu &\geq \left| \sum_{n=0}^{N} \int g_n d\mu - \sum_{n=0}^{N} \int g'_n d\mu \right| \\ &\geq \left| \sum_{n=0}^{\infty} \int g_n d\mu - \sum_{n=0}^{\infty} \int g'_n d\mu \right| - \left| \sum_{n=N}^{\infty} \int g_n d\mu \right| - \left| \sum_{n=N}^{\infty} \int g'_n d\mu \right| \\ &> \varepsilon/2 > \sum_{n=N}^{\infty} \int \left(|g_n| + |g'_n| \right) d\mu. \end{aligned}$$

By the lemma above, there exists $x \in \mathbb{R}$ such that

$$\sum_{n=N}^{\infty} \left(|g_n(x)| + |g'_n(x)| \right) < \left| \sum_{n=0}^{N} g_n(x) - \sum_{n=0}^{N} g'_n(x) \right|.$$

This implies that $x \in D_{(g_n)} \subseteq D_{(f_n)}$ and $x \in D_{(g'_n)} \subseteq D_{(f'_n)}$, and moreover,

$$\begin{aligned} |f(x) - f'(x)| &= \left| \sum_{n=0}^{\infty} g_n(x) - \sum_{n=0}^{\infty} g'_n(x) \right| \\ &\geq \left| \sum_{n=0}^{N} g_n(x) - \sum_{n=0}^{N} g'_n(x) \right| - \sum_{n=N}^{\infty} \left(|g_n(x)| + |g'_n(x)| \right) > 0. \end{aligned}$$

This contradicts the assumption $f =_{L_1} f'$. Therefore, we must have $\sum_{n=0}^{\infty} \int g_n d\mu = \sum_{n=0}^{\infty} \int g'_n d\mu$.

This means that the following definition of the integration of an integrable function is appropriate. It respects the equality relation for the set L_1. (Recall that a representation of f is a witness for $f \in L_1$.)

Definition 6.3. If $f \in L_1$ and (g_n) is a representation of f, define

$$\int f \mathrm{d}\mu \equiv_{df} \sum_{n=0}^{\infty} \int g_n \mathrm{d}\mu.$$

For $f \in C(\mathbb{R})$, let (f_n) be the sequence such that $f_0 = f$, $f_n = 0$ for $n > 0$. Then, $(f_n) \in \Gamma$, $D_{(f_n)} = \mathbb{R}$, and $((f_n), f) \in L_1$. That is, $f \in L_1$ with (f_n) as a representation of f. Note that $dmn(f) = \mathbb{R}$ and $\int f \mathrm{d}\mu$ is just the Riemann integration of f. We will simply say that all continuous functions with compact support are Lebesgue integrable.

From the definition it also follows that if $f_1, ..., f_n \in L_1$ and $a_1, ..., a_n$ are real numbers, then $\sum_{i=1}^{n} a_i f_i \in L_1$ and

$$\int \left(\sum_{i=1}^{n} a_i f_i \right) \mathrm{d}\mu = \sum_{i=1}^{n} a_i \int f_i \mathrm{d}\mu.$$

Moreover, we have

Corollary 6.4. *If $f \in L_1$ with the representation (g_n), then $|f| \in L_1$, and*

$$\left| \int f \mathrm{d}\mu \right| \leq \int |f| \, \mathrm{d}\mu = \lim_{n \to \infty} \int \left| \sum_{i=0}^{n} g_i \right| \mathrm{d}\mu.$$

Proof. The sequence

$$(g_0, -g_0, |g_0|, g_1, -g_1, |g_0 + g_1| - |g_0|, ...),$$

is a representation of $|f|$. The rest follows straightforwardly. □

We define $(f \wedge g)(x) \equiv_{df} \min(f(x), g(x))$, $(f \vee g)(x) \equiv_{df} \max(f(x), g(x))$. Then, if $f \in L_1$ with the representation (g_n), we have $f \wedge n \in L_1$ with the representation

$$(g_0, -g_0, g_0 \wedge n, g_1, -g_1, (g_0 + g_1) \wedge n - g_0 \wedge n, ...).$$

Similarly, if $f_1, ..., f_n \in L_1$, then $f_1 \wedge ... \wedge f_n \in L_1$ and $f_1 \vee ... \vee f_n \in L_1$.

A more interesting example of Lebesgue integrable function is the characteristic function of an interval:

$$\chi_{(0,1)}(x) = \begin{cases} 1, & \text{for } x \in (0,1); \\ 0, & \text{for } x \in (-\infty, 0] \cup [1, \infty). \end{cases}$$

This function is defined on the subset $A = (-\infty, 0] \cup (0,1) \cup [1, \infty) \subseteq \mathbb{R}$. Note that we cannot decide, for an arbitrary real number $x \in \mathbb{R}$, whether $x \in (-\infty, 0]$, or $x \in (0,1)$, or $x \in [1, \infty)$. Therefore, we cannot show that $A = \mathbb{R}$.

To see how $\chi_{(0,1)}$ is Lebesgue integrable, first note that if two continuous functions f, g on $[a,b]$ and $[b,c]$ are such that $f(b) = g(b)$, then we can piece them together into a continuous h on $[a,c]$ such that $h(x) = f(x)$ for $x \in [a,b]$ and $h(x) = g(x)$ for $x \in [b,c]$. Then, let $f_0 = 0$, $f_1 = 0$, and for each $n > 1$, let f_n be a the continuous function such that

$$f_n(x) = \begin{cases} 0, & \text{for } x \in (-\infty, 0] \cup [1, \infty), \\ nx, & \text{for } x \in \left[0, \frac{1}{n}\right], \\ 1, & \text{for } x \in \left[\frac{1}{n}, 1 - \frac{1}{n}\right], \\ n(1-x), & \text{for } x \in \left[1 - \frac{1}{n}, 1\right]. \end{cases}$$

It is easy to see that $(f_n(x))$ is an increasing sequence, and it converges if and only if $x \in A$, and it converges to $\chi_{(0,1)}(x)$ for $x \in A$. Moreover, $\int f_n \mathrm{d}\mu$ converges to 1. Let $g_n = f_{n+1} - f_n$. Then, $(g_n) \in \Gamma$, $D_{(g_n)} = A$, and $\sum_{n=0}^{\infty} g_n(x) = I_{(0,1)}(x)$ for $x \in A$. Therefore, $I_{(0,1)}$ is integrable with (g_n) as a representation.

Then, by linear combinations, all step functions on finite intervals are Lebesgue integrable. Step functions naturally represent some physics quantities, for instance, the potential of a potential well in quantum mechanics. In that case, the exact values of a potential V at positions around two boundary points, e.g. $-a$ and a, are not relevant, and the potential is sufficiently accurately represented by, for instance,

$$V(x) = \begin{cases} 0, & \text{for } x \in (-a, a), \\ V_0, & \text{for } x \in (-\infty, -a] \cup [a, \infty). \end{cases} \tag{6.1}$$

The fact that V is defined on $(-\infty, -a] \cup (-a, a) \cup [a, \infty)$, but not on $\mathbb{R}$, naturally reflects the fact that potential values around the points $-a$ and a are indeterminate. Note that V is not integrable. But it is measurable. See the next section.

Lebesgue integrable functions are thus a more general class of functions. They are basically sequences of continuous functions that converge in some appropriate sense. They allow representing physics quantities that jump at some points.

A subset of $\mathbb{R}$ is a *full* set if it contains a domain of an integrable function. A full set plays the role of the complement of a set with measure zero in the classical theory. We cannot quantify over arbitrary full sets. Assertions about an arbitrary full set are schematic assertions. However, we can make the convention that when we quantify over full sets we always mean a quantification over domains of integrable functions, that is, sets in the family $\mathscr{D}$. For example, '$f = g$ on a full set' is to mean 'there exists $h \in L_1$ such that $f(x) = g(x)$ for $x \in dmn(h)$'. Note that if $f \in L_1$ and X is a full set, then there is always a representation (g_n) of f such that $D_{(g_n)} \subseteq X$. Because, supposing that $D_{(h_n)} \subseteq X$ and (f_n) is any representation of f, we can let

$$(g_n) \equiv (f_0, h_0, -h_0, f_1, h_1, -h_1, \ldots).$$

Then, $D_{(g_n)} \subseteq D_{(h_n)}$ and (g_n) is a representation of f.

Lemma 6.5. *If $\{X_n\}$ is a sequence of full sets, then $\cap_{n=0}^{\infty} X_n$ is also a full set.*

Proof. By the assumption, there exists a sequence (f_n) of integrable functions such that $dmn(f_n) \subseteq X_n$. Let $(g_{n,k})$ be a representation of f_n, $\sum_{k=0}^{\infty} \int |g_{n,k}| \mathrm{d}\mu$ converges. Let $c_n = 2^{-n} \left(\sum_{k=0}^{\infty} \int |g_{n,k}| \mathrm{d}\mu + 1\right)^{-1}$. Then, $\sum_{k=0}^{\infty} \int |c_n g_{n,k}| \mathrm{d}\mu \leq 2^{-n}$. Arrange $(c_n g_{n,k})$ into a sequence (h_j) so that (n_1, k_1) precedes (n_2, k_2) when

$\max(n_1,k_1) < \max(n_2,k_2)$. It is easy to see that $\sum_{j=0}^{\infty} \int |h_j| \, d\mu$ converges, that is, $(h_j) \in \Gamma$. Obviously, $D_{(h_j)} \subseteq \cap_{n=0}^{\infty} X_n$. □

Lemma 6.6. *Suppose that $f, g \in L_1$. If $f \leq g$ on a full set, then $\int f d\mu \leq \int g d\mu$; if $f = g$ on a full set, then $\int f d\mu = \int g d\mu$.*

Proof. It suffices to show that if $f \geq 0$ on full set X, then $\int f d\mu \geq 0$. Suppose that (g_n) is a representation of f such that $D_{(g_n)} \subseteq X$ and suppose that $\sum_{n=0}^{\infty} \int g_n d\mu < 0$. Let $g_n^+ = g_n \vee 0$ and $g_n^- = -g_n \vee 0$. Then, $g_n = g_n^+ - g_n^-$, $|g_n| = g_n^+ + g_n^-$. Since $\sum_{n=0}^{\infty} \int |g_n| \, d\mu$ converges,

$$\sum_{n=0}^{\infty} \int g_n d\mu = \sum_{n=0}^{\infty} \int g_n^+ d\mu - \sum_{n=0}^{\infty} \int g_n^- d\mu.$$

Choose $\varepsilon > 0$ such that $\sum_{n=0}^{\infty} \int g_n d\mu < -\varepsilon$. Then, $\sum_{n=0}^{\infty} \int g_n^+ d\mu < \sum_{n=0}^{\infty} \int g_n^- d\mu - \varepsilon$. Choose N such that $\sum_{n=N}^{\infty} \int |g_n| \, d\mu < \varepsilon/3$. Then,

$$\sum_{n=0}^{\infty} \int g_n^+ d\mu + \sum_{n=N}^{\infty} \int g_n^- d\mu < \sum_{n=0}^{N} \int g_n^- d\mu - \varepsilon/3.$$

By Lemma 6.1, there exists x such that

$$\sum_{n=0}^{\infty} g_n^+(x) + \sum_{n=N}^{\infty} g_n^-(x) < \sum_{n=0}^{N} g_n^-(x).$$

Therefore, $x \in D_{(g_n)}$ and $\sum_{n=0}^{\infty} g_n^+(x) < \sum_{n=0}^{\infty} g_n^-(x)$. That is,

$$f(x) = \sum_{n=0}^{\infty} \left(g_n^+(x) - g_n^-(x)\right) < 0.$$

This contradicts the assumption. □

Now, we can define a new equality relation between members of L_1. From now on, $f = g$ will mean '$f = g$ on a full set'. This is weaker than the old equality relation $=_{L_1}$ of the set L_1. Finite linear combinations, absolute value, the functions $\vee$ and $\wedge$, and integration all respect this new equality relation. From now on, we will consider L_1 a set with this new equality relation. It means that we expect that the functions and concepts defined on L_1 will respect this new equality relation. For example, if Δ is any one of the relations $=, <, >, \leq$, or $\geq$, then $f \Delta g$ is to mean $f(x) \Delta g(x)$ for all x in a full set. Similarly, 'f is bounded' is to mean $|f| \leq c$ on a full set for some $c > 0$.

Lemma 6.7. *If $f \in L_1$ and $\int |f| \, d\mu = 0$, then $f = 0$.*

Proof. Suppose that (g_n) is a representation of f. By Corollary 6.4, $\int |f| \, d\mu = \lim_{N \to \infty} \int \left|\sum_{n=0}^{N} g_n\right| d\mu$. Therefore, $\int |f| \, d\mu = 0$ implies $\lim_{N \to \infty} \int \left|\sum_{n=0}^{N} g_n\right| d\mu = 0$. For each $k \geq 0$, choose N_k such that $\int \left|\sum_{n=0}^{N_k} g_n\right| d\mu < 2^{-k}$. Let $f_k = \sum_{n=0}^{N_k} g_n$. Then,

$\sum_{k=0}^{\infty} \int |f_k| \, d\mu$ converges. If $x \in D_{(f_k)} \cap D_{(g_k)}$, then $\sum_{k=0}^{\infty} |f_k(x)|$ converges. Therefore, $\lim_{k\to\infty} |f_k(x)| = 0$, that is, $\sum_{n=0}^{\infty} g_n(x) = 0$. So, $f(x) = 0$. □

The norm of $f \in L_1$ is defined as $\|f\|_1 \equiv_{df} \int |f| \, d\mu$. Then, $\|f\|_1 = 0$ implies $f = 0$. Define

$$\rho(f,g) \equiv_{df} \|f-g\|_1 .$$

Then, ρ is a metric on L_1. From now on, we will treat L_1 as a metric space.

The following theorem gives the completeness of Lebesgue integration.

Theorem 6.8. *Suppose that (f_n) is a sequence of Lebesgue integrable functions and $\sum_{n=0}^{\infty} \int |f_n| \, d\mu$ converges. Then, there exists a Lebesgue integrable function f such that for $x \in dmn(f)$, $\sum_{n=0}^{\infty} |f_n(x)|$ converges and $f(x) = \sum_{n=0}^{\infty} f_n(x)$, and*

$$\lim_{N\to\infty} \int \left| f - \sum_{n=0}^{N} f_n \right| d\mu = 0.$$

Proof. Let $(g_n^k)_k$ be a representation of f_n. Since $\sum_{k=0}^{\infty} \int |g_n^k| \, d\mu$ converges and $\int |f_n| \, d\mu = \lim_{K\to\infty} \int |\sum_{k=0}^{K} g_n^k| \, d\mu$, we can choose K such that $\sum_{k=K+1}^{\infty} \int |g_n^k| \, d\mu < 2^{-n-1}$ and $\int |\sum_{k=0}^{K} g_n^k| \, d\mu < \int |f_n| \, d\mu + 2^{-n-1}$. By replacing g_n^0 by $\sum_{k=0}^{K} g_n^k$, we can assume that $\sum_{k=1}^{\infty} \int |g_n^k| \, d\mu < 2^{-n-1}$ and $\int |g_n^0| \, d\mu < \int |f_n| \, d\mu + 2^{-n-1}$. Therefore, $\sum_{k=0}^{\infty} \int |g_n^k| \, d\mu < \int |f_n| \, d\mu + 2^{-n}$. Arrange (g_n^k) into a single sequence (h_i) so that g_n^k precedes $g_{n'}^{k'}$ when $\max(n,k) < \max(n',k')$. Since $\sum_{n=0}^{\infty} \int |f_n| \, d\mu$ converges, it is easy to see that $\sum_{i=0}^{\infty} \int |h_i| \, d\mu$ converges. (h_i) is a representation for the function f defined by $f(x) \equiv \sum_{i=0}^{\infty} h_i(x)$ on $D_{(h_i)}$. Note that if $\sum_{i=0}^{\infty} |h_i(x)|$ converges, then for each n, $\sum_{k=0}^{\infty} |g_n^k(x)|$ converges and $f_n(x) = \sum_{k=0}^{\infty} g_n^k(x)$. Therefore, $f(x) = \sum_{n=0}^{\infty} f_n(x)$ on $D_{(h_i)}$. Moreover, since $\int |f| \, d\mu = \lim_{m\to\infty} \int |\sum_{i=0}^{m} h_i(x)| \, d\mu$ by Corollary 6.4, it follows that $\int |f| \, d\mu \le \sum_{n=0}^{\infty} \int |f_n| \, d\mu$.

Now, to show that $\lim_{N\to\infty} \int |f - \sum_{n=0}^{N} f_n| \, d\mu = 0$, we apply the above argument to the sequence $(f_{N+1}, f_{N+2}, \ldots)$. It follows that for some integrable function f^N, $f^N(x) = \sum_{n=N+1}^{\infty} f_n(x)$ on some full set and $\int |f^N| \, d\mu \le \sum_{n=N+1}^{\infty} \int |f_n| \, d\mu$. Therefore, $\lim_{N\to\infty} \int |f^N| \, d\mu = 0$. Since $f(x) = \sum_{n=0}^{\infty} f_n(x)$ on some full set, $f^N = f - \sum_{n=0}^{N} f_n$ on some full set (the intersection of two). Therefore, $\int |f - \sum_{n=0}^{N} f_n| \, d\mu = \int |f^N| \, d\mu$ by Lemma 6.6, and thus $\lim_{N\to\infty} \int |f - \sum_{n=0}^{N} f_n| \, d\mu = 0$. □

Corollary 6.9. *Suppose that (g_n), $g_n \ge 0$, is an increasing sequence of non-negative Lebesgue integrable functions and suppose that $\lim_{n\to\infty} \int g_n d\mu$ exists. Then, there exists a Lebesgue integrable function f such that $\lim_{n\to\infty} g_n(x) = f(x)$ for $x \in dmn(f)$ and $\lim_{n\to\infty} \int |g_n - f| \, d\mu = 0$.*

For any $f \in L_1$, let (f_n) in the theorem be a representation of the integrable function f. Note that it is a sequence of functions in $C(\mathbb{R})$. Then, it follows that

Corollary 6.10. *$C(\mathbb{R})$ is dense in L_1.*

Suppose that (f_n) is a Cauchy sequence in L_1 with a modulus of Cauchyness ω. Consider the sequence

$$\left(f_{\omega(2^0)}, f_{\omega(2^1)} - f_{\omega(2^0)}, f_{\omega(2^2)} - f_{\omega(2^1)}, \ldots\right).$$

Applying the theorem to this sequence we see that (f_n) converges in L_1. Therefore we have

Corollary 6.11. *Suppose that (f_n) is a Cauchy sequence in L_1. Then, there exists $f \in L_1$ and a subsequence $(f_{N(k)})$ of (f_n) such that $\lim_{k\to\infty} f_{N(k)}(x) = f(x)$ for $x \in dmn(f)$, and moreover, $\lim_{n\to\infty} \|f - f_n\|_1 = 0$. In particular, L_1 is complete.*

6.2 Measurable Functions

The potential function (6.1) above is not an integrable function, because it takes a non-zero constant value on infinite intervals $(-\infty, -a]$ and $[a, \infty)$ and its integration would have to be infinite if existed. We must generalize integrable functions to such more general functions. They are measurable functions. Here, we will follow the idea in [6] that measurable functions are those that can be approximated by integrable functions, but we will simplify the definition since we will consider real functions only. Our idea is that a measurable function is a function that can be approximated by continuous functions on any compact interval, where the approximations on a compact interval can further ignore some smaller and smaller sub-intervals of the given compact interval.

Note that we will still consider only partial functions in $\mathscr{F}(\mathscr{D}, \mathbb{R})$, that is, measurable functions also have domains in the family $\mathscr{D}$.

First, we need some definitions and results on intervals. For a finite interval $I \equiv (a,b)$, we define $|I| \equiv b - a$, called the length of I. We will consider only such intervals with $a < b$. In that case, $\mathbb{R} - I$ will be $(-\infty, a] \cup [b, \infty)$. Recall that in the last section we have shown that the characteristic function $\chi_{(a,b)}$ of an open interval is integrable. Therefore, $I \cup (\mathbb{R} - I)$ is a full set.

If (I_n) is a sequence of (possibly mutually overlapping) finite intervals, a *generalized interval* is a union $J \equiv \cup_{n=0}^{\infty} I_n$. We define $|J| \equiv \sum_{n=0}^{\infty} |I_n|$ if it converges. In that case, we say that J is a finite generalized interval. In contrast, we will call $I \equiv (a,b)$ a '*simple interval*', or simply an 'interval'. Note that $x \in J$ requires a witness n such that $x \in I_n$, and $x \in \mathbb{R} - J$ means $x \in \mathbb{R}$ and for each n, $x \in \mathbb{R} - I_n$.

Suppose that $J \equiv \cup_{n=0}^{\infty} I_n$ is a finite generalized interval. Let $f_n = \chi_{I_0} \vee \ldots \vee \chi_{I_n}$. We see that (f_n) is an increasing sequence of non-negative integrable functions. Moreover, for $m > n$, $f_m - f_n \leq \chi_{I_{n+1}} \vee \ldots \vee \chi_{I_m}$. Therefore, $\int |f_m - f_n| \, d\mu \leq \sum_{i=n+1}^{m} |I_i|$. It means that $\lim_{n\to\infty} \int f_n d\mu$ exists. By Corollary 6.9 above, there exists an integrable function g such that for $x \in dmn(g)$, $g(x) = \lim_{n\to\infty} f_n(x)$. Note that $dmn(g) \subseteq \cap_{n=0}^{\infty} dmn(f_n)$, and $f_n(x) = 0$ or $f_n(x) = 1$ for each n and $x \in dmn(g)$. Therefore, $g(x) = 0$ or $g(x) = 1$ for each $x \in dmn(g)$. $g(x) = 0$ implies that $f_n(x) = 0$ for all n, and hence $\chi_{I_n}(x) = 0$ for all n, that is, $x \in \mathbb{R} - J$. Similarly, $g(x) = 1$ implies that $f_n(x) = 1$ for some n, and hence $x \in J$. Therefore, $dmn(g) \subseteq J \cup (\mathbb{R} - J)$ and $J \cup (\mathbb{R} - J)$ is a full set. We define the characteristic function χ_J on $J \cup (\mathbb{R} - J)$ by $\chi_J(x) = 1$ for $x \in J$ and $\chi_J(x) = 0$ for $x \in \mathbb{R} - J$. Then, χ_J is an integrable function and $\chi_J(x) = \lim_{n \to \infty} f_n(x)$ for $x \in dmn(g)$. Note that

$$\int f_n \mathrm{d}\mu \leq \int \chi_{I_0} \mathrm{d}\mu + \ldots + \int \chi_{I_n} \mathrm{d}\mu = |I_0| + \ldots + |I_n|.$$

Therefore, $\int \chi_J \mathrm{d}\mu \leq |J|$.

We will also consider sets in the format $I - J$ for an interval I and a finite generalized interval J. We define

$$\chi_{I-J} \equiv_{df} (\chi_I - \chi_J)^+,$$

whose domain

$$\begin{aligned}&(I \cup (\mathbb{R} - I)) \cap (J \cup (\mathbb{R} - J)) \\ &= (I \cap J) \cup (I \cap (\mathbb{R} - J)) \cup ((\mathbb{R} - I) \cap J) \cup ((\mathbb{R} - I) \cap (\mathbb{R} - J))\end{aligned}$$

is a full set. Clearly, $\chi_{I-J}(x) = 1$ when $x \in I \cap (\mathbb{R} - J)$, and $\chi_{I-J}(x) = 0$ when

$$x \in (I \cap J) \cup ((\mathbb{R} - I) \cap J) \cup ((\mathbb{R} - I) \cap (\mathbb{R} - J)).$$

χ_{I-J} is integrable when both I and J are finite. Since $\chi_{I-J} \leq \chi_I$, $\int \chi_{I-J} \mathrm{d}\mu \leq |I|$. Note that when I is an infinite interval, χ_{I-J} may not be integrable. In particular, $\chi_{\mathbb{R}-J} = 1 - \chi_J$ is not integrable.

We need a few lemmas on the integrability of functions of the format $f\chi_J$ or $f\chi_{I-J}$.

Lemma 6.12. *Suppose that $f \in C(\mathbb{R})$ and I is any interval. Then, $f\chi_I \in L_1$ and $\int f\chi_I \mathrm{d}\mu \leq \int |f| \mathrm{d}\mu$.*

Proof. Similar to the proof of integrability of χ_I in the last section. □

Lemma 6.13. *Suppose that $f \in L_1$ and I is any interval. Then, $f\chi_I \in L_1$ and $\int |f| \chi_I \mathrm{d}\mu \leq \int |f| \mathrm{d}\mu$.*

Proof. We may assume that $f \geq 0$. Let (g_k) be a representation of f. For each k, by the last lemma above, $g_k\chi_I \in L_1$ and $\int |g_k\chi_I| \mathrm{d}\mu \leq \int |g_k| \mathrm{d}\mu$. Therefore, $\sum_{k=0}^{\infty} \int |g_k\chi_I| \mathrm{d}\mu$ converges. Since $f\chi_I = \sum_{k=0}^{\infty} g_k\chi_I$ on $dmn(f) \cap dmn(\chi_I)$, by Theorem 6.8, we have $f\chi_I \in L_1$ and $\int f\chi_I \mathrm{d}\mu = \lim_{N\to\infty} \int \sum_{k=0}^{N} g_k\chi_I \mathrm{d}\mu$. By the last lemma above again, $\int \sum_{k=0}^{N} g_k\chi_I \mathrm{d}\mu \leq \int \left|\sum_{k=0}^{N} g_k\right| \mathrm{d}\mu$. Therefore, $\int f\chi_I \mathrm{d}\mu \leq \int |f| \mathrm{d}\mu$. □

Lemma 6.14. *Suppose that $f \in L_1$. For any $\varepsilon > 0$, there exists finite simple interval I, such that $\int |f| \chi_{\mathbb{R}-I} \mathrm{d}\mu < \varepsilon$.*

Proof. We may assume that $f \geq 0$. Let (g_k) be a representation of f. Choose N such that $\sum_{k=N}^{\infty} \int |g_k| \mathrm{d}\mu < \varepsilon$. Since $\sum_{k=0}^{N} |g_k| \in C(\mathbb{R})$, there exists a finite interval I, such that $\sum_{k=0}^{N} |g_k| = 0$ on $\mathbb{R} - I$, that is, $\int \sum_{k=0}^{N} |g_k| \chi_{\mathbb{R}-I} \mathrm{d}\mu = 0$. Therefore, $\int f\chi_{\mathbb{R}-I} \mathrm{d}\mu \leq \sum_{k=N}^{\infty} \int |g_k| \mathrm{d}\mu < \varepsilon$. □

Lemma 6.15. *Suppose that $J \equiv \cup_{n=0}^{\infty} I_n$ is a finite generalized interval and $f \in C(\mathbb{R})$. Then, $f\chi_J \in L_1$. Moreover, $\int |f|\chi_J \mathrm{d}\mu \leq \int |f| \mathrm{d}\mu$, and if $|f| \leq c$ on J for some constant c, then $\int |f|\chi_J \mathrm{d}\mu \leq c|J|$.*

Proof. We may assume that $f \geq 0$. Since $f \in C(\mathbb{R})$, $|f| \leq c$ for some constant c. Let $\chi_n = \chi_{I_0} \vee ... \vee \chi_{I_n}$. Then, by the lemma above, $f\chi_n = f\chi_{I_0} \vee ... \vee f\chi_{I_n}$ is integrable. Therefore, $(f\chi_n)$ is an increasing sequence of integrable functions. Now, for $m > n$, $\chi_m - \chi_n \leq \chi_{I_m} + ... + \chi_{I_{n+1}}$. Therefore,

$$\int (f\chi_m - f\chi_n) \mathrm{d}\mu \leq \sum_{i=n+1}^{m} \int f\chi_{I_i} \mathrm{d}\mu \leq c \sum_{i=n+1}^{m} |I_i|.$$

Since $\sum_{i=0}^{\infty} |I_i|$ converges, we see that $\lim_{n\to\infty} \int f\chi_n \mathrm{d}\mu$ exists. By Corollary 6.9, there exists an integrable function g such that for $x \in dmn(g)$, $g(x) = \lim_{n\to\infty} f(x)\chi_n(x)$. Obviously, for $x \in J \cup (\mathbb{R} - J)$, $g(x) = f(x)\chi_J(x)$. Therefore, $f\chi_J \in L_1$. The rest follows easily. □

Lemma 6.16. *Suppose that $J \equiv \cup_{n=0}^{\infty} I_n$ is a finite generalized interval and $f \in L_1$. Then, $f\chi_J \in L_1$ and $\int |f|\chi_J \mathrm{d}\mu \leq \int |f| \mathrm{d}\mu$.*

Proof. We may assume that $f \geq 0$. Let (g_k) be a representation of f. Since $g_k \in C(\mathbb{R})$, by the lemma above, $g_k\chi_J \in L_1$ and $\int |g_k\chi_J| \mathrm{d}\mu \leq \int |g_k| \mathrm{d}\mu$. Therefore, $\sum_{n=0}^{\infty} \int |g_k\chi_J| \mathrm{d}\mu$ converges. By Theorem 6.8, there exists an integrable function g such that $g(x) = \sum_{k=0}^{\infty} g_k(x)\chi_J(x)$ for $x \in dmn(g)$. Obviously, $g(x) = f(x)\chi_J(x)$ on a full set. Therefore, $f\chi_J \in L_1$ and $\int f\chi_J \mathrm{d}\mu = \lim_{N\to\infty} \int \sum_{k=0}^{N} g_k\chi_J \mathrm{d}\mu$. Moreover, since $\left|\sum_{k=0}^{N} g_k\right| \in C(\mathbb{R})$, by the lemma above,

$$\int f\chi_J \mathrm{d}\mu \leq \lim_{N\to\infty} \int \left|\sum_{k=0}^{N} g_k\right| \mathrm{d}\mu = \int |f| \mathrm{d}\mu.$$

□

Corollary 6.17. *Suppose that I is any interval and J is a finite generalized interval. Suppose that $f \in L_1$. Then, $f\chi_{I-J} \in L_1$.*

Proof. By the definition above, $f\chi_{I-J} = (f\chi_I - f\chi_J)^+$. Then, this follows from the above lemmas. □

We will denote $\int f\chi_J \mathrm{d}\mu$ as $\int_J f \mathrm{d}\mu$ and denote $\int f\chi_{I-J} \mathrm{d}\mu$ as $\int_{I-J} f \mathrm{d}\mu$.

Lemma 6.18. *Suppose that $f \in L_1$. Then, for any $\varepsilon > 0$, there exists $\delta > 0$, such that if J is a generalized interval and $|J| < \delta$, then $\int_J |f| \mathrm{d}\mu < \varepsilon$.*

Proof. We may assume that $f \geq 0$. Let (g_k) be a representation of f as in the proof of the last lemma. We have

$$\int f\chi_J \mathrm{d}\mu \leq \sum_{k=0}^{\infty} \int |g_k|\chi_J \mathrm{d}\mu.$$

Choose N such that $\sum_{k=N}^{\infty} \int |g_k|\,\mathrm{d}\mu < \varepsilon/2$. Since $\sum_{k=0}^{N} |g_k| \in C(\mathbb{R})$, $\sum_{k=0}^{N} |g_k| \leq C$ for some constant $C > 0$. Let $\delta = \varepsilon/2C$. Then, by Lemma 6.15 above, $\int \sum_{k=0}^{N} |g_k| \chi_J \mathrm{d}\mu < \varepsilon/2$ whenever $|J| < \delta$. Therefore, $\int_J |f|\,\mathrm{d}\mu < \varepsilon$ whenever $|J| < \delta$. □

Lemma 6.19. *Suppose that (J_k) is a sequence of generalized intervals such that $|J_k| \to 0$ as $k \to \infty$. Then, $\cup_k (\mathbb{R} - J_k)$ is a full set.*

Proof. By taking a subsequence, we may assume that $|J_k| < 2^{-k}$ for each k. Let $f_k = k\chi_{J_k}$ for each k. Then, $\int f_k \mathrm{d}\mu \leq k|J_k| < k2^{-k}$. Therefore, $\sum_{k=0}^{\infty} \int |f_k|\,\mathrm{d}\mu$ converges. By Theorem 6.8, there exists $f \in L_1$ such that $f(x) = \sum_{k=0}^{\infty} f_k(x)$ for $x \in dmn(f)$. Obviously, $x \in dmn(f)$ and $f(x) < k$ implies that $x \in \mathbb{R} - J_k$. Therefore, $dmn(f) \subseteq \cup_k (\mathbb{R} - J_k)$. That is, $\cup_k (\mathbb{R} - J_k)$ is a full set. □

Note that $\cup_k (\mathbb{R} - J_k) \subseteq \mathbb{R} - \cap_k J_k$. Therefore, the latter is also a full set. In a special case, we have a finite generalized interval $J \equiv \cup_{n=0}^{\infty} I_n$ and $J_k \equiv \cup_{n=k}^{\infty} I_n$. Then, the condition of the lemma holds. It means that $\mathbb{R} - \cap_{k=0}^{\infty} \cup_{n=k}^{\infty} I_n$ is a full set. This corresponds to the classical conclusion that $\cap_{k=0}^{\infty} \cup_{n=k}^{\infty} I_n$ is of the measure 0.

Now, we can define measurable functions.

Definition 6.20. Let $f \in \mathscr{F}(\mathscr{D}, \mathbb{R})$. f is measurable, if for each $\varepsilon > 0$ and each simple finite interval I, there exists a generalized interval J and $g \in C(\mathbb{R})$, such that $|J| < \varepsilon$ and $|f - g| < \varepsilon$ on $A \cap (I - J)$ for some full set A.

To simplify the presentation, we will simply say "on $I - J$", instead of "on $A \cap (I - J)$ for some full set A".

It directly follows from the definition that any continuous function in $C(\mathbb{R}, \mathbb{R})$ is measurable. It is also easy to see that finite linear combinations of measurable functions are measurable. Similarly, if f and g are measurable, then $|f|$, f^+, f^-, $f \vee g$, $f \wedge g$ are all measurable. Moreover, if f is measurable and $g \in C(\mathbb{R}, \mathbb{R})$, then $g \circ f$ is measurable. Furthermore, any characteristic function of an interval is obviously measurable. Therefore, any step function is measurable, including step functions on infinite sub-intervals.

Since continuous functions in $C(\mathbb{R})$ are bounded, it follows from the definition that if f is measurable, then for any finite simple interval I and any $\varepsilon > 0$, there exists a generalized interval J, such that $|J| < \varepsilon$, and f is bounded on $I - J$.

The measurability of a product of measurable functions needs a little more attention.

Lemma 6.21. *If $f_1, ..., f_n$ is a finite sequence of measurable functions, then $f_1 ... f_n$ is measurable.*

Proof. Given a finite simple interval I and $\varepsilon > 0$, first there exist generalized intervals J_1, ..., J_n and c_1, ..., c_n, such that $|J_i| < \frac{\varepsilon}{2n}$ and $|f_i| \leq c_i$ on $I - J_i$ for $i = 1, ..., n$. Then, again, there exist generalized intervals J_1', ..., J_n' and g_1, ..., $g_n \in C(\mathbb{R})$, such that $|J_i'| < \frac{\varepsilon}{2n}$, and $|f_i - g_i| < \frac{\varepsilon}{n(c_1+1)...(c_n+1)}$ on $I - J_i'$ for $i = 1, ..., n$. Let

$$J = J_1 \cup ... \cup J_n \cup J_1' \cup ... \cup J_n'.$$

Then, $|J| < \varepsilon$. We may assume that $\varepsilon < 1$. Then, we have $|f_i| \leq c_i$ on $I - J$ and $|g_i| \leq c_i + 1$ on $I - J$ for all $i = 1, ..., n$. Therefore, we must have $|f_1...f_n - g_1...g_n| < \varepsilon$ on $I - J$. □

Next, we want to prove that integrable functions are measurable. We need a lemma.

Lemma 6.22. *Suppose that $f \in L_1$, $f \geq 0$, and $\int f \mathrm{d}\mu < 2^{-4}\varepsilon^2$ for some $\varepsilon > 0$. Then, there exists a generalized interval J, such that $|J| < \varepsilon$ and $f < \varepsilon$ on $(\mathbb{R} - J)$.*

Proof. Let (g_n) be a representation for f. By the assumption,

$$\lim_{N\to\infty} \int \left| \sum_{n=0}^{N} g_n \right| \mathrm{d}\mu = \int |f| \mathrm{d}\mu < 2^{-4}\varepsilon^2.$$

Let $N_0 = 0$. Choose N_1 sufficiently large such that $\sum_{n=N_1}^{\infty} \int |g_n| \mathrm{d}\mu < 2^{-6}\varepsilon^2$ and $\int \left| \sum_{n=0}^{N_1} g_n \right| \mathrm{d}\mu < 2^{-4}\varepsilon^2$. For each $k > 1$ there exists N_k such that $\sum_{n=N_k}^{\infty} \int |g_n| \mathrm{d}\mu < 2^{-2k-4}\varepsilon^2$. For $k \geq 0$, let $h_k \equiv \left| \sum_{n=N_k}^{N_{k+1}} g_n \right|$. Then, $\int h_k \mathrm{d}\mu < 2^{-2k-4}\varepsilon^2$ for $k \geq 0$. Note that $h_k \in C(\mathbb{R})$. There exists a partition P_k such that the Riemann sum $S(h_k, P_k) < 2^{-2k-4}\varepsilon^2$ and the size $\delta(P_k)$ of the intervals in the partition is such that $\delta(P_k) < \omega_{h_k}(2^{-k-3}\varepsilon)$, where ω_{h_k} is a modulus of continuity for h_k. For each interval I in the partition, $\max\{h_k(x) : x \in I\}$ is either $< 2^{-k-1}\varepsilon$ or $> 2^{-k-2}\varepsilon$. In the latter case, we have $h_k(x) > 2^{-k-3}\varepsilon$ for all $x \in I$. Let I_1, ..., I_j be all intervals belonging to the later case. We have $S(h_k, P_k) \geq 2^{-k-3}\varepsilon \sum_{i=1}^{j} |I_i|$. Therefore, $\sum_{i=1}^{j} |I_i| < 2^{-k-1}\varepsilon$. Let $J_k \equiv \cup_{i=1}^{j} I_i$. Then, $|J_k| < 2^{-k-1}\varepsilon$ and $h_k(x) < 2^{-k-1}\varepsilon$ on $\mathbb{R} - J_k$. Let $J = \cup_{k=0}^{\infty} J_k$. Then, $|J| < \varepsilon$, and for $x \in D_{(g_n)} \cap (\mathbb{R} - J)$,

$$f(x) = \sum_{n=0}^{\infty} g_n(x) \leq \sum_{k=0}^{\infty} h_k(x) < \varepsilon.$$

□

Lemma 6.23. *Every integrable function f is measurable.*

Proof. Let (g_n) be a representation for f. Then, $\lim_{N\to\infty} \int \left| f - \sum_{n=0}^{N} g_n \right| \mathrm{d}\mu = 0$. Given $\varepsilon > 0$, choose N such that $\int \left| f - \sum_{n=0}^{N} g_n \right| \mathrm{d}\mu < 2^{-4}\varepsilon^2$. The rest easily follows from the last lemma with $g = \sum_{n=0}^{N} g_n$ in the definition of measurability. □

Reversely, we have

Lemma 6.24. *Suppose that f is measurable. Then, for any finite interval I and any $\varepsilon > 0$, there exists generalized interval J, such that $|J| < \varepsilon$ and $f\chi_{I-J}$ is integrable.*

Proof. For each $n > 0$, there exist $g_n \in C(\mathbb{R})$ and a generalized interval J_n, such that $|J_n| < 2^{-n-1}\varepsilon$ and $|f - g_n| < 2^{-n-1}\varepsilon$ on $I - J_n$. Let $J = \cup_{n=0}^{\infty} J_n$. Then, J is also a generalized interval, and $|J| < \varepsilon$, and $|f - g_n| < 2^{-n-1}\varepsilon$ on $I - J$ for all n. Therefore,

$$f\chi_{I-J} = g_0\chi_{I-J} + \sum_{n=0}^{\infty}\left(g_{n+1}\chi_{I-J} - g_n\chi_{I-J}\right)$$

on $dmn\left(\chi_{I-J}\right) \cap dmn(f)$. Moreover, $\left|g_{n+1}\chi_{I-J} - g_n\chi_{I-J}\right| \leq 2^{-n}\varepsilon\chi_{I-J}$ on $dmn\left(\chi_{I-J}\right)$. Therefore,

$$\int \left|g_{n+1}\chi_{I-J} - g_n\chi_{I-J}\right| \mathrm{d}\mu \leq 2^{-n}\varepsilon\,|I|.$$

This implies that $\sum_{n=0}^{\infty}\int\left|g_{n+1}\chi_{I-J} - g_n\chi_{I-J}\right|\mathrm{d}\mu$ converges. Then, by Theorem 6.8, $f\chi_{I-J} \in L_1$. □

Lemma 6.25. *If f is measurable and g is integrable and $|f| \leq g$, then f is also integrable.*

Proof. By Lemma 6.14, for each n, there exists finite interval I_n, such that $\int g\chi_{\mathbb{R}-I_n}\,\mathrm{d}\mu < 2^{-n}$. We may assume that $I_n \subset I_{n+1}$ and $I_n \to \mathbb{R}$ as $n \to \infty$. By Lemma 6.18, for each n, there exists δ_n, such that $\int g\chi_J \mathrm{d}\mu < 2^{-n}$ whenever J is a generalized interval such that $|J| < \delta_n$. We may assume that $\delta_n < 2^{-n}$. By the last lemma, for each n, there exists a generalized interval J_n, such that $|J_n| < \delta_n$ and $f\chi_{I_n-J_n}$ is integrable. Note that

$$\left|\chi_{I_{n+1}-J_{n+1}} - \chi_{I_n-J_n}\right| \leq \chi_{\mathbb{R}-I_n} + \chi_{J_n} + \chi_{J_{n+1}}.$$

Therefore,

$$\int\left|f\chi_{I_{n+1}-J_{n+1}} - f\chi_{I_n-J_n}\right|\mathrm{d}\mu \leq \int g\left(\chi_{\mathbb{R}-I_n} + \chi_{J_n} + \chi_{J_{n+1}}\right)\mathrm{d}\mu < 2^{-n+2}.$$

It means that $\sum_{n=0}^{\infty}\int\left|f\chi_{I_{n+1}-J_{n+1}} - f\chi_{I_n-J_n}\right|\mathrm{d}\mu$ converges. By Theorem 6.8, there exists $h \in L_1$, such that

$$h(x) = f(x)\chi_{I_0-J_0}(x) + \sum_{n=0}^{\infty}\left(f(x)\chi_{I_{n+1}-J_{n+1}}(x) - f(x)\chi_{I_n-J_n}(x)\right)$$

for $x \in dmn(h)$. Note that $J_k^* = \cup_{n\geq k}J_n$ is a generalized interval and

$$|J_k^*| \leq \sum_{n=k}^{\infty}|J_n| < 2^{-k+1}.$$

By Lemma 6.19, $A = \cup_k\left(\mathbb{R} - J_k^*\right)$ is a full set. For $x \in dmn(h) \cap A$, $x \in I_n - J_n$ for all sufficiently large n. Therefore, $h(x) = f(x)$. That is, $f \in L_1$. □

6.3 Convergence

Now we define the familiar notions of convergence for sequences of measurable functions.

Definition 6.26. Let (f_n) be a sequence of measurable functions and f be a function in $\mathscr{F}(\mathscr{D},\mathbb{R})$.

We say that (f_n) converges in measure to f, if for any finite interval I, any $\varepsilon > 0$, there exists N, such that for each $n > N$, there exists generalized interval J, such that $|J| < \varepsilon$ and $|f - f_n| < \varepsilon$ on $I - J$.

We say that (f_n) converges almost everywhere to f, if for any finite interval I, any $\varepsilon > 0$, there exists N and generalized interval J, such that $|J| < \varepsilon$ and for each $n > N$, $|f - f_n| < \varepsilon$ on $I - J$.

We say that (f_n) converges almost uniformly to f, if for any finite interval I, any $\varepsilon > 0$, there exists generalized interval J, such that $|J| < \varepsilon$ and $f_n \to f$ uniformly on $I - J$.

Obviously, almost uniform convergence implies almost everywhere convergence, which in turn implies convergence in measure.

Lemma 6.27. *If (f_n) is a sequence of measurable functions and (f_n) converges in measure to f, then f is measurable.*

Proof. Given any finite interval I any $\varepsilon > 0$, by the assumption, there exists generalized interval J and n, such that $|J| < \varepsilon/2$ and $|f - f_n| < \varepsilon/2$ on $I - J$. Since f_n is measurable, there exists generalized interval J' and $g \in C(\mathbb{R})$ such that $|J'| < \varepsilon/2$ and $|f_n - g| < \varepsilon/2$ on $I - J'$. Let $J'' = J \cup J'$, then $|J''| < \varepsilon$ and $|f - g| < \varepsilon$ on $I - J''$. Therefore, f is measurable. □

Lemma 6.28. *If (f_n) is a sequence of measurable functions converging in measure to f, then there exists a subsequence (f_{N_k}) converging to f point-wise on a full set.*

Proof. Let (I_k) be a sequence of simple intervals such that $I_k \subset I_{k+1}$ for each k and $I_k \to \mathbb{R}$ as $k \to \infty$. For each k, there exists N_k and generalized interval J_k, such that $|J_k| < 2^{-k}$ and $\left|f_{N_k} - f\right| < 2^{-k}$ on $I_k - J_k$. Let

$$A = dmn(f) \cap \cap_k \left(dmn\left(f_{N_k}\right) \cap (I_k \cup (\mathbb{R} - I_k)) \cap (J_k \cup (\mathbb{R} - J_k))\right).$$

A is a full set. By Lemma 6.19, $B = \cup_k (\mathbb{R} - \cup_{n \geq k} J_n)$ is a full set. For $x \in A \cap B$, there exists k such that $x \in I_n$ and $x \in \mathbb{R} - J_n$ for $n \geq k$. Therefore, $|f_{N_n}(x) - f(x)| < 2^{-n}$ for $n \geq k$. That is, $f_{N_n}(x) \to f(x)$ on the full set $A \cap B$. □

Corollary 6.29. *If (f_n) is a sequence of measurable functions and (f_n) converges in measure to both f and g, then $f = g$ on a full set.*

Proof. By the lemma, we have a subsequence converging to f point-wise on a full set and then a subsequence again converging to g point-wise on a full set. The final subsequence will converge to both f and g point-wise on a full set. Therefore, $f = g$ on a full set. □

The following two lemmas connect convergence in measure with convergence in the metric of L_1.

Lemma 6.30. *Suppose that $f \in L_1$ and $f_n \in L_1$ for all n, and suppose that $\int |f_n - f|$ $d\mu \to 0$ as $n \to \infty$. Then, (f_n) converges to f in measure.*

Proof. This follows from Lemma 6.22. □

The following is the dominated convergence theorem.

Lemma 6.31. *Suppose that $g \in L_1$, and $f_n \in L_1$, $|f_n| \leq g$ for all n, and suppose that (f_n) converges to f in measure. Then, $f \in L_1$ and $\int |f_n - f| \mathrm{d}\mu \to 0$ as $n \to \infty$.*

Proof. By Lemma 6.28, a subsequence converges point-wise to f on a full set. Therefore, $|f| \leq g$ on a full set. By Lemma 6.25, $f \in L_1$. To see that $\int |f_n - f| \mathrm{d}\mu \to 0$, given any $\varepsilon > 0$, first by Lemma 6.14, there exists a finite interval I such that $\int_{\mathbb{R}-I} |f| \mathrm{d}\mu < \varepsilon/6$ and $\int_{\mathbb{R}-I} g \mathrm{d}\mu < \varepsilon/6$. Therefore, $\int_{\mathbb{R}-I} |f_n - f| \mathrm{d}\mu < \varepsilon/3$ for all n. By Lemma 6.18, there exists δ such that $\int_J |f| \mathrm{d}\mu < \varepsilon/6$ and $\int_J g \mathrm{d}\mu < \varepsilon/6$, and hence $\int_J |f_n - f| \mathrm{d}\mu < \varepsilon/3$ for all n, whenever J is a generalized interval and $|J| < \delta$. Since (f_n) converges to f in measure, there exists N such that for each $n > N$ there exists J such that $|J| < \delta$ and $|f_n - f| \leq \varepsilon/3|I|$ on $I - J$. Therefore, $|f_n - f| \chi_{I-J} \leq \frac{\varepsilon}{3|I|}\chi_I$ on a full set, and hence

$$\int_{I-J} |f_n - f| \mathrm{d}\mu \leq \int \frac{\varepsilon}{3|I|} \chi_I \mathrm{d}\mu \leq \varepsilon/3.$$

Then, since

$$|f_n - f| \leq |f_n - f| \left(\chi_{\mathbb{R}-I} + \chi_{I-J} + \chi_J\right)$$

on a full set, we have $\int |f_n - f| \mathrm{d}\mu < \varepsilon$ for each $n > N$. □

It turns out that convergence almost everywhere and convergence almost uniformly are equivalent.

Lemma 6.32. *If (f_n) converges almost everywhere to f, then (f_n) converges almost uniformly to f.*

Proof. By the assumption, given any finite interval I and any $\varepsilon > 0$, for each k there exists generalized interval J_k and N_k, such that $|J_k| < 2^{-k-1}\varepsilon$ and $|f_n - f| < 2^{-k-1}\varepsilon$ on $I - J_k$ for $n \geq N_k$. Let $J = \cup_k J_k$. Then, $|J| < \varepsilon$, and $I - J \subseteq I - J_k$ for each k. Therefore, for each k, $|f_n - f| < 2^{-k-1}\varepsilon$ on $I - J$ for $n \geq N_k$, which means that (f_n) converges to f uniformly on $I - J$. That is, (f_n) converges to f almost uniformly. □

For almost everywhere convergence, we have a stronger version of Lemma 6.28:

Lemma 6.33. *If (f_n) is a sequence of measurable functions converging to f almost everywhere, then (f_n) converges to f point-wise on a full set*

Proof. Let (I_k) be a sequence of simple intervals such that $I_k \subset I_{k+1}$ for each k and $I_k \to \mathbb{R}$ as $k \to \infty$. By the assumption, for each k and I_k, there exist generalized interval J_k and N_k, such that $|J_k| < 2^{-k}$ and $|f_n - f| < 2^{-k}$ on $I_k - J_k$ for $n \geq N_k$. Let $A = \cup_n (\mathbb{R} - \cup_{k \geq n} J_k)$. By Lemma 6.19, A is a full set. Then, for each x in the full set

$$A \cap dmn(f) \cap \cap_k (dmn(f_k) \cap (I_k \cup (\mathbb{R} - I_k)) \cap (J_k \cup (\mathbb{R} - J_k))),$$

we have $x \in \mathbb{R} - \cup_{k \geq N} J_k$ and $x \in I_N$ for some N. Therefore, for each $k > N$, $x \in I_k - J_k$. It means that $|f_n(x) - f(x)| < 2^{-k}$ for each $k > N$ and $n > N_k$. That is,

$(f_n(x))$ converges to $f(x)$. Therefore, (f_n) converges to f point-wise on the full set above. □

Now we introduce several notions of Cauchyness corresponding to those notions of convergence.

Definition 6.34. Let (f_n) be a sequence of measurable functions.

(f_n) is Cauchy in measure, if for any finite interval I, any $\varepsilon > 0$, there exists N, such that for each n, $m > N$, there exists generalized interval J, such that $|J| < \varepsilon$ and $|f_m - f_n| < \varepsilon$ on $I - J$.

(f_n) is Cauchy almost everywhere, if for any finite interval I, any $\varepsilon > 0$, there exists N and generalized interval J, such that $|J| < \varepsilon$, and for each m, $n > N$, $|f_m - f_n| < \varepsilon$ on $I - J$.

(f_n) is Cauchy almost uniformly, if for any finite interval I, any $\varepsilon > 0$, there exists generalized interval J, such that $|J| < \varepsilon$, and (f_n) is Cauchy uniformly on $I - J$, that is, for any $\delta > 0$, there exists N, such that $|f_m - f_n| < \delta$ on $I - J$ for all $m, n > N$.

It is obvious that being Cauchy almost uniformly implies being Cauchy almost everywhere and that in turn implies being Cauchy in measure. The following lemmas connect various notions of Cauchyness with the corresponding notions of convergence.

Lemma 6.35. *If (f_n) is Cauchy almost everywhere, then there exists a measurable function f, such that (f_n) converges almost uniformly to f (and hence (f_n) is Cauchy almost uniformly) and converges to f point-wise on a full set.*

Proof. Let (I_k) be a sequence of simple intervals such that $I_k \subset I_{k+1}$ for each k and $I_k \to \mathbb{R}$ as $k \to \infty$. For each k, there exist a generalized interval J_k and a number N_k, such that $|J_k| < 2^{-k}$ and $|f_m - f_n| < 2^{-k}$ on $I_k - J_k$ for $m, n \geq N_k$. Let A be the full set

$$\cap_k (dmn(f_k) \cap (I_k \cup (\mathbb{R} - I_k)) \cap (J_k \cup (\mathbb{R} - J_k))).$$

$F = A \cap \cup_n (\mathbb{R} - \cup_{k \geq n} J_k)$ is a full set. For each $x \in F$, there exists N such that $x \in I_N$ and $x \in \mathbb{R} - \cup_{k \geq N} J_k$. Therefore, for each $k > N$, there exists N_k, such that for any $m, n \geq N_k$, $|f_m(x) - f_n(x)| < 2^{-k}$. This means that $(f_n(x))$ converges. Therefore, we can define a function f on the full set F by $f(x) = \lim_{n \to \infty} f_n(x)$. (f_n) converges to f point-wise on a full set. Moreover, for any finite interval I and any $\varepsilon > 0$, there exists N such that $I \subseteq I_N$ and $|\cup_{k \geq N} J_k| < \varepsilon$. Let $J = \cup_{k \geq N} J_k$. Then, for any $k > N$, $|f_m - f_n| < 2^{-k}$ on $I - J$ for $m, n \geq N_k$. It is obvious that (f_n) converges to f uniformly on $I - J$. Therefore, (f_n) converges almost uniformly to f. □

Lemma 6.36. *If (f_n) is Cauchy in measure, then there exists a measurable function f, such that (f_n) converges to f in measure and a subsequence of (f_n) converges to f almost uniformly.*

Proof. Again, let (I_k) be a sequence of simple intervals such that $I_k \subset I_{k+1}$ for each k and $I_k \to \mathbb{R}$ as $k \to \infty$. For each k, there exists N_k, such that for $m, n \geq N_k$ there exists a generalized interval J, such that $|J| < 2^{-k}$ and $|f_m - f_n| < 2^{-k}$ on $I_k - J$. For each k, let J_k be such that $\left|f_{N_{k+1}} - f_{N_k}\right| < 2^{-k}$ on $I_k - J_k$, and let $J_k^* = \cup_{n \geq k} J_n$. Then,

$\left|J_k^*\right| < 2^{-k+1}$ and $\left|f_{N_l} - f_{N_k}\right| < 2^{-k+1}$ on $I_k - J_k^*$ for all $l \geq k$. It is easy to see that the subsequence $\left(f_{N_k}\right)$ is Cauchy almost everywhere. Therefore, by the last lemma above, $\left(f_{N_k}\right)$ converges almost uniformly to a measurable function f. To see that (f_n) converges to f in measure, let I be any finite interval and $\varepsilon > 0$. Since $\left(f_{N_k}\right)$ converges almost uniformly to f, there exists J such that $|J| < \varepsilon/2$ and $f_{N_k} \to f$ uniformly on $I - J$. Therefore, there exists k_0, such that $\left|f_{N_k} - f\right| < \varepsilon/2$ on $I - J$ for all $k > k_0$. Since (f_n) is Cauchy in measure, there exists N, such that for any $m,n > N$, there exists J', such that $|J'| < \varepsilon/2$ and $|f_m - f_n| < \varepsilon/2$ on $I - J'$. Then, for $n > N$, choose $k > k_0$ such that $N_k > N$. We have $\left|f_{N_k} - f\right| < \varepsilon/2$ on $I - J$. Moreover, $n > N$ and $N_k > N$. Therefore, there exists J', such that $|J'| < \varepsilon/2$ and $\left|f_{N_k} - f_n\right| < \varepsilon/2$ on $I - J'$. That is, $|f - f_n| < \varepsilon$ on $I - (J \cup J')$, where $|J \cup J'| < \varepsilon$. Therefore, (f_n) converges to f in measure. □

6.4 The Space L_2

We construct the space of square-integrable functions in this section. Note that if f is measurable, f^2 is also measurable. Then, $f^2 \in L_1$ and $\int f^2 \mathrm{d}\mu = 0$ imply that $f^2 = 0$ on a full set, and hence $f = 0$ on a full set. Therefore, we can define the space L_2 and the norm on it as follows:

Definition 6.37. $f \in L_2$, if and only if f is measurable and $f^2 \in L_1$. For $f \in L_2$, the norm $\|f\|_2 \equiv_{df} \left(\int f^2 \mathrm{d}\mu\right)^{1/2}$.

We will drop the subscript in $\|f\|_2$ when no confusion will occur. We have some basic inequalities:

Lemma 6.38. *If $f, g \in L_2$, then $fg \in L_1$ and $\int |fg| \mathrm{d}\mu \leq \|f\| \, \|g\|$.*

Proof. For each $n > 0$, let $c_n = \|f\| + \frac{1}{n}$, $d_n = \|g\| + \frac{1}{n}$.

$$\int \left|\frac{f}{c_n} \cdot \frac{g}{d_n}\right| \mathrm{d}\mu \leq \frac{1}{2} \int \left(\frac{f^2}{c_n^2} + \frac{g^2}{d_n^2}\right) \mathrm{d}\mu = \frac{1}{2} \left(\frac{\|f\|^2}{c_n^2} + \frac{\|g\|^2}{d_n^2}\right) \leq 1.$$

Therefore, $\int |fg| \mathrm{d}\mu \leq c_n d_n$. n is arbitrary. It follows that $\int |fg| \mathrm{d}\mu \leq \|f\| \, \|g\|$. □

The following lemma says that L_2 is closed under linear combinations:

Lemma 6.39. *If $f_i \in L_2$ and $a_i \in \mathbb{R}$ for $i = 1, \ldots, n$, then $\sum_{i=1}^n a_i f_i \in L_2$ and $\|\sum_{i=1}^n a_i f_i\| \leq \sum_{i=1}^n |a_i| \, \|f_i\|$.*

Proof. Note that $\left(\sum_{i=1}^n a_i f_i\right)^2 = \sum_{i,j=1}^n a_i a_j f_i f_j$. By the lemma above, $\sum_{i=1}^n a_i f_i \in L_2$ and the inequality also follows. □

A sequence (f_n) of functions in L_2 is a Cauchy sequence in L_2, if for any $\varepsilon > 0$, there exists N, such that whenever $m,n > N$, $\|f_m - f_n\|_2 < \varepsilon$. A sequence (f_n) of functions in L_2 converges to $f \in L_2$ in the norm of L_2, if $\|f_n - f\|_2$ converges to 0.

Theorem 6.40. *L_2 is complete. That is, any Cauchy sequence in L_2 converges to a function in L_2 in the norm of L_2.*

Proof. Suppose that (f_n) is a Cauchy sequence in L_2. By taking a subsequence, we may assume that $f_0 = 0$ and $\|f_m - f_n\| < 2^{-n}$ for $m > n$.

We first show that there exists $g \in L_2$ such that $f_n^2 \le g^2$ for all n. Let

$$g_n \equiv \sum_{k=0}^{n-1} |f_{k+1} - f_k|.$$

Then, by the two lemmas above, $g_n \in L_2$. That is, (g_n^2) is an increasing sequence of integrable functions. Note that for $m > n$,

$$\left|g_m^2 - g_n^2\right| = 2\sum_{i=0}^{n-1}\sum_{j=n}^{m-1} |f_{i+1} - f_i|\left|f_{j+1} - f_j\right| + \sum_{i,j=n}^{m-1} |f_{i+1} - f_i|\left|f_{j+1} - f_j\right|.$$

Therefore, by the lemmas above,

$$\left|\int g_m^2 \mathrm{d}\mu - \int g_n^2 \mathrm{d}\mu\right| \le \int \left|g_m^2 - g_n^2\right| \mathrm{d}\mu$$

$$\le 2\sum_{i=0}^{n-1}\sum_{j=n}^{m-1} \|f_{i+1} - f_i\| \left\|f_{j+1} - f_j\right\| + \sum_{i,j=n}^{m-1} \|f_{i+1} - f_i\| \left\|f_{j+1} - f_j\right\|$$

$$\le 2\sum_{i=0}^{n-1}\sum_{j=n}^{m-1} 2^{-i-j} + \sum_{i,j=n}^{m-1} 2^{-i-j} < 2^{-n+3}.$$

That is, $\int g_n^2 \mathrm{d}\mu$ converges. By Corollary 6.9, there exists an integrable function g^2, such that $\lim_{n\to\infty} g_n^2(x) = g^2(x)$ on a full set. Note that $|f_n| \le g$ for all n. Therefore, $f_n^2 \le g^2$ for all n.

Next, we show that the sequence (f_n) is Cauchy in measure. Given a finite interval I and $\varepsilon > 0$, there exists N, such that for any $m, m > N$, $\|f_m - f_n\| < 2^{-2}\varepsilon^2$. Then, $\int |f_m - f_n|^2 \mathrm{d}\mu < 2^{-4}\varepsilon^4$. By Lemma 6.22, there exists a generalized interval J, such that $|J| < \varepsilon$ and $|f_m - f_n|^2 < \varepsilon^2$ on $(\mathbb{R} - J)$, that is, $|f_m - f_n| < \varepsilon$ on $(\mathbb{R} - J)$. That is, (f_n) is Cauchy in measure.

Now, by Lemma 6.36, (f_n) converges to a measurable function f in measure. By Lemma 6.28, a subsequence (f_{N_k}) converges to f point-wise on a full set. Since by the conclusion above, $f_{N_k}^2 \le g^2$ for all k on a full set, we have $f^2 \le g^2$ on a full set. Therefore, by Lemma 6.25, $f \in L_2$. Note that $(f_n - f)$ converges to 0 in measure. Therefore, $|f_n - f|^2$ also converges to 0 in measure. Moreover,

$$|f_n - f|^2 \le 2\left(f_n^2 + f^2\right) \le 4g^2.$$

Therefore, by the dominated convergence theorem, i.e. Lemma 6.31, $\int |f_n - f|^2$ $d\mu \to 0$. That is, $\|f_n - f\|_2 \to 0$, namely, (f_n) converges to f in the norm of L_2. □

Lemma 6.41. *Suppose that $f \in L_2$. Then, for any $\varepsilon > 0$, there exists $g \in C(\mathbb{R})$, such that $\|f-g\|_2 < \varepsilon$. That is, $C(\mathbb{R})$ is dense in L_2.*

Proof. We call (a,b) a rational interval, if a, b are rational numbers. Similarly, a generalized interval is rational, if it is a union of rational simple intervals. We first prove that every $f \in L_2$ can be approximated, in the norm of L_2, by functions of the format $g\chi_{I-J}$, where $g \in C(\mathbb{R})$, and I is a finite rational interval, and J is a finite and rational generalized interval. Note that we always have $g\chi_{I-J} \in L_2$. Given any $f \in L_2$ and $\varepsilon > 0$, we must find $g\chi_{I-J}$ such that $\|f - g\chi_{I-J}\|_2 < \varepsilon$. First, by Lemma 6.14, we can find a finite simple interval I such that $\int_{\mathbb{R}-I} f^2 \mathrm{d}\mu < \varepsilon^2/3$. Clearly, we can choose I so that it is rational. Then, by Lemma 6.18, we can find $\delta > 0$, such that whenever J is a generalized and $|J| < \delta$, then $\int_J f^2 \mathrm{d}\mu < \varepsilon^2/3$. Since $f \in L_2$, by the definition of L_2, f is measurable. Therefore, we can find a generalized interval J and $g \in C(\mathbb{R})$, such that $|J| < \delta$ and $|f-g| < \frac{\varepsilon}{3\sqrt{(|I|+1)}}$ on $I-J$. By extending each simple interval in J slightly, we can make J a rational generalized interval but still satisfy the condition $|J| < \delta$. Then, we have $\int_J f^2 \mathrm{d}\mu < \varepsilon^2/3$ and $\int_{I-J} |f-g|^2 \mathrm{d}\mu < \varepsilon^2/3$. Therefore,

$$\int |f - g\chi_{I-J}|^2 \mathrm{d}\mu = \int_{\mathbb{R}-I} f^2 \mathrm{d}\mu + \int_J f^2 \mathrm{d}\mu + \int_{I-J} |f-g|^2 \mathrm{d}\mu < \varepsilon^2,$$

and hence $\|f - g\chi_{I-J}\|_2 < \varepsilon$.

Then, we prove that any function $g\chi_{I-J}$ of the format above can be approximated, in the norm of L_2, by functions in $C(\mathbb{R})$. Let $g\chi_{I-J}$ be such a function and let $\varepsilon > 0$. Suppose that $J = \cup_n I_n$. First, since J is finite, we can find N such that $\sum_{n=N+1}^{\infty} |I_n| < \frac{\varepsilon^2}{4(c+1)}$, where c is the supremum of g^2 on its compact support. Let $J_N \equiv \cup_{n=0}^{N} I_n$ and let $J_N^* \equiv \cup_{n=N+1}^{\infty} I_n$. Then, $J = J_N \cup J_N^*$ and $|J_N^*| < \frac{\varepsilon^2}{4(c+1)}$. It is easy to see that $|\chi_{I-J_N} - \chi_{I-J}| \leq \chi_{J_N^*}$ on a full set. Therefore,

$$\int |g\chi_{I-J_N} - g\chi_{I-J}|^2 \mathrm{d}\mu \leq \int g^2 \chi_{J_N^*} \mathrm{d}\mu < \varepsilon^2/4.$$

Since I, J_N are rational, we can compare the end points of the simple intervals in them. Therefore, by merging overlapping intervals, we can express $I - J_N$ as

$$I - J_N = \cup_{i=1}^{m} I_i,$$

where $I_1, \ldots, I_m$ are mutually disjoint (open, closed or half-open) intervals with rational end points. Then, it is easy to see that we can use continuous functions to approximate $g\chi_{I-J_N}$, as in approximating the characteristic function of the interval $(0,1)$ by continuous functions in Sect. 6.1. That is, we can find $h \in C(\mathbb{R})$ such that $\int |h - g\chi_{I-J_N}|^2 \mathrm{d}\mu < \varepsilon^2/4$. Then, we have $\|h - g\chi_{I-J}\|_2 < \varepsilon$. □

Theorem 6.42. *L_2 is separable. That is, there is a sequence (g_n) of functions in L_2 such that for each $f \in L_2$ and $m > 0$, there exists n, such that $\|f - g_n\|_2 < 1/m$.*

Proof. By the lemma above, it suffices to show that $C(\mathbb{R})$ is separable. A function in $C(\mathbb{R})$ is uniformly continuous and vanishes beyond a compact interval. It can be approximated arbitrarily closely, in the metric of L_2, by a function g that is continuous, piecewise linear on finitely many intervals $[q_0, q_1], [q_1, q_2], ..., [q_{n-1}, q_n]$, and that vanishes on $(-\infty, q_0]$ and $[q_n, +\infty)$. Obviously, we can choose $q_0, q_1, ..., q_n$ to be rational numbers and we can also make $g(q_1), ..., g(q_{n-1})$ rational numbers. Such functions g can be enumerated in a sequence. $\square$

Chapter 7
Hilbert Space

This chapter develops the basics of the theory of bounded and unbounded linear operators on Hilbert spaces. We will finally construct the spectral decomposition for an unbounded self-adjoint linear operator on a Hilbert space and prove Stone's Theorem. This shows that strict finitism is in principle sufficient for the basic applications in classical quantum mechanics.

This chapter again takes many ideas from Bishop and Bridges [6], and we have to make many changes as well. In particular, the basic definition of linear space has to be modified to fit into strict finitism. The development of the theory of unbounded linear operators on Hilbert spaces follows the ideas in Ye [40, 41], with necessary improvements to fit into strict finitism.

7.1 Basic Definitions

We let $\mathbb{F}$ be $\mathbb{R}$ or $\mathbb{C}$, and we want to define linear spaces, normed linear spaces and Hilbert spaces over $\mathbb{F}$. A linear space will be a set with some designated operations as the vector sum and scalar product on the set. We must be able to construct an arbitrary finite sum of vectors. Since the set can be of any signature, recursive constructions on elements of the set are not generally available. Therefore, the definition of linear combination needs some special treatment. We need some notations.

For X a set, recall that $X^{<\infty}$ is the set of finite sequences of elements of X. If X, Y are sets and $f : X \to Y$, let f^* be the function from $X^{<\infty}$ to $Y^{<\infty}$ defined by

$$f^* \equiv_{df} \lambda u.\lambda \mathbf{x}_1 \ldots .\lambda \mathbf{x}_n.\langle f((u)_0)(\mathbf{x}_1)\ldots(\mathbf{x}_n), \ldots\ldots, f\left((u)_{lh(u)-1}\right)(\mathbf{x}_1)\ldots(\mathbf{x}_n)\rangle.$$

That is, for $u = \langle u_1, \ldots, u_k \rangle \in X^{<\infty}$, $f^*(u) \simeq \langle f(u_1), \ldots\ldots, f(u_k)\rangle$. We will simply write $f^*(u)$ as $f(u)$. Similarly, if $g : \mathbb{F} \times X \to Y$, then for $a \in \mathbb{F}$, $x \in X$, $r = \langle r_1, \ldots, r_k \rangle \in \mathbb{F}^{<\infty}$, and $u = \langle u_1, \ldots, u_l \rangle \in X^{<\infty}$, we let

F. Ye, *Strict Finitism and the Logic of Mathematical Applications*, Synthese Library 355,
DOI 10.1007/978-94-007-1347-5_7,

$$g(a,u) \equiv_{df} \langle g(a,u_1), \ldots\ldots, g(a,u_l)\rangle,$$
$$g(r,x) \equiv_{df} \langle g(r_1,x), \ldots\ldots, g(r_k,x)\rangle,$$
$$g(r,u) \equiv_{df} \langle g(r_1,u_1), \ldots\ldots, g\left(r_{\min(k,l)}, u_{\min(k,l)}\right)\rangle.$$

For a finite sequence $r = \langle r_0, \ldots, r_{k-1}\rangle \in \mathbb{F}^{<\infty}$, we denote $\Sigma r \equiv_{df} \sum_{i=0}^{k-1} r_i$.

A permutation of length k is (the code of) a sequence of natural numbers that is a permutation of $0, \ldots, k-1$. If π is a permutation of length k, $u \in X^{<\infty}$, and $lh(u) = k$, we let $\pi(u)$ denote the permutation of elements of u obtained in the way π permutes the sequence $\langle 0, 1, \ldots, k-1\rangle$:

$$\pi(u) \equiv_{df} \lambda \mathbf{x}_1 \ldots . \lambda \mathbf{x}_n . \langle (u)_{(\pi)_0}(\mathbf{x}_1) \ldots (\mathbf{x}_n), \ldots\ldots, (u)_{(\pi)_{k-1}}(\mathbf{x}_1) \ldots (\mathbf{x}_n)\rangle.$$

$\pi(u)$ can be constructed as a term containing π and u. If $lh(u) > k$, we also use $\pi(u)$ to denote the result of permuting the first k elements of u.

A grouping of a sequence of length k is the code of a sequence of natural numbers, such that each item in the sequence is again a code of a sequence of natural numbers and all the numbers are exactly $0, \ldots, k-1$ in that order, that is,

$$\langle\langle 0, \ldots, k_1 - 1\rangle, \langle k_1, \ldots, k_2 - 1\rangle, \ldots, \langle k_h, \ldots, k-1\rangle\rangle.$$

If γ is a grouping of a sequence of length k, and $u \in X^{<\infty}$, $lh(u) \geq k$, then we let $\gamma(u) \in (X^{<\infty})^{<\infty}$ be the sequence obtained by grouping u as γ groups $\langle 0, 1, 2, \ldots, k-1\rangle$. $\gamma(u)$ can also be constructed as a term containing γ and u.

We make the convention that when the length of π or γ is greater than the length of u, $\pi(u) = \gamma(u) = u$.

Then, the definition of linear space can be stated as follows:

Definition 7.1. A linear space consists of a set X, a function $(a,x) \mapsto ax$ from $\mathbb{F} \times X$ to X, a function Σ (for finite sum) from $X^{<\infty}$ to X, and a zero element $0 \in X$, such that for $x \in X$, $a, b \in \mathbb{F}$, $r \in \mathbb{F}^{<\infty}$, $u \in X^{<\infty}$, π a permutation, and γ a grouping, we have

$$\Sigma(\langle x, 0\rangle) = x,\ 1x = x,\ 0x = a0 = 0,\ a(bx) = (ab)x,$$
$$(\Sigma r)x = \Sigma(rx),\ a\Sigma(u) = \Sigma(au),$$
$$\Sigma(\pi(u)) = \Sigma u,\ \Sigma(\Sigma^*(\gamma(u))) = \Sigma u.$$

We will call the members of X 'vectors'. The last four equations are the distributive law for number addition, the distributive law for vector addition, the commutative law, and the associative law. For easy reading we will write $\Sigma(u)$ as $\sum_{i=0}^{n-1} u_i$ where $n \equiv lh(u)$ and $u_i \equiv (u)_i$, and we write $\Sigma(\langle x, y\rangle)$ as $x + y$. Then, the ordinary distributive, commutative and associative laws follow. Notice that the extensionality condition in the definition of sets is extensively utilized here.

The common properties of addition and scalar multiplication on linear spaces are easy to prove. For instance, 0 is the unique additive zero, and $-x = (-1)x$ is the unique additive inverse of x, and $(-a)x = -ax$. Moreover, if $a \neq 0$ and $ax = 0$, then $x = 0$. However, from $ax = 0$ we cannot generally derive $a = 0 \vee x = 0$.

It is easy to verify that ordinary linear spaces do satisfy these conditions. For instance, the spaces $\mathbb{R}^n$, $\mathbb{C}^n$, and the ordinary spaces of real or complex value functions are linear spaces with the usual scalar multiplication and vector addition.

In the last chapter, L_2 is defined as a space of real value functions. A complex value function is a pair of real value functions. The definition and results on L_2 in the last section for real value functions can be easily carried over for complex value functions. By Lemma 6.39, the set L_2 with the ordinary finite sum and scalar product operations is a linear space.

A linear subset of a linear space is a subset that is closed under the operations ax and Σu. Given a subset $A \subseteq X$, the linear subset M spanned by A is

$$x \in M \equiv_{df} \exists r \in \mathbb{F}^{<\infty} \exists u \in A^{<\infty} (x = \Sigma ru).$$

It is easy to verify that M is a linear subset. If $e = \langle e_1, ..., e_n \rangle$ is a finite sequence of vectors, the linear subset $[e]$ spanned by e is

$$x \in [e] \equiv_{df} \exists \langle r_1, ..., r_n \rangle \in \mathbb{F}^{<\infty} \left(x = \sum_{i=1}^{n} r_i e_i \right).$$

Similarly, if $e = (e_n)$ is an infinite sequence of vectors, the linear subset $[e]$ spanned by e is the linear subset spanned by the subset $\{e_n : n \in \mathbb{N}\}$. A subset $B \subseteq X$ is linearly independent, if for any $u = \langle u_1, ..., u_k \rangle \in B^{<\infty}$ and $r = \langle r_1, ..., r_k \rangle \in \mathbb{F}^{<\infty}$, $\Sigma (ru) = 0$ implies that $r_i = 0$ for $i = 1, ..., k$.

Definition 7.2. A norm $\|\cdot\|$ on a linear space X over $\mathbb{F}$ is a function from X to the set $\mathbb{R}^{+0}$ of non-negative numbers, such that for $x \in X$, $a \in \mathbb{F}$, and $u = \langle u_1, ..., u_k \rangle \in X^{<\infty}$, we have

$$\|x\| = 0 \to x = 0, \ \|ax\| = |a| \, \|x\|, \ \|\Sigma u\| \leq \sum_{i=1}^{k} \|u_i\|.$$

Given a norm, we will call the linear space a normed linear space. It becomes a metric space with the metric defined as $\rho(x,y) \equiv_{df} \|x-y\|$. Moreover, it then has the standard inequality: $x \neq y$ if and only if $\rho(x,y) > 0$. With the metric available, notions such as continuity, convergence and so on are available.

By Lemma 6.39, $\|\cdot\|_2$ is a norm on L_2. We consider this the standard norm on L_2 and consider L_2 a normed linear space. The standard norm on $\mathbb{F}^n$ is

$$\|\langle a_1, ..., a_n \rangle\| \equiv_{df} \left(\sum |a_i|^2 \right)^{1/2}.$$

Definition 7.3. A normed linear space is a Banach space if it is separable and complete as a metric space.

$\mathbb{F}^n$ is obviously a Banach space. By Theorem 6.40 and 6.42, we have

Theorem 7.4. *L_2 is a Banach space.*

The definition of inner product also needs some special attention on finite sum.

Definition 7.5. An inner product on a linear space X is a function $(\cdot,\cdot)$ from $X \times X$ to $\mathbb{F}$, such that for any $x, y \in X$, $u \in X^{<\infty}$, $a \in \mathbb{F}$,

$$(x,y) = (y,x)^*,\ (ax,y) = a\,(x,y),\ (\Sigma u, y) = \Sigma\,(u,y),$$
$$(x,x) \in \mathbb{R},\ (x,x) \geq 0,\ (x,x) = 0 \text{ iff } x = 0.$$

A linear space with an inner product is called an inner product space.

For the space $\mathbb{F}^n$, the standard inner product is, for $x \equiv \langle a_1, ..., a_n \rangle$, $y \equiv \langle b_1, ..., b_n \rangle$,

$$(x,y) \equiv_{df} \sum_{i=1}^{n} a_i b_i^*.$$

For the space L_2, the standard inner product is,

$$(f,g) \equiv_{df} \int f g^* \mathrm{d}\mu.$$

It follows from Lemma 6.7 and 6.38 that this is an inner product.

Given an inner product (x,y) on a linear space, define

$$\|x\| \equiv_{df} (x,x)^{1/2}.$$

A direct calculation gives

$$\|x+y\|^2 + \|x-y\|^2 = 2\,\|x\|^2 + 2\,\|y\|^2. \tag{7.1}$$

We have

Lemma 7.6. *If (x,y) is an inner product on a linear space, then $|(x,y)| \leq \|x\|\,\|y\|$.*

Proof. For any x, y and any $\varepsilon > 0$, let $a = (y,x)$, $b = \|x\|^2 + \varepsilon$. Note that

$$(ax,y) = (y,ax) = |(x,y)|^2.$$

We have

$$0 \leq (ax - by, ax - by) = |(x,y)|^2\,\|x\|^2 - 2b\,|(x,y)|^2 + b^2\,\|y\|^2.$$

Therefore,

$$b\,\|y\|^2 \geq |(x,y)|^2 \left(2 - \frac{\|x\|^2}{b}\right) \geq |(x,y)|^2.$$

Since ε is arbitrary. We see that $\|x\|^2\,\|y\|^2 \geq |(x,y)|^2$. □

This means that for fixed y, $x \mapsto (x,y)$ is a continuous function from X to $\mathbb{F}$.

Lemma 7.7. *If (x,y) is an inner product on a linear space, then $\|x\|$ is a norm. Therefore, an inner product space is a normed linear space.*

Proof. It suffices to prove that $\left\|\sum_{i=1}^{n} x_i\right\| \le \sum_{i=1}^{n} \|x_i\|$. We have

$$\left\|\sum_{i=1}^{n} x_i\right\|^2 = \left(\sum_{i=1}^{n} x_i, \sum_{i=1}^{n} x_i\right) = \sum_{i,j=1}^{n} (x_i, x_j) \le \sum_{i,j=1}^{n} |(x_i, x_j)|$$
$$\le \sum_{i,j=1}^{n} \|x_i\| \, \|x_j\| = \left(\sum_{i=1}^{n} \|x_i\|\right)^2.$$

□

Note that our definitions of linear space and inner product space allow this to be proved without using any recursive construction on higher type entities. Finally, we can define Hilbert spaces.

Definition 7.8. A Hilbert space is a complete, separable inner product space.

It is obvious that $\mathbb{F}^n$ and L_2 are Hilbert spaces.

7.2 Linear Operators

A linear transformation T from a linear space X to another linear space Y is a function $T : X \to Y$ such that

$$T(\Sigma(ru)) = \Sigma(rT(u))$$

for any $u \in X^{<\infty}$ and $r \in \mathbb{F}^{<\infty}$. For any linear space X, we use I to denote the identity linear transformation from X to itself, that is, $I(x) = x$, for $x \in X$.

For a linear transformation T from X to another normed linear space Y, T is bounded if there exists $c \ge 0$ such that $\|T(x)\| \le c\|x\|$ for all $x \in X$. We say that T is bounded by c. We will denote this fact by

$$\|T\| \le c.$$

Suppose that a linear transformation T is bounded on the unit ball $Sc(0,1)$, that is, for some c, $\|T(x)\| \le c$ for all x such that $\|x\| \le 1$. Then, $\left\|T\left(\frac{x}{\|x\|+\varepsilon}\right)\right\| \le c$ for all $x \in X$ and $\varepsilon > 0$. Therefore, $\|T(x)\| \le c\|x\|$ for all $x \in X$. Conversely, when T is bounded, T is obviously bounded on the unit ball. Therefore, T is bounded if and only if T is bounded on the unit ball.

When T is bounded, we say that T is normable, if

$$\|T\| = \sup\{\|T(x)\| : x \in X, \|x\| \le 1\}$$

exists. Obviously, if T is normable, then $\|T(x)\| \le \|T\| \, \|x\|$. Note that we will use the notation $\|T\| \le c$ even if we don't know if T is normable.

Recall that a function from a metric space to another metric space is continuous if it is uniformly continuous on every closed ball. If f is bounded, then obviously f

is uniformly continuous. Conversely, if f is continuous, then it must be uniformly continuous and thus bounded on the unit ball. Therefore, we have

Lemma 7.9. *Suppose that f is a linear function from X to another normed linear space. Then, the following are equivalent:*

(1) f is bounded;
(2) f is bounded on the unit ball;
(3) f is continuous;
(4) f is uniformly continuous.

A linear transformation from X into itself is called a linear operator (or simply operator) on X. If A, B are linear operators on X, then the product AB can be defined. Obviously, if A and B are bounded by c, d respectively, then AB is bounded by cd.

We use $Hom(X)$ to denote the set of all bounded linear operators on X. Since we cannot generally prove that all bounded operators are normable, we do not have a norm on $Hom(X)$. However, $Hom(X)$ is complete in the following sense.

Lemma 7.10. *Suppose that X is a Hilbert space and (A_n) is a sequence in $Hom(X)$ such that for any $\varepsilon > 0$, there exists N, such that for any $m,n \geq N$, $\|A_m - A_n\| \leq \varepsilon$. Then, there exists $A \in Hom(X)$ such that $\|A_n - A\| \to 0$, that is, for any $\varepsilon > 0$, there exists N, such that for any $n \geq N$, $\|A_n - A\| \leq \varepsilon$.*

Proof. For any x, $(A_n x)$ is a Cauchy sequence in X. Therefore, there exists y such that $A_n x \to y$. Define $Ax \equiv y$. It is easy to verify that A is linear. Moreover, given any $\varepsilon > 0$, there exists N, such that for any $m,n \geq N$, $\|A_m - A_n\| \leq \varepsilon/2$. For any x with $\|x\| \leq 1$, choose $m \geq N$ such that $\|A_m x - Ax\| \leq \varepsilon/2$. Then, for any $n \geq N$, $\|A_n x - Ax\| \leq \varepsilon$. Therefore, $\|A_n - A\| \leq \varepsilon$ for any $n \geq N$. So, $\|A_n - A\| \to 0$. □

When the condition of the lemma holds, we say that (A_n) is a Cauchy sequence in $Hom(X)$ and (A_n) uniformly converges to A in $Hom(X)$. We denote this as $A = \lim^u_{n\to\infty} A_n$. We say that (A_n) strongly converges to A in $Hom(X)$, if for any $x \in X$, $A_n x \to Ax$ in X, and all A_n are bounded by a common bound c. We denote this fact as $A = \lim_{n\to\infty} A_n$. Apparently, uniform convergence implies strong convergence.

Lemma 7.11. *Suppose that (A_n) is a sequence in $Hom(X)$ with a common bound $c > 0$, and suppose that there is a dense subset $M \subseteq X$ such that $(A_n x)$ converges for any $x \in M$. Then, (A_n) strongly converges to an operator A in $Hom(X)$.*

Proof. Let $x \in X$. For any $\varepsilon > 0$, choose $y \in M$ such that $\|y - x\| \leq \varepsilon/4c$. Since $(A_n y)$ converges, there exists N, such that for $m,n \geq N$, $\|A_m y - A_n y\| \leq \varepsilon/2$. Therefore, for $m,n \geq N$,

$$\begin{aligned}\|A_m x - A_n x\| &\leq \|A_m x - A_m y\| + \|A_m y - A_n y\| + \|A_n y - A_n x\| \\ &\leq \varepsilon/4 + \varepsilon/2 + \varepsilon/4 = \varepsilon.\end{aligned}$$

So, $(A_n x)$ converges to some x'. Define $Ax \equiv x'$. It is easy to verify that A is linear and bounded. □

Two linear operators A and B on a Hilbert space X are called adjoint, if for any $x, y \in X$,

$$(Ax, y) = (x, By).$$

It is easy to verify that if A and B' are also adjoint, then $B = B'$. Therefore, in case A and B are adjoint, we denote B as A^* and call it the adjoint of A. A is *self-adjoint* or *Hermitian*, if $A^* = A$, that is, for any $x \in X$,

$$(Ax, y) = (x, Ay).$$

In that case, (Ax, x) must be a real number. It is easy to see that if (A_n) strongly converges to A in $Hom(X)$ and all A_n are self-adjoint, then A is also self-adjoint.

An operator U is a *unitary* operator, if its adjoint U^* exists and $UU^* = U^*U = I$. This implies that $\|Ux\| = \|x\|$ for any x. Therefore, $\|U\| = 1$ if the space is non-zero.

7.3 Subspace and Base

Subspaces are required to be located ([6], p. 307).

Definition 7.12. A subset M of a Hilbert space X is a subspace, if it is a linear subset and is further closed and located. A subspace M is non-zero, if there exists $x \in M$ such that $x \neq 0$.

Two vectors x, y are orthogonal if $(x, y) = 0$. We denote it as $x \perp y$. If $x \perp y$, then

$$\|x + y\|^2 = \|x\|^2 + \|y\|^2.$$

Let $e = \langle e_1, ..., e_n \rangle$ be a finite sequence of vectors in a Hilbert space X. e is orthogonal, if $(e_i, e_j) = 0$ for $i, j = 1, ..., n$, $i \neq j$. e is further orthonormal, if it is orthogonal and $\|e_i\| = 1$ or $\|e_i\| = 0$ for $i = 1, ..., n$. Since it is decidable whether $\|e_i\| < 1$ or $\|e_i\| > 0$, given an orthonormal sequence $\langle e_1, ..., e_n \rangle$, for each $i = 1, ..., n$ it is decidable if $\|e_i\| = 1$ or $\|e_i\| = 0$. Then, given an expansion $x = \sum_{i=1}^{n} r_i e_i$, we can construct another expansion $x = \sum_{i=1}^{n} r'_i e_i$, such that $r'_i = r_i$ if $\|e_i\| = 1$, and $r'_i = 0$ if $\|e_i\| = 0$. Such an expansion $x = \sum_{i=1}^{n} r'_i e_i$ is called a normalized expansion on the orthonormal sequence $\langle e_1, ..., e_n \rangle$. In case $x = \sum_{i=1}^{n} r_i e_i$ is normalized, we have

$$\left\| \sum_{i=1}^{n} r_i e_i \right\|^2 = \sum_{i=1}^{n} |r_i|^2.$$

An infinite orthonormal sequence (e_n) of vectors is defined similarly. Note that we have to allow the zero vector 0 to appear in an orthonormal sequence. The reason will be explained later. If (e_n) is an infinite orthonormal sequence of vectors and (r_n) is a sequence of numbers in $\mathbb{F}$ such that $\sum_{n=0}^{\infty} |r_n|^2$ converges, then $\sum_{n=0}^{\infty} r_n e_n$ converges. We can similarly normalize a sequence (r_n) with respect to an orthonormal sequence (e_n). That is, we can make sure that $r_n = 0$ when $e_n = 0$. From now on, whenever we write an expansion $x = \sum_{i=1}^{n} r_i e_i$ or $x = \sum_{i=1}^{\infty} r_i e_i$, we always tacitly assume that it has been normalized. It is easy to see that if (e_n) is an orthonormal sequence, then the set $\{e_n : n \in \mathbb{N} \wedge \|e_n\| = 1\}$ is linearly independent.

A vector x is orthogonal to a subset A, if $x \perp y$ for each $y \in A$. We denote this as $x \perp A$. We have

Lemma 7.13. *Suppose that A is a linear subset, and x is any vector, and $y \in A$. Then, $(x-y) \perp A$, if and only if $\rho(x,A) = \|x-y\|$. Moreover, the vector y satisfying this condition is unique in A (if it exists).*

Proof. First, suppose that $(x-y) \perp A$. Given any $y' \in A$, we have $y - y' \in A$. Therefore, $(x-y) \perp (y-y')$. Hence,

$$\|x-y'\|^2 = \|x-y\|^2 + \|y-y'\|^2 \geq \|x-y\|^2 .$$

So, $\rho(x,A) = \|x-y\|$.

Next, suppose that $\rho(x,A) = \|x-y\|$. Then, for any $z \in A$ and any $r \in \mathbb{F}$,

$$(x-y-rz, x-y-rz) \geq (x-y, x-y) .$$

Therefore, $|r|^2 \|z\|^2 \geq (x-y, rz) + (rz, x-y)$. Let $r = (x-y, z)$. Then,

$$|(x-y,z)|^2 \|z\|^2 \geq 2\,|(x-y,z)|^2 .$$

Replace z by $z' = \frac{z}{\|z\|+1}$. Then, $\|z'\| < 1$. We see that $(x-y, z') = 0$. Therefore, $(x-y, z) = 0$. That is, $(x-y) \perp A$.

The uniqueness of y follows from the equation $\|x-y'\|^2 = \|x-y\|^2 + \|y-y'\|^2$. □

If M is a subspace and x is any vector and $y \in M$ is such that $(x-y) \perp M$, then y is called the projection of x onto M. The following lemma shows that the projection always exists uniquely.

Lemma 7.14. *If M is a subspace and x is any vector, then there exists a unique $y \in M$ such that $\rho(x,M) = \|x-y\|$ and thus $(x-y) \perp M$.*

Proof. By definition, $\rho(x,M)$ exists. So, there is a sequence (y_n) of vectors in M such that $\|y_n - x\| \to \rho(x,M)$. Then, using the equation (7.1) above, we have

$$\begin{aligned} \|y_n - y_m\|^2 &= \|(y_n - x) - (y_m - x)\|^2 \\ &= 2\|y_n - x\|^2 + 2\|y_m - x\|^2 - 4\left\|\frac{1}{2}(y_n + y_m) - x\right\|^2 . \end{aligned}$$

Since $\frac{1}{2}(y_n + y_m) \in M$, $\left\|\frac{1}{2}(y_n + y_m) - x\right\| \geq \rho(x,M)$. Therefore,

$$\|y_n - y_m\|^2 \leq 2\left(\|y_n - x\|^2 - \rho(x,M)^2\right) + 2\left(\|y_m - x\|^2 - \rho(x,M)^2\right) .$$

It means that (y_n) is a Cauchy sequence. Since X is complete and M is closed, (y_n) converges to some $y \in M$. Then, $\rho(x,M) = \|x-y\|$. □

Therefore, given a subspace, we can define a function $P_M : X \to M$ such that for any x, $(x - P_M x) \perp M$. Then, $x = P_M x + (x - P_M x)$ is an orthogonal decomposition of x. Suppose that $x = \sum_{i=1}^n r_i x_i$. Then,

$$x - \sum_{i=1}^{n} r_i P_M x_i = \sum_{i=1}^{n} r_i (x_i - P_M x_i),$$

which is orthogonal to M. Therefore, $P_M x = \sum_{i=1}^n r_i P_M x_i$. That is, P_M is a linear operator. Since $\|x\| = \|P_M x\| + \|x - P_M x\|$ and $P_M x = x$ for $x \in M$, we see that if M is non-zero, then $\|P_M\| = 1$ and $P_M^2 = P_M$. P_M is called the projection onto the subspace M.

Note that $P_M x \perp (y - P_M y)$. Therefore,

$$(P_M x, y) = (P_M x, P_M y) + (P_M x, y - P_M y) = (P_M x, P_M y).$$

Similarly, $(x, P_M y) = (P_M x, P_M y)$. So, $(P_M x, y) = (x, P_M y)$ and P_M is self-adjoint. On the other side, we have

Lemma 7.15. *Suppose that P is a self-adjoint operator on a Hilbert space X and $P^2 = P$. Then, there exists a subspace M such that $P = P_M$.*

Proof. Let $M \equiv \{x : Px = x\}$. M is a closed linear subset. For any $x \in M$ and any vector $y \in X$,

$$(y - Py, x) = (y,x) - (Py,x) = (y,x) - (y,Px) = 0.$$

So, $(y - Py) \perp M$. Since $P^2 = P$, $Py \in M$. Therefore, $\|y - Py\| = \rho(y,M)$, that is, M is located and is thus a subspace. $(y - Py) \perp M$ also implies that $Py = P_M y$. That is, P is the projection onto M. □

Given a subspace, we define $M^{\perp}$ as the subset of vectors y such that $y \perp M$. It is easy to verify that $M^{\perp}$ is a linear subset and is closed. Moreover, for any x, $y = (x - P_M x) \in M^{\perp}$ and $x - y = P_M x \perp M^{\perp}$. Therefore, $\rho(x, M^{\perp}) = \|x - y\|$ exists. That is, $M^{\perp}$ is located and is a subspace. Therefore, we have the orthogonal decomposition

$$X = M \oplus M^{\perp}.$$

As an example of subspace, we have

Lemma 7.16. *If $e = \langle e_1, \ldots, e_n \rangle$ is an orthonormal sequence of vectors, then the linear subset $[e]$ spanned by e is a subspace.*

Proof. To see that $[e]$ is closed, suppose that (x_k), $x_k \equiv \sum_{i=1}^n r_{k,i} e_i$, is a sequence of vectors in $[e]$ and (x_k) converges to x. Note that we assume that the expansion of each x_k is normalized. Therefore, $\|x_k - x_m\|^2 = \sum_{i=1}^n |r_{k,i} - r_{m,i}|^2$. We see that each $(r_{k,i})_k$ is a Cauchy sequence and therefore converges to some r_i. Then, (x_k) converges to $\sum_{i=1}^n r_i e_i$. Therefore, $x = \sum_{i=1}^n r_i e_i \in [e]$. So, $[e]$ is closed.

To see that $[e]$ is located, let x be any vector, and let $y \equiv \sum_{i=1}^{n} (x, e_i) e_i$. It is easy to see that $(x-y) \perp e_i$ for each i. So, $(x-y) \perp [e]$. By Lemma 7.13 above, $\rho(x,[e]) = \|x-y\|$ exists. Therefore, $[e]$ is located. □

Definition 7.17. An orthonormal sequence (e_n) is a basis of a Hilbert space X, if for any vector x,

$$x = \sum_{i=0}^{\infty} (x, e_i) e_i.$$

We consider the Gram-Schmidt orthogonalization process for constructing a basis for a Hilbert space X. By the definition of Hilbert space, X is separable. That is, there exists a sequence of vectors (y_n) that is dense in the Hilbert space X. By repeating each item in the sequence (y_n) infinitely, we may assume that for each $y \in X$ and any $\varepsilon > 0$ there exists arbitrarily large n such that $|y - y_n| < \varepsilon$. We want to construct an orthonormal basis (e_n).

Informally, (e_n) is constructed as follows ([6], pp. 368–369): If $\|y_0\| > 1$ let $e_0 = \|y_0\|^{-1} y_0$, and if $\|y_0\| < 2$, let $e_0 = 0$. Suppose that $e_0, \ldots, e_{n-1}$ have been constructed. Decide if

$$\left\| y_n - \sum_{i=0}^{n-1} (y_n, e_i) e_i \right\| > \frac{1}{2n} \quad \text{or} \quad \left\| y_n - \sum_{i=0}^{n-1} (y_n, e_i) e_i \right\| < \frac{1}{n}.$$

In the former case, let

$$e_n = \left\| y_n - \sum_{i=0}^{n-1} (y_n, e_i) e_i \right\|^{-1} \left(y_n - \sum_{i=0}^{n-1} (y_n, e_i) e_i \right), \tag{7.2}$$

and in the latter, let $e_n = 0$.

We must revise this construction to avoid any recursion on higher type entities. We can express e_n as

$$e_n \equiv \sum_{i=0}^{n} r_i^n y_i.$$

So we can instead construct the sequence $r_1^1, r_1^2, r_2^2, r_1^3, r_2^3, r_3^3 \ldots$. Denote $b_{i,j} \equiv (y_i, y_j)$; then we should have

$$\begin{aligned} \sum_{i=0}^{n-1} (y_n, e_i) e_i &= \sum_{i=0}^{n-1} \left(y_n, \sum_{j'=0}^{i} r_{j'}^i y_{j'} \right) \sum_{j=0}^{i} r_j^i y_j \\ &= \sum_{j=0}^{n-1} \left(\sum_{i=j}^{n-1} \sum_{j'=0}^{i} b_{n,j'} \left(r_{j'}^i \right)^* r_j^i \right) y_j. \end{aligned}$$

Denote $c_{n,j} \equiv \sum_{i=j}^{n-1} \sum_{j'=0}^{i} b_{n,j'} \left(r_{j'}^i \right)^* r_j^i$ for $j = 0, \ldots, n-1$, and $c_{n,n} = -1$. Then,

$$y_n - \sum_{i=0}^{n-1} (y_n, e_i)\, e_i = -\sum_{j=0}^{n} c_{n,j} y_j,$$

$$\left\| y_n - \sum_{i=0}^{n-1} (y_n, e_i)\, e_i \right\|^2 = \sum_{j,j'=0}^{n} c_{n,j} c_{n,j'}^* b_{j,j'}.$$

Denote $d_n \equiv \sum_{j,j'=0}^{n} c_{n,j} c_{n,j'}^* b_{j,j'}$. Then, when $d_n > \left(\frac{1}{2n}\right)^2$,

$$r_j^n \equiv -\frac{c_{n,j}}{\sqrt{d_n}}, \text{ for } j = 0, \ldots, n;$$

and when $d_n < \left(\frac{1}{n}\right)^2$,

$$r_j^n \equiv 0, \text{ for } j = 0, \ldots, n.$$

Note that $c_{n,j}$ and d_n are constructed using product and finite sum from r_j^i for $i = 0, \ldots, n-1$ and $j \le i$. Therefore, the construction of r_j^i is iteratable.

Let $e_n \equiv \sum_{j=1}^{n} r_j^n y_j$. Then, either $e_n = 0$, or e_n is given by (7.2). In the latter case, $\|e_n\| = 1$, and in both cases $(e_n, e_j) = 0$ for $j = 0, \ldots, n-1$. Therefore, (e_n) is an orthonormal sequence.

Note that we may not be able to delete the zero vector 0 from the sequence (e_n) and get another orthonormal sequence (e_k') such that $\|e_k'\| = 1$ for all k. To do that, we will need a function f such that $e_k' = e_{f(k)}$, but this function f could be beyond elementary recursive functions. (Given a function f that grows faster than all elementary recursive functions, from the sequence (e_k'), we can actually construct the corresponding sequence (e_n), by inserting the zero vector 0 sufficiently many times. Note that the sequence (e_n) itself can still be an elementary recursive sequence, because to decide if e_n should be 0, we only need to compute $f(0), f(1), \ldots$ with values bounded by n.)

Let M_n denote the subspace spanned by $e_0, \ldots, e_n$. For any $y \in X$ and $\varepsilon > 0$, choose $n > 2/\varepsilon$ such that $\|y - y_n\| < \varepsilon/2$. If $e_n = 0$, then

$$\left\| y_n - \sum_{i=0}^{n-1} (y_n, e_i)\, e_i \right\| < \frac{1}{n} < \varepsilon/2,$$

and hence we must have $\rho(y, M_n) < \varepsilon$. If e_n is given by (7.2) then $y_n \in M_n$, and hence $\rho(y, M_n) < \varepsilon$. It means that in the decomposition

$$y = P_{M_n} y + (y - P_{M_n} y),$$

we have $\|y - P_{M_n} y\| = \rho(y, M_n) < \varepsilon$ whenever $n > 2/\varepsilon$. Therefore, the sequence $(P_{M_n} y)$ converges to y. That is, $y = \sum_{i=0}^{\infty} (y, e_i)\, e_i$. Therefore, (e_n) is a basis. So, we have

Theorem 7.18. *Every Hilbert space has a orthonormal basis. If (e_n) is a basis, $x = \sum_{i=0}^{\infty} a_i e_i$ and $y = \sum_{i=0}^{\infty} b_i e_i$ (are normalized expansions), then $a_i = (x, e_i)$, and*

$$\|x\|^2 = \sum_{i=0}^{\infty} |a_i|^2, \ (x,y) = \sum_{i=0}^{\infty} a_i b_i^*.$$

A Hilbert space X is finite dimensional, if it has a finite orthonormal sequence of vectors $e = \langle e_1, ..., e_n \rangle$ as a basis. In that case, X is the space $[e]$ spanned by e. The Gram-Schmidt orthogonalization process can be applied to a finite dimensional space as well. It will give some results about finite dimensional spaces. First, we need a lemma:

Lemma 7.19. *Suppose that M is a subspace of a Hilbert space X such that the unit ball $Sc(0,1) \cap M$ of M is totally bounded, and suppose that (e_n) is an orthonormal basis of X. Then, there exists N such that for all $n > N$, if $\|e_n\| = 1$, then $\rho(e_n, M) > 0$.*

Proof. Suppose that $\{x_1, ..., x_K\}$ is a $\frac{1}{4}$ approximation to $Sc(0,1) \cap M$. For each $i = 1, ..., K$, the theorem above implies that $(x_i, e_n) \to 0$ as $n \to \infty$. Therefore, there exists N, such that $|(x_i, e_n)| < \frac{3}{8}$ for $n > N$, for all $i = 1, ..., K$. Then, for $n > N$, supposing that $\|e_n\| = 1$, we have

$$\|e_n - x_i\|^2 = \|e_n\|^2 - 2\mathrm{Re}((x_i, e_n)) + \|x_i\|^2 \geq 1 - 2|(x_i, e_n)| > \frac{1}{4}.$$

Now, $\rho(e_n, M) = \|e_n - P_M e_n\|$. Since $\|P_M e_n\| \leq 1$. Then, $P_M e_n \in Sc(0,1) \cap M$. Therefore, there exists $i = 1, ..., K$ such that $\|P_M e_n - x_i\| \leq \frac{1}{4}$. Then,

$$\|e_n - P_M e_n\| \geq \|e_n - x_i\| - \|P_M e_n - x_i\| > \frac{1}{2} - \frac{1}{4} = \frac{1}{4}.$$

□

Corollary 7.20. *Suppose that M is a finite dimensional subspace of a Hilbert space X and (e_n) is an orthonormal basis of X. Then, there exists N such that for all $n > N$, if $\|e_n\| = 1$, then $\rho(e_n, M) > 0$.*

Proof. Suppose that $e = \langle e'_1, ..., e'_m \rangle$ is a finite orthonormal basis for M. Then, every $x \in M$ can be expressed as $x = \sum_{i=1}^m r_i e'_i$, with $\|x\| = \left(\sum_{i=1}^m |r_i|^2\right)^{\frac{1}{2}}$, and for $y = \sum_{i=1}^m s_i e'_i$, we have $\|x - y\| = \left(\sum_{i=1}^m |r_i - s_i|^2\right)^{\frac{1}{2}}$. Therefore, by approximating the unit ball $Sc(0,1)$ in $\mathbb{F}^m$, we can see that the unit ball $Sc(0,1) \cap M$ of M is totally bounded. Then, the conclusion follows from the lemma above. □

Corollary 7.21. *If X is finite dimensional and (e_n) is an orthonormal basis of X, then for some N, $e_n = 0$ for all $n > N$.*

Proof. Let M be X in the last corollary. □

Corollary 7.22. *Any subspace of a finite dimensional space is finite dimensional.*

Proof. Suppose that M is a subspace of a finite dimensional space Y. The unit ball $Sc(0,1)$ of Y is totally bounded. If $\{x_1,...,x_K\}$ is an ε approximation to $Sc(0,1)$, it is easy to see that $\{P_M x_1,...,P_M x_K\}$ is an ε approximation to $Sc(0,1)\cap M$. Therefore, the unit ball $Sc(0,1)\cap M$ of M is totally bounded. Let (e_n) be an orthonormal basis of M. Applying the lemma with $X = M$, we see that $e_n = 0$ for all sufficiently large n. Therefore, M is finite dimensional. □

We say that a Hilbert space X is infinite dimensional, if for any finite dimensional subspace M of X, there exists $x \in X$ such that $\rho(x,M) > 0$. If X has an orthonormal basis (e_n) such that $\|e_n\| = 1$ for arbitrarily large n, from the lemma above it easily follows that X is infinite dimensional. Conversely, we have

Lemma 7.23. *Suppose that (y_n) is a sequence of vectors dense in X, and there exists an iteratable function h, such that for each $n > 0$, there exists $m \le h(n)$ such that $\|y_m - y\| > 1/n$ for all $y \in [y_0,...,y_n]$. Then, X is infinite dimensional and there exists an orthonormal basis (e_n) for X such that $\|e_n\| = 1$ for all n.*

Proof. First, we show that for the orthonormal basis (e_n) constructed in the Gram-Schmidt orthogonalization process above for the given (y_n), and for each $n > 0$, there exists m, $n < m \le h(n)$ such that $e_m \ne 0$. Note that $[e_0,...,e_n] \subseteq [y_0,...,y_n]$. By the assumption, there exists $m \le h(n)$ such that $\rho(y_m,[e_0,...,e_n]) \ge 1/n$. For each $i = n+1,...,h(n)$, $\|e_i\| = 1$ or 0 is decidable. If $\|e_i\| = 0$ for all $i = n+1,...,h(n)$, then

$$\rho(y_m,[e_0,...,e_{m-1}]) = \rho(y_m,[e_0,...,e_n]) \ge 1/n \ge 1/(m-1).$$

Therefore, by the Gram-Schmidt orthogonalization process, we should have $\|e_m\| = 1$, a contradiction. Therefore, there exists $m \le h(n)$ such that $e_m \ne 0$.

Note that by replacing e_0 with the minimum $m \le h(0)$ such that $\|e_m\| = 1$, we may assume that $\|e_0\| = 1$. Then, we can recursively define a function g:

$$g(0) = 0,$$
$$g(n+1) = \mu m \le h(g(n))\,(m > g(n) \wedge \|e_m\| = 1).$$

Since h is iteratable, this is a bounded primitive recursion. The sequence $(e'_n) \equiv (e_{g(n)})$ will be an orthonormal basis such that $\|e'_n\| = 1$ for all n. □

An orthonormal basis (e_n) such that $\|e_n\| = 1$ for all n is called a non-zero orthonormal basis.

Theorem 7.24. *There exists a non-zero orthonormal basis for L_2.*

Proof. Recall that a sequence of vectors dense in L_2 consists of continuous functions f in $C(\mathbb{R})$ of this format: f is linear on some rational intervals $[p_i, p_{i+1}]$, $p_i \le p_{i+1}$, $i = 0,...,k-1$; f vanishes on $(-\infty, p_0]$ and $[p_k,+\infty)$; and $q_i = f(p_i)$ are all rational numbers for $i = 1,...,k-1$. We need to arrange such functions f into a sequence (f_n) so that for each n, there exists $m \le h(n)$ such that $\int |f_m - g|^2 d\mu > 1/n$ for any $g \in [f_0,...,f_n]$, and so that the operation h is iteratable.

The function f above can be represented by the sequence

$$\#(f) \equiv \langle p_0,...,p_k,q_1,...,q_{k-1}\rangle \tag{7.3}$$

of rational numbers. A rational number is encoded as a sequence $\langle \delta, m_1, m_2 \rangle$ with $\delta = 0$ or 1 and $m_2 \neq 0$. We can arrange them into a sequence such that $\langle \delta, m_1, m_2 \rangle$ precedes $\langle \delta', m_1', m_2' \rangle$ if $\max(m_1, m_2) < \max(m_1', m_2')$. We call $\max(m_1, m_2)$ the size of the rational number $\langle \delta, m_1, m_2 \rangle$. Then, there are $2N(N+1)$ codes of rational numbers with size$\leq N$. Note that if (the code of) a rational number p has a size$\leq N$, then $|p| \leq N$. For f in (7.3), let $size(f)$ denote the maximum size of $p_0, \ldots, p_k, q_1, \ldots, q_{k-1}$, and let $length(f) = k$. If $size(f) \leq N$, then f vanishes on $(-\infty, -N]$ and $[N, +\infty)$.

We can arrange the finite sequences (7.3) into an infinite sequence (f_n) as follows: For each $N > 1$, at the step N, we arrange the sequences (7.3) with $k = length(f) = N$ and $size(f) \leq N$; and then we proceed to the step $N+1$. There will be many repetitions. We do not consider the order of $p_0, \ldots, p_k$. If the order is not correct, it is considered a repetition. Then, there are $(2N)^{2N(N+1)}$ items at the step N, and there are less than $N(2N)^{2N(N+1)}$ items before the step $N+1$. This will actually cover all the functions represented in the format (7.3), because if a sequence in the format (7.3) is such that $size(f) > length(f)$, then the function represented by the sequence must also be represented by another sequence of the format (7.3) arranged at some step $N > size(f)$, with some p_i repeated among $p_0, \ldots, p_N$.

Now, given any f_n, let $N = length(f_n)$. That is, f_n is arranged at the step N. Since there are $(2(N-1))^{2(N-1)N}$ items at the step $N-1$, we have

$$n \geq (2(N-1))^{2(N-1)N}.$$

For any $i \leq n$, $size(f_i) \leq N$. Therefore, any function in $[f_0, \ldots, f_n]$ vanishes on $(-\infty, -N]$ and $[N, +\infty)$. Let f be the function that vanishes on $(-\infty, N]$ and $[N+3, +\infty)$ but equals to 1 on $[N+1, N+2]$. Then, for any $g \in [f_0, \ldots, f_n]$, $\int |f-g|^2 d\mu > 1$. Note that f can be represented by a sequence f_m in the format (7.3) with $length(f_m) = N+3$ and $size(f_m) = N+3$. Therefore, f_m must be arranged in the sequence at the step $N+3$. Then,

$$m \leq (N+3)(2(N+3))^{2(N+3)(N+4)}.$$

It is easy to see that $m < h(n) \equiv n^{l_0}$ for some constant l_0. h is iteratable. □

Suppose that A is a bounded linear operator on a Hilbert space X, and (e_i) is an orthonormal basis for X, and $x = \sum_{i=0}^{\infty} r_i e_i$. Since A is continuous, $Ax = \sum_{i=0}^{\infty} r_i A e_i$. Since the inner product is continuous, $(Ax, e_j) = \sum_{i=0}^{\infty} r_i (Ae_i, e_j)$. Denote $a_{i,j} = (Ae_i, e_j)$, we have

$$Ae_i = \sum_{j=0}^{\infty} a_{i,j} e_j, \; Ax = \sum_{j=0}^{\infty} \left(\sum_{i=0}^{\infty} r_i a_{i,j} \right) e_j.$$

$(a_{i,j})$ represents A on the basis (e_j). If B is another bounded linear operator and $b_{j,k} \equiv (Be_j, e_k)$, then $(BAe_i, e_k) = \sum_{j=0}^{\infty} a_{i,j} b_{j,k}$.

7.4 The Spectral Decomposition of a Unitary Operator

In this section, we construct the spectral decomposition of a unitary operator, following the classical proof in Riesz and Sz.-Nagy [31], p. 281. This will be used in the next section to construct the spectral decomposition of an unbounded self-adjoint operator.

Suppose that U is a unitary operator on a Hilbert space H. Let $p(z) \equiv \sum_{k=-n}^{n} c_k z^k$ be a complex polynomial in variables z and z^{-1}. We want to construct $p(U)$. Let $\mathscr{P}$ denote the algebra of complex formal polynomials, $\sum_{k=-n}^{n} c_k z^k$, in variables z and z^{-1}. So, we want to construct a mapping from $\mathscr{P}$ into $Hom(H)$, the set of bounded operators defined on the space H. Then, we will extend it into a mapping from $C(S^1)$, the set of continuous real functions on $S^1 = \{z : |z| = 1,\, z \in \mathbb{C}\}$, into $Hom(H)$. After that, one can easily follow the proof of the spectral theorem for bounded self-adjoint operators on pp. 378–379 of [6].

First, note that we cannot always construct arbitrary products of bounded operators. For instance, suppose that (e_n) is a non-zero orthonormal basis for H, and suppose that $U(e_{2i}) = e_{2^i}$, for $i = 0,1,2,\ldots$, and $U(e_{2i+1}) = e_{\pi(2i+1)}$, where π maps odd numbers one-one onto all numbers that are not a power of 2. Then, U is unitary, but $U^n(e_i)$ will require iterating the power function n times.

Suppose that $A_0,\ldots,A_n$ are mutually commuting operators. That is, $A_iA_j = A_jA_i$ for $i,j = 0,\ldots,n$. We say that the operators $A_0,\ldots,A_n$ are positively multiplicable, if we can construct operators B_u indexed by an arbitrary finite sequence $u = \langle i_1,\ldots,i_k\rangle$, $i_1 \le n$, ..., $i_k \le n$, such that for any sequences u, v, any $i = 0,\ldots,n$, and any permutation π,

$$B_{<>} = I,\; B_{\langle i\rangle} = A_i,\; B_{u*v} = B_uB_v,\; B_u = B_{\pi(u)}.$$

Suppose that an operator A has an inverse A^{-1}, that is, $AA^{-1} = A^{-1}A = I$. We say that A is multiplicable, if we can construct a mapping $n \longmapsto A^n$, from integers to operators, such that $A^1 = A$, $A^0 = I$, A^{-1} is the inverse, and for all integers m,n, $A^{m+n} = A^mA^n$. Then, it follows that $(A^n)^{-1} = A^{-n}$. We define $(A^m)^n \equiv_{df} A^{mn}$. Then, $A^{-n} = \left(A^{-1}\right)^n$. Note that for a constant numeral $\overline{n}$, we have

$$(A^m)^{\overline{n}} = A^m \ldots A^m \text{ (n times)}.$$

Recall that for a unitary operator U, $U^{-1} = U^*$. In the rest of this section, we assume that U is a multiplicable unitary operator on H. Then, for a polynomial $p(z) = \sum_{k=-n}^{n} c_k z^k \in \mathscr{P}$, $p(U)$ is naturally defined:

$$p(U) \equiv_{df} \sum_{k=-n}^{n} c_k U^k.$$

Moreover, $p \mapsto p(U) = \sum_{k=-n}^{n} c_k U^k$ defines a linear, multiplicative mapping from $\mathscr{P}$ into $Hom(H)$. Note that

$$\left(U^{n-i}x, U^{-i}y\right) = \left(U^{n-(i+1)}x, U^{-(i+1)}y\right).$$

By the finite transitivity of equality between real numbers, we have $(U^n)^* = U^{-n}$. Then, we have $p(U)^* = p^*(U)$, where $p^*(z) = \sum c_k^* z^{-k}$. Similarly, we have $\|U^n x\| \leq \|x\|$, and $\|p(U)x\| \leq \sum_{k=-n}^{n} |c_k| \, \|x\|$.

We say that $\sum_{k=-n}^{n} c_k z^k$ is *n-degree*. So n-degree polynomials are also m-degree for $m \geq n$. Suppose that $p(z) = \sum_{k=-n}^{n} c_k z^k \in \mathscr{P}$ and $p(z) = 0$ for all $z \in S^1$. Then, by Lemma 5.6, Corollary 5.9, and Corollary 5.14, we have $2\pi i c_{-n} = \int_{S^1} p(z) z^{n-1} dz = 0$, $2\pi i c_{-(n-1)} = \int_{S^1} p(z) z^{n-2} dz = 0$, ... So $p = 0$. Therefore, $p_1(z) = p_2(z)$ on S^1 implies $p_1 = p_2$.

Note that for $z \in S^1$, $z^{-1} = z^*$. Suppose that p is real on S^1. Then, for $z \in S^1$,

$$p(z) = p(z)^* = \sum_{k=-n}^{n} c_k^* z^{-k} = \sum_{k=-n}^{n} c_{-k}^* z^k.$$

Therefore, $c_k^* = c_{-k}$, and hence $p(U)$ is self-adjoint.

We say that an operator A on H is *positive* (denoted as $A \geq 0$), if $(Ax, x) \geq 0$ for all $x \in H$. We want to prove that the mapping $p \mapsto p(U)$ preserves positivity and bounds, but first we need a lemma, a finitistic version of the lemma on Riesz and Sz.-Nagy [31], p. 118. For $s = \sum_{k=-m}^{m} b_k z^k$, define

$$Coef(s) \equiv \max\{|b_{-m}|, \ldots, |b_m|\}.$$

Lemma 7.25. *If p is m-degree and $p(z) \geq \varepsilon > 0$ on S^1, then for any $\delta > 0$, there exist m-degree q and s such that $p = q^* q + s$ and $Coef(s) \leq \delta$.*

Proof. Let $p(z) = \sum_{k=-m}^{m} c_k z^k$. As p is real on S^1, $c_k^* = c_{-k}$. We assume that $\delta < \frac{\varepsilon}{4m}$. For each k, $|c_k| > 0$ or $|c_k| < \delta$. We may assume that there is $n \geq 0$ such that $|c_n| > 0$ and $|c_k| < \delta$ for $k = \pm(n+1), \ldots, \pm m$. Let $p'(z) = \sum_{k=-n}^{n} c_k z^k$. Then for $z \in S^1$, $p'(z)$ is also real and $p'(z) \geq \varepsilon/2$. Now we show that p' can be expressed as $p' = q^* q$.

The case $n = 0$ is trivial. Suppose that $n > 0$. Let

$$r(z) \equiv z^n p'(z) = \sum_{k=-n}^{n} c_k z^{n+k}.$$

Then $|r(z)| \geq \frac{\varepsilon}{2}$ on S^1, and hence $r(z) = 0$ implies $||z| - 1| = d(z, S^1) \geq \frac{1}{2}\omega\left(\frac{\varepsilon}{2}\right) > 0$, where ω is the modulus of continuity for $r(z)$. This means that the zeros of r are either inside or outside the unit circle. Similarly, since $|r(0)| = |c_{-n}| > 0$, $r(z) = 0$ implies $|z| \geq \frac{1}{2}\omega(|c_{-n}|)$. So the zeros of r are away from 0. Since $|c_n| > 0$, by the fundamental theorem of algebra, there exits z_1 such that $r(z_1) = 0$. Since $c_k^* = c_{-k}$,

$$r(z) = z^{2n} r\left(\frac{1}{z^*}\right)^* \quad \text{for } |z| > 0, \tag{7.4}$$

and therefore $r\left(\frac{1}{z_1^*}\right) = 0$ as we have $|z_1| > 0$. Since either $|z_1| > 1$ or $|z_1| < 1$, we have $\left|\frac{1}{z_1^*}\right| < 1$ or $\left|\frac{1}{z_1^*}\right| > 1$ correspondingly. So $z_1 \neq \frac{1}{z_1^*}$ are different zeros of r. Two successive polynomial divisions give

$$r(z) = (z - z_1)(z_1^* z - 1)\, r_1(z) \tag{7.5}$$

for some $r_1(z)$. Write $r_1(z)$ as

$$r_1(z) \equiv \sum_{k=-(n-1)}^{n-1} d_k z^{(n-1)+k}.$$

By expanding the right hand side of (7.5) and comparing the coefficients, and by the fact that $c_k^* = c_{-k}$ for $k = 0, ..., n$, we see that $d_k^* = d_{-k}$ for $k = 0, ..., n-1$, and $d_{(n-1)} = \frac{c_n}{z_1^*} \neq 0$. That is, the construction can be repeated for $r_1(z)$. Moreover, each d_k can be constructed by applying finite sum and product on c_k, z_1, z_1^*, $\frac{1}{z_1}$, and $\frac{1}{z_1^*}$, and the construction from r to r_1 is iteratable. Therefore, by a bounded recursion, we have

$$r(z) = \frac{c_n}{z_1^* ... z_n^*}(z - z_1)(z_1^* z - 1) ... (z - z_n)(z_n^* z - 1).$$

For $z \in S^1$, we have

$$p'(z) = z^{-n} r(z) = b(z - z_1) ... (z - z_n)(z^* - z_1^*) ... (z^* - z_n^*)$$

for some constant b. Since $p'(z) \geq \frac{\varepsilon}{2}$, we must have $b > 0$. So, $p' = qq^*$ for $q = \sqrt{b}(z - z_1) ... (z - z_n)$. □

Remember that we use $\|A\| \leq M$ to mean that M is a bound of A, without considering if A is normable or not.

Lemma 7.26. *(a) If $p(z) \geq 0$ on S^1, then $p(U) \geq 0$;*
(b) If $|p(z)| \leq M$ on S^1, then $\|p(U)\| \leq M$.

Proof. (a) By the previous lemma, for any $\varepsilon > 0$ and $\delta > 0$, we can find q, s such that $p + \varepsilon = q^* q + s$ and $Coef(s) < \delta$. So we have

$$|(s(U)x, x)| \leq \|s(U)x\|\, \|x\| \leq \delta\,(2n+1)\,\|x\|^2,$$

assuming that p is n-degree. Therefore,

$$((p(U) + \varepsilon)x, x) = (q(U)x, q(U)x) + (s(U)x, x) \geq -\delta(2n+1)\,\|x\|^2.$$

δ and ε are arbitrary. So, $(p(U)x, x) \geq 0$.

(b) Since $M^2 - p^* p \geq 0$, this follows from (a). □

Next, we extend the mapping $\mathscr{P} \to Hom(H)$ to $C(S^1)$. Let

$$\mathscr{RP} \equiv \{p \in \mathscr{P} : \ p(z) \text{ is real on } S^1\} \subset C(S^1).$$

$\mathscr{RP}$ is a subspace of the metric space $C(S^1)$, and it is closed under finite product and finite sum, and $p(U)$ is self-adjoint for p in $\mathscr{RP}$. It is easy to see that $\mathscr{RP}$ is separating in $C(S^1)$ (see Sect. 4.4). By Stone-Weierstrass Theorem, $\mathscr{RP}$ is dense in $C(S^1)$. Then, given any $f \in C(S^1)$, there exists a sequence (p_n) in $\mathscr{RP}$ such that

$p_n \to f$ in the norm of $C(S^1)$. Then, Lemma 7.26 implies that $p_n(U)$ is a Cauchy sequence in $Hom(H)$. Therefore, by Lemma 7.10, there exists $A \in Hom(H)$ such that $p_n(U)$ converges uniformly to A. We define $f(U) \equiv A$. That is, we have

Lemma 7.27. *The mapping $\mathscr{RP} \to Hom(H)$, $p \mapsto p(U)$, can be extended into a linear, multiplicative mapping $C(S^1) \to Hom(H)$, $f \mapsto f(U)$, such that $f(U)$ is self-adjoint, and (i) $f(U) \geq 0$ whenever $f(z) \geq 0$ on S^1, and (ii) $\|f(U)\| \leq M$ whenever $|f(z)| \leq M$ on S^1.*

Proof. Suppose that for each $k = 1,\ldots,K$, $p_{k,n} \to f_k$ as $n \to \infty$ in the norm of $C(S^1)$. Then, $\sum_{k=1}^{K} r_k p_{k,n} \to \sum_{k=1}^{K} r_k f_k$ in the norm of $C(S^1)$. Therefore,

$$\left(\sum_{k=1}^{K} r_k f_k\right)(U) = \overset{u}{\lim_{n\to\infty}} \left(\sum_{k=1}^{K} r_k p_{k,n}\right)(U) = \overset{u}{\lim_{n\to\infty}} \sum_{k=1}^{K} r_k p_{k,n}(U) = \sum_{k=1}^{K} r_k f_k(U).$$

That is, the mapping $C(S^1) \to Hom(H)$ is linear.

To see that it is multiplicative, note that if $p_{k,n} \to f_k$ in $C(S^1)$ for $k = 1,\ldots,K$, then $\prod_{k=1}^{K} p_{k,n} \to \prod_{k=1}^{K} f_k$. Therefore,

$$\left(\prod_{k=1}^{K} f_k\right)(U) = \overset{u}{\lim_{n\to\infty}} \left(\prod_{k=1}^{K} p_{k,n}\right)(U) = \overset{u}{\lim_{n\to\infty}} \prod_{k=1}^{K} p_{k,n}(U) = \prod_{k=1}^{K} f_k(U).$$

$f(U)$ is self-adjoint as the uniform limit of self-adjoint operators. If $f(z) \geq 0$ on S^1, then for any $\varepsilon > 0$, we can find a sequence (p_n) in $\mathscr{RP}$ such that $p_n \to f + \varepsilon$ and $p_n > 0$ on S^1 for all n. Therefore, $f(U) + \varepsilon I \geq 0$ for any $\varepsilon > 0$. Then, we have $f(U) \geq 0$. The conclusion (ii) is similar. □

Next, we want to extend the mapping $C(S^1) \to Hom(H)$, $f \longmapsto f(U)$, to include the characteristic functions of some arc intervals on S^1. We use the ideas in [6], pp. 237–251. Choose an orthonormal basis (e_k) of H. For convenience, we assume that $e_0 = 0$. Define $\mu : C(S^1) \to \mathbb{R}$ by

$$\mu(f) \equiv_{df} \sum_{k=1}^{\infty} 2^{-k} (f(U)e_k, e_k).$$

Since $f \in C(S^1)$ is bounded, $f(U)$ is bounded by Lemma 7.27, and hence μ is well defined. Note that $0 < \mu(1) \leq 1$ (assuming that H is non-zero).

From the linearity and positivity of the mapping $f \mapsto f(U)$, we have

Lemma 7.28. *μ is linear and positive, that is, for any finite sequence $f_1,\ldots,f_n$ of functions in $C(S^1)$ and any finite sequence $a_1,\ldots,a_n$ of real numbers, we have*

$$\mu\left(\sum_{i=1}^{n} a_i f_i\right) = \sum_{i=1}^{n} a_i \mu(f_i),$$

and for any $f \in C(S^1)$ such that $f(z) \geq 0$ on S^1, we have $\mu(f) \geq 0$

Such a function μ is called a positive measure on S^1. As a corollary, we have

Corollary 7.29. *If* $f,g \in C(S^1)$ *and* $f \geq g$ *on* S^1, *then* $\mu(f) \geq \mu(g)$.

We need to define and locate smooth points on S^1 with respect to that positive measure. We need some notations. For $z,w \in S^1$, $|z-w| > 0$, let (z,w) denote the open arc from z to w along the positive (counterclockwise) direction on S^1. This description is geometrical, but it can be easily translated into an analytical definition. We define closed and semi-closed arcs $[z,w]$, $(z,w]$, $[z,w)$ similarly. Note that these are totally bounded subsets and are therefore located. We use $lh(z,w)$ to denote the length of the arcs $(z,w), [z,w]$ and so on.

For two subsets A,B of a metric space, we define the weak complement

$$A \sim B \equiv_{df} \{x \in A : \rho(x,y) > 0 \text{ for all } y \in B\}.$$

Then, it is easy to see that for $z \neq w$, the complement $S^1 \sim (z,w) = [w,z]$ and $S^1 \sim [z,w] = (w,z)$. Moreover, when $(z,w) \subset (z',w')$, $z \neq z'$, $w \neq w'$, we have

$$(z',w') \sim (z,w) = (z',z] \cup [w,w'),$$
$$((z',w') \sim (z',z)) \sim (w,w') = [z,w].$$

For $z \in S^1$, $\varepsilon > 0$, let $z\tilde{+}\varepsilon$ denote the point on S^1 with ε arc length from z along the positive (counterclockwise) direction on S^1. Therefore, $(z,z\tilde{+}\varepsilon)$ is an arc of the length ε. $z\tilde{-}\varepsilon$ or $z\tilde{+}(-\varepsilon)$ similarly denotes the point on S^1 with ε arc length from z along the negative direction on S^1.

For $z \neq w$ on S^1 and $\delta > 0$, let $\chi^{\delta}_{[z,w]}$ denote the function in $C(S^1)$ such that $\chi^{\delta}_{[z,w]}(u) = 1$ for $u \in [z,w]$, and $\chi^{\delta}_{[z,w]}(u) = 0$ for $u \in [w\tilde{+}\delta, z\tilde{-}\delta]$, and $\chi^{\delta}_{[z,w]}(u)$ is linear (on the arc length parameter) on $[z\tilde{-}\delta, z]$ and $[w, w\tilde{+}\delta]$. $\chi^{\delta}_{[z,w]}$ is the outer characteristic function of $[z,w]$ with the precision δ. We will write $\chi^{1/n}_{[z,w]}$ as $\chi^{n}_{[z,w]}$, and write $\chi^{1/n}_{[z,z]}$ as χ^n_z. Note that $\left(\chi^n_{[z,w]}\right)_n$ is a decreasing sequence of non-negative functions and therefore $\left(\mu\left(\chi^n_{[z,w]}\right)\right)$ is a decreasing sequence of non-negative real numbers.

The following definition is adapted from [6], p. 237:

Definition 7.30. Let $z \in S^1$. For $\varepsilon > 0$, we say that $[z,w]$ has a profile lower than ε, denoted as $[z,w] \ll \varepsilon$, if there exists n such that $\mu\left(\chi^n_{[z,w]}\right) < \varepsilon$. We say that z is *smooth*, if $[z,z]$ has arbitrarily low profiles, that is, $\mu(\chi^n_z) \to 0$ as $n \to \infty$.

The following lemmas follow the ideas on [6], pp. 237–241. We need some more notations. For $t \in [0,1]$, let

$$z_t \equiv_{df} \cos 2\pi t + \mathrm{i} \sin 2\pi t.$$

Then, when t ranges from 0 to 1, z_t ranges from 1 to 1 on S^1. For $r,s,t \in [0,1]$, $s < t$, and $\delta > 0$, $n > 0$, we define

$$\chi^{\delta}_{[s,t]}(r) \equiv_{df} \chi^{\delta}_{[z_s,z_t]}(z_r),$$

and we similarly denote $\chi^{n}_{[s,t]} \equiv_{df} \chi^{1/n}_{[s,t]}$. Moreover, we use $[s,t] \ll \varepsilon$ to mean $[z_s,z_t] \ll \varepsilon$, and we similarly say 'a number $t \in [0,1]$ is smooth', meaning that z_t is smooth.

Lemma 7.31. *If $z,w,u \in S^1$, $w \in [z,u]$, $[z,w] \ll \varepsilon$ and $[w,u] \ll \varepsilon$, then $[z,u] \ll 2\varepsilon$.*

Proof. We may assume that for some n, $\mu\left(\chi^{n}_{[z,w]}\right) < \varepsilon$ and $\mu\left(\chi^{n}_{[w,u]}\right) < \varepsilon$. Note that $\chi^{n}_{[z,u]} \leq \chi^{n}_{[z,w]} + \chi^{n}_{[w,u]}$ on S^1. Therefore,

$$\mu\left(\chi^{n}_{[z,u]}\right) \leq \mu\left(\chi^{n}_{[z,w]}\right) + \mu\left(\chi^{n}_{[w,u]}\right) < 2\varepsilon.$$

□

Lemma 7.32. *Suppose that $s,t \in [0,1]$, and $s < t$, and s,t are rational numbers, and $M \geq 0$, $K > 1$, $[s,t] \ll \frac{M+1}{K}$. Then, there exists a rational number r_0,*

$$s + \frac{1}{3}(t-s) < r_0 < s + \frac{2}{3}(t-s),$$

and there exist non-negative integers A,B, such that $M = A+B$, $[s,r_0] \ll \frac{A+1}{K}$, and $[r_0,t] \ll \frac{B+1}{K}$.

Proof. Suppose that $L,n > 0$ are such that $\mu\left(\chi^{n}_{[s,t]}\right) + \frac{1}{L} < \frac{M+1}{K}$ for some n. Choose $N > \frac{4(M+1)L}{K}$. Divide the interval $\left[s + \frac{1}{3}(t-s), s + \frac{2}{3}(t-s)\right]$ into N equal rational subintervals

$$p_0 = s + \frac{1}{3}(t-s) < p_1 < \ldots < p_N = s + \frac{2}{3}(t-s).$$

Choose $m > n$ such that $\frac{1}{m} < \frac{p_{i+1}-p_i}{4}$, that is, $m > \frac{12N}{t-s}$. Then, for any $r \in [0,1]$,

$$\sum_{i=0}^{N-1} \chi^{m}_{[p_i,p_{i+1}]}(r) \leq 2\chi^{n}_{[s,t]}(r).$$

Therefore,

$$\sum_{i=0}^{N-1} \mu\left(\chi^{m}_{[p_i,p_{i+1}]}\right) \leq 2\mu\left(\chi^{n}_{[s,t]}\right) < \frac{2(M+1)}{K} - \frac{2}{L}.$$

Then, there exists i such that $\mu\left(\chi^{m}_{[p_i,p_{i+1}]}\right) < \frac{2(M+1)}{NK} < \frac{1}{2L}$. Note that to determine i, we only need to estimate each $\mu\left(\chi^{m}_{[p_i,p_{i+1}]}\right)$ up to the precision of $\frac{1}{NL}$. Let $r_0 \equiv \frac{p_i+p_{i+1}}{2}$. Note that for any $r \in [0,1]$,

$$\chi^{m}_{[s,r_0]}(r) + \chi^{m}_{[r_0,t]}(r) \leq \chi^{n}_{[s,t]}(r) + \chi^{m}_{[p_i,p_{i+1}]}(r).$$

Therefore,

$$\mu\left(\chi^m_{[s,r_0]}\right)+\mu\left(\chi^m_{[r_0,t]}\right)<\frac{M+1}{K}-\frac{1}{L}+\frac{1}{2L}=\frac{M+1}{K}-\frac{1}{2L}.$$

That is,

$$\mu\left(\chi^m_{[s,r_0]}\right)K+\mu\left(\chi^m_{[r_0,t]}\right)K+\frac{K}{2L}<M+1.$$

We can choose a rational number r such that

$$\mu\left(\chi^m_{[s,r_0]}\right)K<r-\frac{K}{8L}<r+\frac{K}{8L}<M+1-\mu\left(\chi^n_{[r_0,t]}\right)K.$$

Note that to determine r we only need to approximate $\mu\left(\chi^n_{[s,r_0]}\right)K$ up to the precision of $\frac{1}{8L}$. Let A be the integer such that $r-1\le A<r$. Since $r>0$, $A\ge 0$. Then, $\mu\left(\chi^m_{[s,r_0]}\right)K+\frac{K}{8L}<r\le A+1$, that is, $\mu\left(\chi^m_{[s,r_0]}\right)+\frac{1}{8L}<\frac{A+1}{K}$, which implies $[s,r_0]\ll\frac{A+1}{K}$. Similarly, let $B=M-A\ge 0$. Then,

$$\mu\left(\chi^n_{[r_0,t]}\right)K+\frac{K}{8L}<M+1-r<M+1-A=B+1,$$

that is, $\mu\left(\chi^m_{[r_0,t]}\right)+\frac{1}{8L}<\frac{B+1}{K}$, which implies $[r_0,t]\ll\frac{B+1}{K}$. Finally, note that the construction of r_0 and the witnesses for $[s,r_0]\ll\frac{A+1}{K}$ and $[r_0,t]\ll\frac{B+1}{K}$, that is, the construction of i, m and $\frac{1}{8L}$ from n and $\frac{1}{L}$, is by iteratable operations. □

Lemma 7.33. *For any $K>1$, there is a finite sequence $z_1,\ldots,z_M$ of points on S^1, such that $[z,z]\ll 1/K$ whenever $z\neq z_i$, for all $i=1,\ldots,M$.*

Proof. First, there exists M such that $[0,1]\ll\frac{M+1}{2K}$. Given s,t,M,K in the last lemma, let $R(s,t,M,K)$, $A(s,t,M,K)$, $B(s,t,M,K)$ denote the constructed r_0, A, B in the lemma respectively. For any finite sequence e of 0 and 1, we construct rational numbers a_e, b_e and integer M_e such that

$$\begin{aligned}
a_e&=0,\, b_e=1,\, M_e=M,\ \text{ for } e=\langle\,\rangle,\\
a_{e*\langle 0\rangle}&=a_e,\ b_{e*\langle 0\rangle}=R(a_e,b_e,M_e,2K),\, M_{e*\langle 0\rangle}=A(a_e,b_e,M_e,2K),\\
a_{e*\langle 1\rangle}&=R(a_e,b_e,M_e,2K),\, b_{e*\langle 1\rangle}=b_e,\ M_{e*\langle 1\rangle}=B(a_e,b_e,M_e,2K).
\end{aligned}$$

In the proof of the last lemma, we already make sure that the operations used in the constructions are iteratable. Therefore, this can be achieved by a bounded primitive recursion. The constructed are all rational numbers. Therefore, quantifier-free induction is enough to prove some basic properties of the constructed sequences. It is easy to see that we have:

(i) $a_e=a_{e*\langle 0\rangle}<b_{e*\langle 0\rangle}=a_{e*\langle 1\rangle}<b_{e*\langle 1\rangle}=b_e$, and $M_{e*\langle 0\rangle}+M_{e*\langle 1\rangle}=M_e$.
(ii) For each $n>0$, $\sum_{\{e:lh(e)=n\}}M_e=M$.
(iii) $[a_e,b_e]\ll\frac{M_e+1}{2K}$ (with rational numbers as witnesses).

(iv) $|b_e - a_e| \leq \left(\frac{2}{3}\right)^{lh(e)}$.
(v) If $lh(e) = lh(e')$ and $e \neq e'$, then the intervals $[a_e, b_e]$ and $[a_{e'}, b_{e'}]$ do not overlap.
(vi) For each $n > 0$, the intervals $[a_e, b_e]$ with $lh(e) = n$ constitute a partition of $[0, 1]$.

For each e, let $x_e = \frac{a_e + b_e}{2}$. For each $k = 1, ..., M$, we construct a sequence $s_k \equiv (x_{k,n})_{n=0}^{\infty}$ of rational numbers such that:

(I) Each $x_{k,n} = x_e$ for some unique e with $lh(e) = n$.
(II) For each e with $lh(e) = n$, there are exactly M_e values of k such that $x_{k,n} = x_e$.
(III) For each $x_{k,n}$, if $x_{k,n} = x_e$ for some e, then $x_{k,n+1} = x_{e*\langle 0\rangle}$ or $x_{e*\langle 1\rangle}$.

$x_{k,n}$ are constructed as follows:

(1) $x_{k,0} = x_{\langle\rangle}$ for $k = 1, ..., M$.
(2) Suppose that $x_{k,n}$, $k = 1, ..., M$, have been constructed to satisfy (I) and (II). Then, for each e with $lh(e) = n$, there are M_e values of k such that $x_{k,n} = x_e$. Recall that $M_e = M_{e*\langle 0\rangle} + M_{e*\langle 1\rangle}$. For $M_{e*\langle 0\rangle}$ of those values of k, we let $x_{k,n+1} = x_{e*\langle 0\rangle}$, and for the rest $M_{e*\langle 1\rangle}$ of those values of k, we let $x_{k,n+1} = x_{e*\langle 1\rangle}$.

Here, we are again constructing rational numbers. Therefore, with the quantifier-free induction, we can derive the basic properties (I), (II) and (III) of the sequences.

Then, by (III) and (i), (iv) above, for each $x_{k,n} = x_e$, we have $x_{k,n'} \in [a_e, b_e]$ for all $n' \geq n$. Therefore, $|x_{k,n'} - x_{k,n}| \leq |b_e - a_e| \leq \left(\frac{2}{3}\right)^n$. It is easy to see that $(x_{k,n})_n$ is a Cauchy sequence and converges to some x_k. Moreover, if $x_{k,n} = x_e$, we have $x_k \in [a_e, b_e]$ and $|x_{k,n} - x_k| \leq \left(\frac{2}{3}\right)^n$. Therefore, we get M points $x_1, ..., x_k \in [0, 1]$.

Now, let $z_k \equiv z_{x_k}$ for $k = 1, ..., M$. Suppose that $z \in S^1$ and $z \neq z_k$ for $k = 1, ..., M$. For each ε, $0 \leq \varepsilon \leq 1$, let $arclh(\varepsilon)$ be the length of the arc on S^1 corresponding to an interval of the length ε in $[0, 1]$. We can find N such that $lh(z, z_k) > 4arclh\left(\left(\frac{2}{3}\right)^N\right)$ for all $k = 1, ..., M$. Then, for each k, if $x_{k,N} = x_e$, then $|x_e - x_k| \leq \left(\frac{2}{3}\right)^N$, and therefore $arclh(z_{x_e}, z_k) \leq arclh\left(\left(\frac{2}{3}\right)^N\right)$. Then, $lh(z, z_{x_e}) > 3arclh\left(\left(\frac{2}{3}\right)^N\right)$ and therefore $z \notin [z_{a_e}, z_{b_e}]$ and $\widetilde{\rho}(z, [z_{a_e}, z_{b_e}]) > 2arclh\left(\left(\frac{2}{3}\right)^N\right)$, where $\widetilde{\rho}$ denotes the arc distance on S^1. Note that for each e of the length N, there exists k with $x_{k,N} = x_e$ if and only if $M_e > 0$. Therefore, the above inequality holds for each e of the length N such that $M_e > 0$. That is, z is at least two blocks away from any arc interval $[z_{a_e}, z_{b_e}]$ on S^1 with $M_e > 0$. By (vi) above and by estimating z up to the precision of $\frac{1}{4}arclh\left(\left(\frac{2}{3}\right)^N\right)$, we can find two adjacent arc intervals $[z_{a_e}, z_{b_e}]$, $\left[z_{a_{e'}}, z_{b_{e'}}\right]$ such that $z \in \left[z_{a_e}, z_{b_{e'}}\right]$ and either $b_e = a_{e'}$ or $b_e = 1$ and $a_{e'} = 0$. We must have $M_e = M_{e'} = 0$. Therefore, by (iii) above, $[z_{a_e}, z_{b_e}] \ll \frac{1}{2K}$ and $\left[z_{a_{e'}}, z_{b_{e'}}\right] \ll \frac{1}{2K}$. Therefore, by Lemma 7.31 above, $\left[z_{a_e}, z_{b_{e'}}\right] \ll \frac{1}{K}$. Then, $[z, z] \ll \frac{1}{K}$. □

Lemma 7.34. *There is a sequence (z_n) of points on S^1 such that if $z \in S^1$ and $z \neq z_n$ for each n, then z is a smooth point.*

Proof. By the lemma above, for each $K > 0$, there exists a finite sequence $z_{K,1}, ...,$ z_{K,M_K} of points on S^1, such that if $z \in S^1$ and $z \neq z_{K,i}$ for each i, then $[z,z] \ll 1/K$. Arrange all $z_{K,i}$ into a single sequence (z_n). If $z \in S^1$ and $z \neq z_n$ for all n, then $[z,z] \ll 1/K$ for all $K > 0$. Therefore, z is smooth. □

Let $\rho_s(U)$ denote the set of all smooth points on S^1. By a construction similar to the proof for the Cantor's theorem for real numbers (i.e. Theorem 3.2), we see that $\rho_s(U)$ is dense in S^1. Note that for $z, w \in S^1$ and $m > n$, we have

$$0 \leq \chi^n_{[z,w]} - \chi^m_{[z,w]} \leq \chi^n_z + \chi^n_w.$$

Therefore, we have

Lemma 7.35. *If $z, w \in \rho_s(U)$, then $\mu\left(\chi^n_{[z,w]}\right)$ converges.*

Now, we can extend the mapping $C(S^1) \rightarrow Hom(H)$, $f \longmapsto f(U)$, to include the characteristic functions of arc intervals on S^1 with smooth end points. First, we need some lemmas.

Lemma 7.36. *Suppose that $f \in C(S^1)$. If $\mu(f^2) \leq \varepsilon$, then for each k, $\|f(U)e_k\| \leq \sqrt{2^k \varepsilon}$.*

Proof. By definition, $\mu(f^2) \leq \varepsilon$ means that

$$\sum_{k=1}^{\infty} 2^{-k}\left(f^2(U)e_k, e_k\right) = \sum_{k=1}^{\infty} 2^{-k}\|f(U)e_k\|^2 \leq \varepsilon.$$

Therefore, $\|f(U)e_k\| \leq \sqrt{2^k \varepsilon}$. □

Lemma 7.37. *Suppose that (f_n) is a sequence in $C(S^1)$, and for any $\varepsilon > 0$, there exists N, such that $\mu(|f_m - f_n|) < \varepsilon$ whenever $m, n \geq N$, and moreover for all n, $|f_n| \leq C$ on S^1. Then, $(f_n(U))$ strongly converges to some operator in $Hom(H)$. Moreover, if (g_n) is another sequence in $C(S^1)$ with a common bound, and $\mu(|g_n - f_n|) \rightarrow 0$ as $n \rightarrow \infty$, then $(g_n(U))$ strongly converges to the same operator in $Hom(H)$.*

Proof. Note that $(f_m - f_n)^2 \leq 2C|f_m - f_n|$ on S^1, and hence

$$\mu\left((f_m - f_n)^2\right) \leq 2C\mu(|f_m - f_n|).$$

For each $k > 0$, by the lemma above and the assumptions, for any $\varepsilon > 0$, there exists N, such that $\|(f_m - f_n)(U)e_k\| < \sqrt{2^k 2C\varepsilon}$ whenever $m, n > N$. It means that for each k, $(f_n(U)e_k)_n$ converges in H. Then, for any finite linear combination x of e_k, $k = 0, 1, 2, ...$, $(f_n(U)x)$ also converges. Since such x constitute a dense subset of H, by Lemma 7.11, $f_n(U)$ strongly converges to an operator in $Hom(H)$.

By the same argument, we have $\|(g_n - f_n)(U)e_k\| \rightarrow 0$ for each e_k. Then, it is easy to see that $g_n(U)$ and $f_n(U)$ strongly converges to the same operator. □

Let $g_n = g$ for some $g \in C(S^1)$ in the lemma. We have

Corollary 7.38. *Let (f_n) be as in the lemma. If $g \in C(S^1)$ and $\mu(|f_n - g|) \to 0$ as $n \to \infty$, then $(f_n(U))$ strongly converges to $g(U)$.*

Then, for any $z, w \in \rho_s(U)$, by Lemma 7.35 above, $\left(\chi^n_{[z,w]}(U)\right)_n$ strongly converges to an operator in $Hom(H)$. We denote the limit as

$$E_{[z,w]} \equiv_{df} \chi_{[z,w]}(U) \equiv_{df} \lim_{n\to\infty} \chi^n_{[z,w]}(U).$$

$\{E_{[z_1,z_2]} : z_1, z_2 \in \rho_s(U), z_1 \neq z_2\}$ is called the spectral family associated with U. It has the common properties of a spectral family:

Lemma 7.39. *Let $z, u, w, z_n \in \rho_s(U)$, $z \neq u$.*

(i) $E_{[z,w]}$ is a projection.
(ii) $E_{[z,z_n]} \to 0$ if $z_{n+1} \in (z, z_n)$ for all n and $z_n \to z$; $E_{[z,z_n]} \to I$ if $z_{n+1} \in (z_n, z)$ for all n and $z_n \to z$.
(iii) $E_{[z,z_n]} \to E_{[z,u]}$ if $z_n \to u$.
(iv) $E_{[u,z]} = I - E_{[z,u]}$; $E_{[z,w]} = E_{[z,u]}E_{[z,w]} = E_{[z,w]}E_{[z,u]}$ if $w \in (z, u)$.
(v) Suppose that $z = z_0, z_1, \ldots, z_n = u$ are separated points along the positive direction from z to u. Then, $E_{[z,u]} = \sum_{i=0}^{n-1} E_{[z_i,z_{i+1}]}$.

Proof. (i) Since all $\chi^n_{[z,w]}(U)$ are self-adjoint, their strong limit $E_{[z,w]}$ is also self-adjoint. To see that $E^2_{[z,w]} = E_{[z,w]}$, note that the mapping from $C(S^1)$ to $Hom(H)$ is multiplicative. Therefore,

$$\left(\chi^n_{[z,w]}\right)^2(U) = \chi^n_{[z,w]}(U)\,\chi^n_{[z,w]}(U).$$

Apparently, $\chi^n_{[z,w]}(U)\,\chi^n_{[z,w]}(U) \to E^2_{[z,w]}$, since the strong limit is multiplicative. On the other side, $\left|\left(\chi^n_{[z,w]}\right)^2 - \chi^n_{[z,w]}\right| \leq \chi^n_z + \chi^n_w$. Therefore,

$$\mu\left(\left|\left(\chi^n_{[z,w]}\right)^2 - \chi^n_{[z,w]}\right|\right) \to 0.$$

So, by the lemma above, $\left(\chi^n_{[z,w]}\right)^2(U) \to E_{[z,w]}$.

(ii) For the first half, it suffices to show that for each e_k, $\left\|E_{[z,z_n]}e_k\right\| \to 0$. Given any $\varepsilon > 0$, first we can find N, such that $\mu\left(\chi^N_z\right) < \varepsilon$. For all sufficiently large n, we have $z_n \in \left[z \tilde{-} \frac{1}{2N}, z \tilde{+} \frac{1}{2N}\right]$. Then, for any $m > 2N$, $\chi^m_{[z,z_n]} < 2\chi^N_z$. Therefore,

$$\mu\left(\left(\chi^m_{[z,z_n]}\right)^2\right) \leq \mu\left(\chi^m_{[z,z_n]}\right) < 2\varepsilon.$$

Therefore, by Lemma 7.36, $\left\|\chi^m_{[z,z_n]}e_k\right\| < \sqrt{2^k 2\varepsilon}$. Since m is arbitrary, we have $\left\|E_{[z,z_n]}e_k\right\| < \sqrt{2^{k+1}\varepsilon}$ for all sufficiently large n. Therefore, $\left\|E_{[z,z_n]}e_k\right\| \to 0$. The second half of (ii) follows from (iv) and the first half.

(iii) can be proved similarly, by noting that given any $\varepsilon > 0$, for sufficiently large n, $\mu\left(\left|\chi^m_{[z,z_n]} - \chi^m_{[z,u]}\right|^2\right) < \varepsilon$ will hold for all sufficiently large m.

(iv) To see that $E_{[u,z]} = I - E_{[z,u]}$, similarly note that

$$\left|\chi^n_{[u,z]} + \chi^n_{[z,u]} - 1\right| \le \chi^n_z + \chi^n_u.$$

From there we have $\left\|\chi^n_{[u,z]}(e_k) + \chi^n_{[z,u]}(e_k) - e_k\right\| \to 0$ for each e_k. Therefore, $E_{[u,z]} + E_{[z,u]} - I = 0$.

The rest of (iv) and (v) are similar. □

Suppose that z_0, z_1, ..., $z_{n+1} = z_0$ are distinct smooth points on S^1, chosen along the positive direction. Then, $P = (z_0, z_1, ..., z_n)$ is a partition of S^1. Let

$$Mesh(P) \equiv_{df} \max\{|z_i - z_{i+1}| : i = 0, ..., n\}.$$

Another partition P' is a refinement of P if $P' \supset P$. For $g \in C(S^1)$, we define

$$g_P(U) \equiv_{df} \sum_{i=1}^{n} g(z_i) E_{[z_i, z_{i+1}]}.$$

Now, suppose that $P_m = \left(z_0^{(m)}, z_1^{(m)}, \ldots\right)$ is a sequence of partitions such that P_{m+1} is a refinement of P_m and $Mesh(P_m) \to 0$. We want to show that $g_{P_m}(U)$ strongly converges to $g(U)$ as $m \to \infty$. Suppose that $C > 0$ and $|g| \le C$ on S^1. Let $\varepsilon > 0$. Let $\delta > 0$ be such that $|g(z) - g(w)| < \varepsilon$ whenever $|z - w| < \delta$, $z, w \in S^1$. Let m be any sufficiently large number such that $Mesh(P_m) < \delta/2$. Since $z_0^{(m)}, z_1^{(m)}, \ldots$ are finitely many smooth points on S^1, we can choose n sufficiently large so that $\sum_i \mu\left(\chi^n_{z_i^{(m)}}\right) < \varepsilon$. We may also assume that $\frac{1}{n} < \frac{1}{4} lh\left(z_i^{(m)}, z_{i+1}^{(m)}\right)$ for all i. Then,

$$\mu\left(\left|g - \sum_i g\chi^n_{\left[z_i^{(m)}, z_{i+1}^{(m)}\right]}\right|\right) \le C\mu\left(\left|1 - \sum_i \chi^n_{\left[z_i^{(m)}, z_{i+1}^{(m)}\right]}\right|\right)$$
$$\le C\mu\left(\sum_i \chi^n_{z_i^{(m)}}\right) \le C\varepsilon.$$

Note that for $z \in \left[z_i^{(m)} \tilde{-} \frac{1}{n}, z_{i+1}^{(m)} \tilde{+} \frac{1}{n}\right]$, $\left|z - z_i^{(m)}\right| < \delta$. So, $\left|g(z) - g\left(z_i^{(m)}\right)\right| < \varepsilon$. Then,

$$\mu\left(\left|\sum_i g\chi^n_{\left[z_i^{(m)}, z_{i+1}^{(m)}\right]} - \sum_i g\left(z_i^{(m)}\right)\chi^n_{\left[z_i^{(m)}, z_{i+1}^{(m)}\right]}\right|\right)$$
$$= \mu\left(\left|\sum_i \left(g - g\left(z_i^{(m)}\right)\right)\chi^n_{\left[z_i^{(m)}, z_{i+1}^{(m)}\right]}\right|\right) \le \varepsilon\mu\left(\sum_i \chi^n_{\left[z_i^{(m)}, z_{i+1}^{(m)}\right]}\right) \le 2\varepsilon.$$

Combining the last two inequalities, we have

$$\mu\left(\left|g-\sum_i g\left(z_i^{(m)}\right)\chi^n_{\left[z_i^{(m)},z_{i+1}^{(m)}\right]}\right|\right)\leq (C+2)\,\varepsilon.$$

Note that $\left|g-\sum_i g\left(z_i^{(m)}\right)\chi^n_{\left[z_i^{(m)},z_{i+1}^{(m)}\right]}\right|\leq 3C$. Therefore,

$$\mu\left(\left|g-\sum_i g\left(z_i^{(m)}\right)\chi^n_{\left[z_i^{(m)},z_{i+1}^{(m)}\right]}\right|^2\right)\leq 3C\,(C+2)\,\varepsilon.$$

So, by Lemma 7.36

$$\left\|\left(g(U)-\sum_i g\left(z_i^{(m)}\right)\chi^n_{\left[z_i^{(m)},z_{i+1}^{(m)}\right]}(U)\right)e_k\right\|\leq\sqrt{2^k 3C\,(C+2)\,\varepsilon}.$$

Since n can be arbitrarily large, we have

$$\left\|\left(g(U)-g_{P_m}(U)\right)e_k\right\|\leq\sqrt{2^k 3C\,(C+2)\,\varepsilon}.$$

Therefore, $g_{P_m}(U)\,e_k\to g(U)\,e_k$ for each k. So, $g_{P_m}(U)\to g(U)$.

We will express $g_{P_m}(U)\to g(U)$ as

$$g(U)=\int_{S^1} g(z)\,\mathrm{d}E_z,$$

where the integration is understood as the strong limit above.

For $g=g_1+\mathrm{i}g_2$, we define $g(U)\equiv g_1(U)+\mathrm{i}g_2(U)$. Then the same expression holds as well. So we have the spectral decompositions:

Theorem 7.40. *If U is a multiplicable unitary operator on H, and $g\in C\left(S^1,\mathbb{C}\right)$, and $\left\{E_{(z_1,z_2]}:z_1,z_2\in\rho_s(U),z_1\neq z_2\right\}$ is the spectral family associated with U, then*

$$g(U)=\int_{S^1} g(z)\,\mathrm{d}E_z.$$

In particular,

$$U=\int_{S^1} z\,\mathrm{d}E_z.$$

Furthermore, if $A\in Hom(H)$, then $AU=UA$ iff $AE_{[z_1,z_2]}=E_{[z_1,z_2]}A$ for all $z_1\neq z_2$ in $\rho_s(U)$.

Proof. The first half has been proved above. If $AU=UA$, then $Ap(U)=p(U)A$ for any $p\in\mathscr{R}\mathscr{P}$, and then $Af(U)=f(U)A$ for any $f\in C\left(S^1\right)$. From there, we have $AE_{[z_1,z_2]}=E_{[z_1,z_2]}A$. Conversely, if $AE_{[z_1,z_2]}=E_{[z_1,z_2]}A$ for all $z_1\neq z_2$ in $\rho(U)$, then $Ag_{P_m}(U)=g_{P_m}(U)A$ for $g(z)\equiv z$. Since $g_{P_m}(U)\to U$, we have $AU=UA$. □

Finally we prove a corollary which will be used later.

Corollary 7.41. *If $w \in \rho_s(U)$, then $(U-w)x=0$ implies $x=0$.*

Proof. Suppose that $(U-w)x=0$. Then $((U^*-w^*)(U-w)x,x)=0$. Note that

$$(U^*-w)(U-w)x = \lim_{Mesh(P)\to 0}\sum_{i=0}^{n}|z_i-w|^2 E_{[z_i,z_{i+1}]}x$$
$$= \lim_{Mesh(P)\to 0}\sum_{i=0}^{n}|z_i-w|^2 E^2_{[z_i,z_{i+1}]}x.$$

Therefore,

$$0=((U^*-w^*)(U-w)x,x) = \lim_{Mesh(P)\to 0}\sum_{i=0}^{n}|z_i-w|^2\left\|E_{[z_i,z_{i+1}]}x\right\|^2.$$

We write this as $\int_{S^1}|z-w|^2 d\|E_z x\|^2=0$. Let $u,v\in\rho_s(U)$, $w\in(u,v)$. Then we should have $\int_{[v,u]}|z-w|^2 d\|E_z x\|^2=0$. Suppose that $|u-w|>\varepsilon$ and $|v-w|>\varepsilon$ for some $\varepsilon>0$. Note that for any smooth points $z_0,\ldots,z_n$ along the positive direction on S^1,

$$\sum_{i=0}^{n-1}\left\|E_{[z_i,z_{i+1}]}x\right\|^2=\sum_{i=0}^{n-1}\left(E_{[z_i,z_{i+1}]}x,x\right)=\left(E_{[z_0,z_n]}x,x\right)=\left\|E_{[z_0,z_n]}x\right\|^2.$$

Therefore, $\int_{[v,u]}d\|E_z x\|^2=\left\|E_{[v,u]}x\right\|^2$. Then,

$$\int_{[v,u]}|z-w|^2 d\|E_z x\|^2\geq\varepsilon\int_{[v,u]}d\|E_z x\|^2=\varepsilon\left\|E_{[v,u]}x\right\|^2.$$

So $E_{[v,u]}x=0$. Let $u\to w$, $v\to w$. Then $E_{[v,u]}\to I$. So $x=0$. □

7.5 Unbounded Operators

So far, we have considered only bounded operators defined on the entire Hilbert space. In this section, we present basic notions and facts about unbounded operators, which are defined on the linear subsets of a Hilbert space H. The basic definition is:

Definition 7.42. A *linear operator* A on a Hilbert space H is a pair consisting of a specification of a domain $D(A)$, a linear subset of H, and a linear mapping $A: D(A)\to H$.

From now on, 'linear operator' will be used in this general sense and we will explicitly say 'bounded operator' when we mean bounded operators defined on the entire space. Note that an unbounded operator comes with a subset of the Hilbert space as its domain. Therefore, we cannot generally quantify over all unbounded operators. Instead, we can consider a family $\{A_i : i \in I\}$ of unbounded operators with a parameter and quantify over such a family. In that case, we assume that their domains are a family $\{D(A_i) : i \in I\}$ of linear subsets.

The range and null space of A are defined as

$$R(A) \equiv_{df} \{Ax : x \in D(A)\},$$
$$N(A) \equiv_{df} \{x \in D(A) : Ax = 0\}.$$

A is called injective, if $Ax = Ay$ implies $x = y$ for any $x, y \in D(A)$. If A is injective, we can define A^{-1} with $D(A^{-1}) = R(A)$: For $x \in R(A)$, there exists y such that $Ay = x$. We define $A^{-1}x \equiv y$. The sum $A + B$ and product BA of two operators A and B are defined as:

$$D(A+B) \equiv_{df} D(A) \cap D(B), \quad (A+B)x \equiv_{df} Ax + Bx;$$
$$D(BA) \equiv_{df} \{x \in D(A) : Ax \in D(B)\}, \quad (BA)x \equiv_{df} B(Ax).$$

We say that an operator A is included in another operator B, denoted as $A \subseteq B$, if $D(A) \subseteq D(B)$ and $Ax = Bx$ for $x \in D(A)$. Then, some familiar equalities or inclusion relations hold, for instance (cf. Riesz and Sz.-Nagy [31], p. 299),

$$(B+C)A = BA + CA, \; A(B+C) \supseteq AB + AC.$$

Here is an example of an unbounded linear operator on L_2:

$$D\left(\widehat{X}\right) \equiv_{df} \{f \in L_2 : xf \in L_2\},$$
$$\widehat{X}f \equiv_{df} xf, \text{ for } f \in D\left(\widehat{X}\right).$$

$\widehat{X}$ is the position operator in quantum mechanics. Recall that $C(\mathbb{F})$ is dense in L_2. Obviously, $C(\mathbb{F}) \subseteq D\left(\widehat{X}\right)$. Therefore, $D\left(\widehat{X}\right)$ is dense in L_2.

When $D(A)$ is dense in H, we can define the adjoint A^* of A as follows:

Definition 7.43. If A is an operator on the Hilbert space H, and $D(A)$ is dense in H, then A^* is:

$$D(A^*) = \{y : \text{for some } y^*, \; (Ax, y) = (x, y^*) \text{ for all } x \in D(A)\},$$
$$A^*y = y^* \text{ above}, \quad \text{for } y \in D(A^*).$$

y^* is uniquely determined by the fact that $D(A)$ is dense in H. Note that $D(A) = H$ does not imply $D(A^*) = H$. Therefore, A^* is defined for any operator A with $D(A)$ dense in H, including bounded operators. Some familiar relations for adjoint still hold:

Lemma 7.44. *For any operators A, B,*

$$(cI)^* = c^*I, \; (A+B)^* \supseteq A^* + B^*, \; (AB)^* \supseteq B^*A^*,$$

$$(AB)^* = B^*A^*, \text{ if } D(A) = D(A^*) = H,$$
$$(A+B)^* = A^* + B^*, \text{ if } D(A) = D(A^*) = H,$$
$$B^* \supseteq A^*, \text{ if } A \supseteq B.$$

Proof. Straightforward from the definitions. □

A is called closed if whenever $x_n \in D(A)$, $x_n \to x$, $Ax_n \to y$, we have $x \in D(A)$ and $Ax = y$. Suppose that A is such that whenever $x_n, x'_n \in D(A)$, $x_n \to x$, $x'_n \to x$, $Ax_n \to y$, and $Ax'_n \to y'$, we have $y = y'$. Then, we say that A is closable and its closure $\overline{A}$ is defined as

$$D\left(\overline{A}\right) = \{x : \text{ for some } x_n \in D(A), \text{ and some } y, x_n \to x, \text{ and } Ax_n \to y\}$$
$$\overline{A}x = y \text{ above, for } x \in D\left(\overline{A}\right).$$

If B is closed and $A \subset B$, then A must be closable and $\overline{A} \subset B$. If A is closed, then $\overline{A} = A$. Moreover, we have

Lemma 7.45. *If $D(A)$ is dense, then A^* is always closed.*

Proof. Suppose that $x_n \in D(A^*)$, $x_n \to x$, and $A^*x_n \to y$. By definition, we have a sequence (x_n^*) such that $(Az, x_n) = (z, x_n^*)$ for all $z \in D(A)$. Moreover, $x_n \to x$, $x_n^* = A^*x_n \to y$. Therefore, $(Az, x) = (z, y)$ for any $z \in D(A)$. Then, by the definition again, we have $x \in D(A^*)$ and $A^*x = y$. □

Definition 7.46. A is called symmetric if $D(A)$ is dense and $A \subseteq A^*$.

It is easy to see that when A is symmetric, A is closable, and $\overline{A}$ is also symmetric. The position operator $\widehat{X}$ is clearly symmetric.

Lemma 7.47. *If A is symmetric, then for $x \in D(A)$ and real number r,*

$$\|(A + ri)x\|^2 = \|Ax\|^2 + r^2 \|x\|^2. \tag{7.6}$$

This implies that $A + ri$ is injective if $r \neq 0$. Note that unlike what is in classical mathematics, in strict finitism, this does not imply that $(A + ri)^{-1}$ can be constructed. However, if $R(A + ri) = H$, then for any $y \in H$, there exists unique x such that $(A + ri)x = y$. Therefore, we can define $(A + ri)^{-1} y \equiv x$. So, $(A + ri)^{-1}$ exists and $D\left((A + ri)^{-1}\right) = H$. Conversely, $D\left((A + ri)^{-1}\right) = H$ implies that $R(A + ri) = H$. Moreover, in that case, the equation (7.6) above implies that $\left\|(A + ri)^{-1}\right\| \leq r^{-1}$.

Then, we can define self-adjointness in strict finitism [41], which is different from the common classical definition.

Definition 7.48. A is *self-adjoint* if A is symmetric, and for any real number $r \neq 0$, $R(A + ri) = H$. A is *essentially self-adjoint*, if it is closable and $\overline{A}$ is self-adjoint. A is strongly self-adjoint, if A is self-adjoint and $(A + ri)^{-1}$ is positively multiplicable for any real number $r \neq 0$.

Consider the position operator $\widehat{X}$. Note that $\left(\widehat{X}+ri\right)f=(x+ri)f$ for $f\in D\left(\widehat{X}+ri\right)$. Since $\left|(x+ri)^{-1}g\right|^2\leq|r|^{-2}|g|^2$, by Lemma 6.25, we have $(x+ri)^{-1}g\in L_2$ for $g\in L_2$. Therefore, $(x+ri)^{-1}g\in D\left(\widehat{X}+ri\right)$. That is, $D\left(\left(\widehat{X}+ri\right)^{-1}\right)=H$ and $\left(\widehat{X}+ri\right)^{-1}g=(x+ri)^{-1}g$ for any $g\in L_2$. Clearly, $\left(\widehat{X}+ri\right)^{-n}g=(x+ri)^{-n}g$. That is, $\left(\widehat{X}+ri\right)^{-1}$ is positively multiplicable. Therefore, $\widehat{X}$ is strongly self-adjoint.

Lemma 7.49. *Suppose that $D(A)$ is dense.*

(a) $R(A)^{\perp}=N(A^*)$.
(b) If $A^*=A$, then $R(A+r\mathrm{i})^{\perp}=\{0\}$ for any real number $r\neq 0$.
(c) If A is self-adjoint, then $A^*=A$.
(d) If $A\subset B$, and A is self-adjoint, and B is symmetric, then $A=B$.
(e) If A is symmetric and for any real number $r\neq 0$, $R(A+r\mathrm{i})$ is dense in H, then A is essentially self-adjoint.

Proof. (a) is trivial by definition. (b) follows from (a) and the equation (7.6).

(c) We must prove $D(A^*)\subset D(A)$. Let $x\in D(A^*)$. Since $D\left((A-\mathrm{i})^{-1}\right)=H$, there exists $y\in D(A)$ such that $(A-\mathrm{i})y=(A^*-\mathrm{i})x$. So $(A^*-\mathrm{i})(x-y)=0$. Since $R(A+\mathrm{i})=H$, by (a),

$$x-y\in N(A^*-\mathrm{i})=R(A+\mathrm{i})^{\perp}=\{0\}.$$

Hence $x=y\in D(A)$.

(d) follows from (c) directly.

(e) A is closable, since it is symmetric. By the assumption, for any $y\in H$ there exists $x_n\in H$ such that $(A+r\mathrm{i})x_n\to y$. By the equation (7.6),

$$\|(A+r\mathrm{i})x_n-(A+r\mathrm{i})x_m\|^2=\|Ax_n-Ax_m\|^2+r^2\|x_n-x_m\|^2.$$

Therefore, both (x_n) and (Ax_n) are Cauchy sequences. So, $x_n\to x$ and $Ax_n\to y'$ for some x and y'. So, $x\in D\left(\overline{A}\right)$ and $\left(\overline{A}+r\mathrm{i}\right)x=y$. Hence $R\left(\overline{A}+r\mathrm{i}\right)=H$. □

(c) justifies the term 'self-adjoint' in Definition 7.48. In classical analysis, for A closed and symmetric, $R(A\pm\mathrm{i})^{\perp}=\{0\}$ is equivalent to $R(A\pm\mathrm{i})=H$, but we don't have a finitistic proof of this. So $R(A\pm\mathrm{i})=H$ is perhaps stronger than $A^*=A$.

We still need to show that for bounded operators defined on the entire space H, Definition 7.48 is equivalent to the original definition of self-adjointness. We need a lemma.

Lemma 7.50. *Suppose that $D(A)=H$, and $\|A\|\leq 1-\varepsilon$ for some $\varepsilon>0$, and A is positively multiplicable. Then, $(1-A)^{-1}$ exists, and it is bounded by ε^{-1}, and $D\left((1-A)^{-1}\right)=R(1-A)=H$,*

$$(1-A)^{-1} = 1 + A + A^2 + ...,$$

where the infinite sum is understood as a strong limit. Moreover, $(1-A)^{-1}$ *is positively multiplicable.*

Proof. Since A is multiplicable, and since $\|A\| \leq 1-\varepsilon$, by the finite transitivity of inequality between real numbers, we see that $\|A^n\| \leq (1-\varepsilon)^n$ and $\sum_{n=0}^{\infty} A^n x$ converges for any x. Then, it is easy to verify that $\|\sum_{n=0}^{\infty} A^n x\| \leq \varepsilon^{-1} \|x\|$ and

$$(1-A)\sum_{n=0}^{\infty} A^n x = \sum_{n=0}^{\infty} A^n (1-A) x = x.$$

Moreover, $(1-A)^{-k}$ can be constructed as

$$\sum_{n=0}^{\infty} \left(\sum_{\substack{i_1+\ldots+i_k=n, \\ i_1\geq 0,\ldots,i_k\geq 0}} 1 \right) A^n.$$

Therefore, $(1-A)^{-1}$ is also positively multiplicable. □

Then we can prove the lemma which guarantees the equivalence.

Lemma 7.51. *If* $D(A) = H$, *and* A *is bounded and positively multiplicable, and* $A^* = A$, *then* A *is strongly self-adjoint.*

Proof. Suppose that r is a real number and $r \neq 0$. We may assume that $r > 0$. The case for $r < 0$ is similar. Choose $M > r$ such that $\|A\| \leq M - \varepsilon$ for some $\varepsilon > 0$. We assume that $M > 1$. By Lemma 7.50, $(A+M\mathrm{i})^{-1} = (M\mathrm{i})^{-1}\left((M\mathrm{i})^{-1}A+1\right)^{-1}$ exists and $D\left((A+M\mathrm{i})^{-1}\right) = H$ and $(A+M\mathrm{i})^{-1}$ is positively multiplicable. Recall that $\left\|(A+M\mathrm{i})^{-1}\right\| \leq M^{-1}$. We have

$$\begin{aligned} A + r\mathrm{i} &= \left(1-(M-r)\mathrm{i}(A+M\mathrm{i})^{-1}\right)(A+M\mathrm{i}) \\ &= (A+M\mathrm{i})\left(1-(M-r)\mathrm{i}(A+M\mathrm{i})^{-1}\right). \end{aligned}$$

Now $\left\|(M-r)\mathrm{i}(A+M\mathrm{i})^{-1}\right\| \leq (M-r)M^{-1} < 1$. By Lemma 7.50 again, we have

$$\begin{aligned} (A+r\mathrm{i})^{-1} &= \left(1-(M-r)\mathrm{i}(A+M\mathrm{i})^{-1}\right)^{-1}(A+M\mathrm{i})^{-1} \\ &= (A+M\mathrm{i})^{-1}\left(1-(M-r)\mathrm{i}(A+M\mathrm{i})^{-1}\right)^{-1}. \end{aligned}$$

Therefore, $D\left((A+r\mathrm{i})^{-1}\right) = H$ and $(A+r\mathrm{i})^{-1}$ is positively multiplicable. □

Definition 7.52. The *resolvent set* of an operator A is

$$\rho(A) \equiv \left\{ \begin{array}{l} z \in \mathbb{C} : z - A \text{ is injective, and } (z-A)^{-1} \text{ is bounded} \\ \text{and positively multiplicable with } D\left((z-A)^{-1}\right) = H \end{array} \right\},$$

and the *spectrum* of A is

$$\sigma(A) = \mathbb{C} \sim \rho(A).$$

We need the boundedness and multiplicability of $(z-A)^{-1}$ to be explicitly stated in the definition of resolvent set, because we don't have the closed graph theorem, and not all bounded operators are positively multiplicable.

Lemma 7.53. *If $z_0 \in \rho(A)$, $\left\|(z_0-A)^{-1}\right\| \le M$, and $|z| < M^{-1}$, then $z_0 + z \in \rho(A)$.*

Proof. Note that

$$z_0 + z - A = \left(I + z(z_0-A)^{-1}\right)(z_0 - A).$$

By Lemma 7.50, $\left(I + z(z_0-A)^{-1}\right)^{-1}$ exists, and it is multiplicable, bounded and defined on the whole H. Therefore,

$$(z_0 + z - A)^{-1} = (z_0-A)^{-1}\left(I + z(z_0-A)^{-1}\right)^{-1}$$

is bounded, multiplicable and defined on the whole H. □

We have a characterization of self-adjointness similar to that in the classical analysis. (cf. Weidmann [37], p. 108, Theorem 5.23)

Lemma 7.54. *Suppose that A is symmetric. The following are equivalent:*

(1) A is strongly self-adjoint;
(2) For any $N > 0$, there exists real number $r > N$ such that $\pm r\mathrm{i} \in \rho(A)$;
(3) $\rho(A) \supset \mathbb{C} \sim \mathbb{R}$.

Proof. It is trivial that (1) implies (2) and (3) implies (1). To see that (2) implies (3), first note that for any $r \neq 0$, we have $\left\|(r\mathrm{i}-A)^{-1}\right\| \le |r|^{-1}$, and therefore by the lemma above, $r\mathrm{i} \in \rho(A)$ implies that $r\mathrm{i} + z \in \rho(A)$ for any z such that $|z| < |r|$. Now, let $z' \subset \mathbb{C} \sim \mathbb{R}$. We may assume that z' belongs to the upper half of the complex plane. We can find sufficiently large r such that $z' \in S(r\mathrm{i}, r')$ for some $r' < r$. Then, $|r\mathrm{i} - z'| < r$. Therefore, $z' = r\mathrm{i} + (z' - r\mathrm{i}) \in \rho(A)$. □

7.6 The Spectral Theorem

We prove the Spectral Theorem for unbounded self-adjoint operators in this section. We will use the Cayley transformation method. See, for instance, Riesz and Sz.-Nagy [31], for the classical proof.

Lemma 7.55. *If A is self-adjoint, then the Cayley transformation of A,*

$$U = (A - \mathrm{i})(A + \mathrm{i})^{-1},$$

is a unitary operator on H. Moreover, if A is strongly self-adjoint, then U is multiplicable. Besides, $1 - U$ is injective, and $D(A) = R(1 - U)$, and

$$A = \mathrm{i}(1 + U)(1 - U)^{-1}.$$

Proof. It is easy to see that $U^{-1} = (A + \mathrm{i})(A - \mathrm{i})^{-1}$. To see that U is unitary, we need to show that for any x, y,

$$\left((A - \mathrm{i})(A + \mathrm{i})^{-1} x, y\right) = \left(x, (A + \mathrm{i})(A - \mathrm{i})^{-1} y\right).$$

First, there exist $x', y' \in D(A)$ such that $x = (A + \mathrm{i})x'$, and $y = (A - \mathrm{i})y'$. Then, the equation is reduced to

$$((A - \mathrm{i})x', (A - \mathrm{i})y') = ((A + \mathrm{i})x', (A + \mathrm{i})y').$$

This is obvious, because by the self-adjointness of A, $(Ax', -\mathrm{i}y') = (\mathrm{i}x', Ay')$ and $(-\mathrm{i}x', Ay') = (Ax', \mathrm{i}y')$. Therefore, U is unitary. Moreover, note that

$$U = (A + \mathrm{i} - 2\mathrm{i})(A + \mathrm{i})^{-1} = 1 - 2\mathrm{i}(A + \mathrm{i})^{-1},$$
$$U^{-1} = (A - \mathrm{i} + 2\mathrm{i})(A - \mathrm{i})^{-1} = 1 + 2\mathrm{i}(A - \mathrm{i})^{-1}.$$

Therefore, U is multiplicable if $(A + \mathrm{i})^{-1}$ and $(A - \mathrm{i})^{-1}$ are, that is, if A is strongly self-adjoint.

For any x, let $x' \in D(A)$ be such that $x = (A + \mathrm{i})x'$. Then, $(1 - U)x = 2\mathrm{i}x'$. Therefore, $(1 - U)x = 0$ implies that $x' = 0$, and that in turn implies that $x = 0$. That is, $(1 - U)$ is injective. It also follows that $R(1 - U) = D(A)$.

For any $x' \in D(A)$, let $x = (A + \mathrm{i})x'$. Then, $(A + \mathrm{i})^{-1} x = x'$ and $(1 - U)^{-1} x' = \frac{1}{2\mathrm{i}} x$. By the definition of U, $Ux = (A - \mathrm{i})x'$. Therefore,

$$\begin{aligned} \mathrm{i}(1 + U)(1 - U)^{-1} x' &= \mathrm{i}(1 + U)\frac{1}{2\mathrm{i}} x = \frac{1}{2}(x + Ux) \\ &= \frac{1}{2}\left((A + \mathrm{i})x' + (A - \mathrm{i})x'\right) = Ax'. \end{aligned}$$

So, $A = \mathrm{i}(1 + U)(1 - U)^{-1}$. □

Lemma 7.56. *If U is the Cayley transformation of a strongly self-adjoint operator A, then $1 \in \rho_s(U)$.*

Proof. Since $\rho_s(U)$ is dense in S^1, we can choose $w_n, u_n \in \rho_s(U)$ such that $w_n \to 1$, $u_n \to 1$, and $1 \in [w_n, u_n] \subset (w_{n-1}, u_{n-1})$. Take $f_n \in C(S^1)$, $f_n = 1$ on $[w_n, u_n]$, $f_n = 0$ on $[u_{n-1}, w_{n-1}]$, and $0 \le f_n \le 1$ on S^1. We will prove that $f_n(U) \to 0$ strongly.

If $x \in R(1-U)$, then $x = (1-U)y$ for some y. $f_n(U)x = f_n(U)(1-U)y$. It is obvious that $f_n(z)(1-z) \to 0$ uniformly on S^1. So, $f_n(U)(1-U) \to 0$ strongly. Hence $f_n(U)x \to 0$. Since $R(1-U) = D(A)$ is dense in H and $f_n(U)$ is uniformly bounded, we must have $f_n(U) \to 0$ strongly. □

In the following, we assume that A is strongly self-adjoint. $\varphi \mapsto \mathrm{e}^{\mathrm{i}\varphi}$ is a continuous and invertible mapping from $(0, 2\pi)$ onto $S^1 - \{1\}$, and $\lambda \mapsto 2\operatorname{arccot}(-\lambda)$ is a continuous and invertible mapping from $(-\infty, +\infty)$ onto $(0, 2\pi)$, so $\lambda \mapsto \tau(\lambda) = \mathrm{e}^{2\operatorname{arccot}(-\lambda)\mathrm{i}}$ is a continuous and invertible mapping from $(-\infty, +\infty)$ onto $S^1 - \{1\}$, such that when λ increases, $\tau(\lambda)$ moves along the positive direction on $S^1 - \{1\}$. Let

$$\rho_s(A) \equiv_{df} \{\lambda \in \mathbb{R} : \tau(\lambda) \in \rho_s(U)\}.$$

We also call points in $\rho_s(A)$ the smooth points of A. Note that $\rho_s(A)$ is dense in $\mathbb{R}$. Then, we can construct the spectral family on $\rho_s(A)$ corresponding to the spectral family on $\rho_s(U)$:

$$E_\lambda \equiv_{df} E_{[1,\tau(\lambda)]}, \text{ for } \lambda \in \rho_s(A).$$

By Lemma 7.56 and the properties of $E_{[z_1,z_2]}$, we can easily verify that E_λ, $\lambda \in \rho_s(A)$, have the common properties for a spectral family:

(i) $E_\lambda \to 0$ as $\lambda \to -\infty$, $E_\lambda \to I$ as $\lambda \to +\infty$;
(ii) $E_{\lambda'} \to E_\lambda$ as $\lambda' \to \lambda$;
(iii) $E_{\lambda'} E_\lambda = E_\lambda E_{\lambda'} = E_\lambda$ when $\lambda < \lambda'$.

Then, for g a continuous function on $\mathbb{R}$, similar to the definition of $\int_{S^1} g(z)\,\mathrm{d}E_z$, we can define $\int_{\lambda_1}^{\lambda_2} g(\lambda)\,\mathrm{d}E_\lambda$ for $\lambda_1 < \lambda_2$, as a strong limit.

We first construct the decomposition of A restricted to $R(E_{\lambda_2} - E_{\lambda_1})$.

Lemma 7.57. *For $\lambda_1, \lambda_2 \in \rho_s(A)$, $\lambda_1 < \lambda_2$,*

(a) $R(E_{\lambda_2} - E_{\lambda_1}) \subset D(A)$, $A(E_{\lambda_2} - E_{\lambda_1}) \supset (E_{\lambda_2} - E_{\lambda_1})A$;
(b) $A(E_{\lambda_2} - E_{\lambda_1})$ is bounded and self-adjoint, and $D(A(E_{\lambda_2} - E_{\lambda_1})) = H$,

$$A(E_{\lambda_2} - E_{\lambda_1}) = \int_{\lambda_1}^{\lambda_2} \lambda\,\mathrm{d}E_\lambda,$$

where the integration is understood as a strong limit.

Proof. Consider $z_1, z_2 \in \rho_s(U)$, $[z_1, z_2] \subset S^1 - \{1\}$. Let

$$f_n(z) \equiv \frac{\chi^n_{[z_1,z_2]}}{(1-z)(1-z^*)} \in C(S^1).$$

We have

$$(1-U)(1-U^*)f_n(U) = (1-U^*)f_n(U)(1-U) = \chi^n_{[z_1,z_2]}(U).$$

Note that $|f_m(z) - f_n(z)| \le c\left|\chi^m_{[z_1,z_2]}(z) - \chi^n_{[z_1,z_2]}(z)\right|$ for some constant $c > 0$. Since z_1, z_2 are smooth points, by Lemma 7.37, $f_n(U)$ strongly converges to some bounded operator B on H. Therefore,

$$(1-U)(1-U^*)B = (1-U^*)B(1-U) = E_{[z_1,z_2]}.$$

So, $R\left(E_{[z_1,z_2]}\right) \subset R(1-U) = D(A)$, and $(1-U)^{-1}E_{[z_1,z_2]} \supset E_{[z_1,z_2]}(1-U)^{-1}$. By Lemma 7.55, $A = \mathrm{i}(1+U)(1-U)^{-1}$. Since $UE_{[z_1,z_2]} = E_{[z_1,z_2]}U$, we have $AE_{[z_1,z_2]} \supset E_{[z_1,z_2]}A$. (a) follows when $\tau(\lambda_1) = z_1$, $\tau(\lambda_2) = z_2$.

Take

$$g_n(z) \equiv \mathrm{i}\frac{1+z}{1-z}\chi^n_{[z_1,z_2]}(z) \in C\left(S^1\right).$$

Again, $g_n(U)$ strongly converges to some bounded, self-adjoint operator C defined on the whole space H, and $(1-U)C = \mathrm{i}(1+U)E_{[z_1,z_2]}$. On the other hand,

$$(1-U)AE_{[z_1,z_2]} = (1-U)\mathrm{i}(1+U)(1-U)^{-1}E_{[z_1,z_2]} = \mathrm{i}(1+U)E_{[z_1,z_2]}.$$

Since $(1-U)$ is injective, we have $AE_{[z_1,z_2]} = C$. So, $AE_{[z_1,z_2]}$ is bounded, self-adjoint, and defined on the whole space H. Following the proof of Theorem 7.40, it can be seen that $\int_{[z_1,z_2]}\mathrm{i}\frac{1+z}{1-z}\mathrm{d}E_z$ exists as a strong limit and $C = \int_{[z_1,z_2]}\mathrm{i}\frac{1+z}{1-z}dE_z$. By the familiar trigonometrical equations, $\mathrm{i}\frac{1+\tau(\lambda)}{1-\tau(\lambda)} = \lambda$. Hence (b) follows when $\tau(\lambda_1) = z_1$, $\tau(\lambda_2) = z_2$. □

To generalize the decomposition we must define $\int_{-\infty}^{+\infty}\lambda\mathrm{d}E_\lambda$.

Definition 7.58. For $g \in C(\mathbb{R},\mathbb{C})$, the set of continuous functions from $\mathbb{R}$ to $\mathbb{C}$, we define

$$D\left(\int_{-\infty}^{+\infty} g(\lambda)\mathrm{d}E_\lambda\right) \equiv \left\{x : \lim_{\substack{\lambda_1\to-\infty\\ \lambda_2\to+\infty}} \int_{\lambda_1}^{\lambda_2} g(\lambda)\mathrm{d}E_\lambda x \text{ exists}\right\}.$$

The limit is considered for smooth points λ_1, λ_2. For $x \in D\left(\int_{-\infty}^{+\infty} g(\lambda)\mathrm{d}E_\lambda\right)$, we define $\int_{-\infty}^{+\infty} g(\lambda)\mathrm{d}E_\lambda x$ as the limit.

Note that if $\lambda_1 < \lambda_2 \le \lambda_3 < \lambda_4$ are smooth points, then

$$(E_{\lambda_2}-E_{\lambda_1})x \perp (E_{\lambda_4}-E_{\lambda_3})x, \quad (E_{\lambda_2}-E_{\lambda_1})x \perp E_{\lambda_1}x$$

for any x, and hence

$$\begin{aligned}\left\|(E_{\lambda_2}-E_{\lambda_1})x + (E_{\lambda_4}-E_{\lambda_3})x\right\|^2 &= \left\|(E_{\lambda_2}-E_{\lambda_1})x\right\|^2 + \left\|(E_{\lambda_4}-E_{\lambda_3})x\right\|^2,\\ \left\|(E_{\lambda_2}-E_{\lambda_1})x\right\|^2 &= \left\|E_{\lambda_2}x\right\|^2 - \left\|E_{\lambda_1}x\right\|^2.\end{aligned}$$

Therefore, we have

Corollary 7.59. *(a) $x \in D\left(\int_{-\infty}^{+\infty} g(\lambda)\mathrm{d}E_\lambda\right)$ iff $\int_{-\infty}^{+\infty}|g(\lambda)|^2\mathrm{d}\|E_\lambda x\|^2$ exists, and in that case,*

$$\left\|\int_{-\infty}^{+\infty} g(\lambda)\mathrm{d}E_\lambda x\right\|^2 = \int_{-\infty}^{+\infty}|g(\lambda)|^2\mathrm{d}\|E_\lambda x\|^2,$$

where the right hand side of the equation is understood as the limit of a Riemann-Stieltjes integration, $\lim_{\substack{\lambda_1\to-\infty\\ \lambda_2\to+\infty}} \int_{\lambda_1}^{\lambda_2} |g(\lambda)|^2 \mathrm{d}\|E_\lambda x\|^2$.

(b) If g is bounded and $\|g\| < M$, *then* $D\left(\int_{-\infty}^{+\infty} g(\lambda)\,\mathrm{d}E_\lambda\right) = H$ *and* $\left\|\int_{-\infty}^{+\infty} g(\lambda)\,\mathrm{d}E_\lambda\right\| \leq M$, *and if* $h \in C(\mathbb{R},\mathbb{C})$ *is also bounded, then*

$$\left(\int_{-\infty}^{+\infty} g(\lambda)\,\mathrm{d}E_\lambda\right)\left(\int_{-\infty}^{+\infty} h(\lambda)\,\mathrm{d}E_\lambda\right) = \int_{-\infty}^{+\infty} g(\lambda)h(\lambda)\,\mathrm{d}E_\lambda.$$

Then we have the Spectral Theorem.

Theorem 7.60. *Suppose that A is strongly self-adjoin and* E_λ *is the spectral family on* $\rho_s(A)$ *constructed above. Then,*

$$A = \int_{-\infty}^{+\infty} \lambda\,\mathrm{d}E_\lambda,$$

and for any $B \in Hom(H)$, $BA \subset AB$ *if and only if* $BE_\lambda = E_\lambda B$ *for all* E_λ.

Proof. Choose $\lambda_m \in \rho_s(A)$, $m = 0, \pm1, \pm2, \ldots$, such that $\lambda_m < \lambda_n$ for $m < n$, and $\lambda_m \to +\infty$ as $m \to +\infty$, and $\lambda_m \to -\infty$ as $m \to -\infty$. For any x, let $x_m = \left(E_{\lambda_m} - E_{\lambda_{-m}}\right)x$, $m = 1, 2, \ldots$. Then $x_m \to x$. By Lemma 7.57,

$$Ax_m = A\left(E_{\lambda_m} - E_{\lambda_{-m}}\right)x = \int_{\lambda_{-m}}^{\lambda_m} \lambda\,\mathrm{d}E_\lambda x.$$

Now, if $x \in D\left(\int_{-\infty}^{+\infty} \lambda\,\mathrm{d}E_\lambda x\right)$, $\lim_{m\to\infty} Ax_m$ exists. Since A is closed, $x \in D(A)$ and

$$Ax = \lim_{m\to\infty} \int_{\lambda_{-m}}^{\lambda_m} \lambda\,\mathrm{d}E_\lambda x = \int_{-\infty}^{+\infty} \lambda\,\mathrm{d}E_\lambda x.$$

Thus $A \supset \int_{-\infty}^{+\infty} \lambda\,\mathrm{d}E_\lambda x$. On the other hand, if $x \in D(A)$, by Lemma 7.57,

$$\int_{\lambda_{-m}}^{\lambda_m} \lambda\,\mathrm{d}E_\lambda x = A\left(E_{\lambda_m} - E_{\lambda_{-m}}\right)x = \left(E_{\lambda_m} - E_{\lambda_{-m}}\right)Ax.$$

So, $\lim_{m\to\infty} \int_{\lambda_{-m}}^{\lambda_m} \lambda\,\mathrm{d}E_\lambda x$ exists. That means $x \in D\left(\int_{-\infty}^{+\infty} \lambda\,\mathrm{d}E_\lambda x\right)$. So, $A = \int_{-\infty}^{+\infty} \lambda\,\mathrm{d}E_\lambda$.

Now suppose that $BA \subset AB$. For each x there exists y such that $x - (A + \mathrm{i})y$. Let U be the Cayley transformation of A. Then, $Ux = (A - \mathrm{i})y$ and $BUx = (A - \mathrm{i})By = UBx$. So $UB = BU$. By Theorem 7.40, $BE_{[1,z]} = E_{[1,z]}B$ for all $z \in \rho_s(U)$. That is, $BE_\lambda = E_\lambda B$. Conversely, suppose that $BE_\lambda = E_\lambda B$ for all E_λ. We can also infer by Theorem 7.40 that $BU = UB$. Then, for $x \in D(A)$, $x = (1 - U)y$ for some y and $Ax = \mathrm{i}(1 + U)y$. Then, $ABx = \mathrm{i}(1 + U)By = BAx$. That is, $BA \subset AB$. □

As a corollary we have the decompositions of bounded operators.

Corollary 7.61. *If A is strongly self-adjoint,* E_λ *is the spectral family for A, and* $\|A\| < M$, *then* $(-\infty, -M) \cup (M, +\infty) \subset \rho_s(A)$, $E_\lambda = 0$ *for* $\lambda < -M$, $E_\lambda = I$ *for* $\lambda > M$, *and* $A = \int_{-M''}^{M'} \lambda\,\mathrm{d}E_\lambda$ *for any* $M', M'' > M$.

Proof. We need only to prove that $E_{\lambda_2} - E_{\lambda_1} = 0$ for $M < \lambda_1 < \lambda_2$ or $\lambda_1 < \lambda_2 < -M$. Consider the first case. Let $x \in R\left(E_{\lambda_2} - E_{\lambda_1}\right)$. Then, $x \in D(A)$, $Ax = \int_{\lambda_1}^{\lambda_2} \lambda \mathrm{d}E_\lambda x$, and

$$\|Ax\|^2 = \int_{\lambda_1}^{\lambda_2} \lambda^2 \mathrm{d}\,\|E_\lambda x\|^2 \geq \lambda_1^2 \|x\|^2 .$$

So, $M^2 \|x\|^2 \geq \lambda_1^2 \|x\|^2$. Since $M < \lambda_1$, we must have $x = 0$. That is, $E_{\lambda_2} - E_{\lambda_1} = 0$. □

As an application of the Spectral theorem, we discuss the spectrum of an operator. As in the classical case, we can characterize $\rho(A)$ by the spectral family E_λ, $\lambda \in \rho_s(A)$ (cf. Weidmann [37], p. 200, Theorem 7.22).

Lemma 7.62. *Suppose that A is strongly self-adjoint and E_λ is the spectral family of A. Then, for each $\lambda \in \mathbb{R}$, $\lambda \in \rho(A)$ if and only if there exist $\lambda_1, \lambda_2 \in \rho_s(A)$ such that $\lambda_1 < \lambda < \lambda_2$ and $E_{\lambda_1} = E_{\lambda_2}$.*

Proof. First, suppose that $\lambda_0 \in \rho(A)$ and $\left\|(\lambda_0 - A)^{-1}\right\| \leq M$. Then, $\|(\lambda_0 - A)x\| \geq \frac{1}{M}\|x\|$ for any $x \in D(A)$. Choose $\lambda_1, \lambda_2 \in \rho_s(A)$ such that $\lambda_1 < \lambda_0 < \lambda_2$ and $\lambda_2 - \lambda_1 < \frac{1}{2M}$. We need to prove $R\left(E_{\lambda_2} - E_{\lambda_1}\right) = \{0\}$. Let $x \in R\left(E_{\lambda_2} - E_{\lambda_1}\right)$. Then, $x \in D(A)$ and

$$\|(\lambda_0 - A)x\|^2 = \int_{\lambda_1}^{\lambda_2} (\lambda_0 - \lambda)^2 \mathrm{d}\,\|E_\lambda x\|^2 \leq \left(\frac{1}{2M}\right)^2 \|x\|^2 .$$

So we have $\frac{1}{M}\|x\| \leq \frac{1}{2M}\|x\|$. Hence $x = 0$. Conversely, suppose that $\lambda_1, \lambda_2 \in \rho_s(A)$, $\lambda_1 < \lambda_0 < \lambda_2$, and $E_{\lambda_2} = E_{\lambda_1}$. Let $\varepsilon = \min\{\lambda_2 - \lambda_0, \lambda_0 - \lambda_1\}$ and $z = \lambda_0 + \varepsilon i$. By Lemma 7.54, $z \in \rho(A)$. For $\lambda \leq \lambda_1$ or $\lambda \geq \lambda_2$, $|z - \lambda|^2 \geq 2\varepsilon^2$. We have

$$\|(z - A)x\|^2 = \int_{-\infty}^{+\infty} |z - \lambda|^2 \mathrm{d}\,\|E_\lambda x\|^2 \geq 2\varepsilon^2 \|x\|^2 ,$$

since $E_{\lambda_2} = E_{\lambda_1}$. So, $\left\|(z - A)^{-1}\right\| \leq \frac{1}{\sqrt{2}\varepsilon}$. Now, $|\varepsilon i| < \sqrt{2}\varepsilon$. By Lemma 7.53, we have $\lambda_0 = z - \varepsilon i \in \rho(A)$. □

Lemma 7.63. *Suppose that A is strongly self-adjoint and E_λ is the spectral family of A, and suppose that $\lambda_1, \lambda_2 \in \rho_s(A)$, $\lambda_1 < \lambda_2$, and $[\lambda_1, \lambda_2] \subseteq \rho(A)$, and suppose that for some M, $\left\|(\lambda - A)^{-1}\right\| \leq M$ for all $\lambda \in [\lambda_1, \lambda_2]$. Then, $E_{\lambda_1} = E_{\lambda_2}$.*

Proof. From the first half of the proof of the last lemma, we see that we can cover the interval $[\lambda_1, \lambda_2]$ by finitely many small intervals $\left[\lambda_i', \lambda_{i+1}'\right]$, $i = 0, \ldots, n-1$, such that $\frac{1}{4M} < \lambda_{i+1}' - \lambda_i' < \frac{1}{2M}$ and $E_{\lambda_i'} = E_{\lambda_{i+1}'}$ for $i = 0, \ldots, n-1$. Note that the assumption that $\left\|(\lambda - A)^{-1}\right\|$ has a uniform bound on $[\lambda_1, \lambda_2]$ is critical here, although it is redundant in the classical theory. Then the conclusion follows. □

As an example, we compute the spectrum of a self-adjoint operator on a finite-dimensional space H. Let A be a self-adjoint operator on H. Suppose that H is n-dimensional and choose a non-zero orthonormal basis $\{e_1,...,e_n\}$ of H. A is represented by an $n\times n$ matrix on this basis, also denoted by A. Let $|B|$ denote the determinant of an arbitrary matrix B. Then, as in the classical case, we have

Theorem 7.64. *(i) $\lambda\in\rho(A)$ if and only if $|\lambda I-A|\neq 0$; (ii) $\lambda\in\sigma(A)$ if and only if $|\lambda I-A|=0$.*

Proof. First, by a pure algebraic computation, we see that when $|\lambda I-A|\neq 0$, the matrix $(\lambda I-A)$ has an inverse, and hence the operator $(\lambda I-A)$ has a bounded inverse defined on the whole space H, that is, $\lambda\in\rho(A)$. On the other hand, when $\lambda\in\rho(A)$, the operator $(\lambda I-A)^{-1}$ is represented by a matrix B. It is easy to verify that the products of operators are represented by the products of corresponding matrices. So we have $(\lambda I-A)B=I$. Finally, $|(\lambda I-A)B|=|(\lambda I-A)||B|$ can be proved by pure algebraic computations. So we must have $|(\lambda I-A)|\neq 0$. Hence we have proved that $\lambda\in\rho(A)$ iff $|(\lambda I-A)|\neq 0$. The second conclusion follows from the first and the continuity of $|(\lambda I-A)|$ as a function of λ. □

Since $\sigma(A)\subset\mathbb{R}$, the zeros of the equation $|(\lambda I-A)|=0$ are all real. By the Fundamental Theorem of Algebra, we can find n real numbers $\lambda_1,...,\lambda_n$ such that

$$|(\lambda I-A)|=(\lambda-\lambda_1)\cdots(\lambda-\lambda_n).$$

Then, we have another characterization of $\sigma(A)$:

Theorem 7.65. *$\lambda\in\sigma(A)$ if and only if for any $\varepsilon>0$, there exists a λ_i such that $|\lambda-\lambda_i|<\varepsilon$; or equivalently,*

$$\sigma(A)=\cap_{k=1}^{\infty}\cup_{i=1}^{n}\left[\lambda_i-\frac{1}{k},\,\lambda_i+\frac{1}{k}\right].$$

Proof. The sufficiency is clear by the continuity of $|(\lambda I-A)|$. Conversely, when $\lambda\in\sigma(A)$, for each i, we can decide if $|\lambda-\lambda_i|<\varepsilon$ or $>\frac{\varepsilon}{2}$. If $|\lambda-\lambda_i|>\frac{\varepsilon}{2}$ for all $i=1,...,n$, we would have $|(\lambda I-A)|\neq 0$, a contradiction. So we can find one λ_i with $|\lambda-\lambda_i|<\varepsilon$. □

Certainly $\{\lambda_1,...,\lambda_n\}\subset\sigma(A)$. However, we cannot claim that $\sigma(A)=\{\lambda_1,...,\lambda_n\}$. Similarly, we cannot generally claim that the eigen-space of the eigen-value λ_i exists. Indeed, for any $\varepsilon>0$, $i=1,...,n$, we can find $\lambda,\lambda'\in\rho(A)$, $\lambda<\lambda_i<\lambda'$, $\lambda'-\lambda<\varepsilon$, and an orthonormal basis for the subspace $R(E_{\lambda'}-E_\lambda)$, which is the eigen-space of the eigen-values in (λ,λ') in the classical sense. But we cannot guarantee that $E_{\lambda'}-E_\lambda$ converges as $\varepsilon\to 0$. Note that $N(\lambda_i-A)$ is a linear subset of H, but we may not be able to find a basis for it, for it may not be located and hence may fail to be a subspace. The following theorem shows that this is due to our inability to decide $\lambda_i=\lambda_j\vee\lambda_i\neq\lambda_j$, $i,j=1,...,n$, because in case the λ_i's appeared in the eigen-equation

$$|(\lambda I-A)|=(\lambda-\lambda_1)\cdots(\lambda-\lambda_n)=0$$

are mutually distinguishable, everything will be the same as the classical case.

Theorem 7.66. *If A is a self-adjoint operator on a finite-dimensional space H, and*

$$|\lambda I - A| = (\lambda - \lambda_1)\cdots(\lambda - \lambda_n),$$

then

(a) *if for each $j = 1,...,n$, either $\lambda_i = \lambda_j$ or $\lambda_i \neq \lambda_j$, then $N(\lambda_i - A) = R\big(E_{\lambda_i+\varepsilon} - E_{\lambda_i-\varepsilon}\big)$ for some $\varepsilon > 0$, and $N(\lambda_i - A)$ is a finite-dimensional subspace of H with a dimension > 0;*

(b) *if for all $i, j = 1,...,n$, either $\lambda_i = \lambda_j$ or $\lambda_i \neq \lambda_j$, then for any λ, $\lambda \in \sigma(A)$ iff $\lambda = \lambda_i$ for some $i = 1,...,n$, and if $\lambda_{i_1},...,\lambda_{i_m}$ are the mutually distinct representatives from $\lambda_1,...,\lambda_n$, then*

$$H = N(\lambda_{i_1} - A) \oplus \cdots \oplus N(\lambda_{i_m} - A),$$
$$A = \lambda_{i_i} P_1 + \cdots \lambda_{i_m} P_m,$$

where P_j is the projection onto $N\left(\lambda_{i_j} - A\right)$.

Proof. (a) If the condition holds, we can find $\varepsilon > 0$ such that $\lambda_j = \lambda_i$ or $\left|\lambda_j - \lambda_i\right| > 2\varepsilon$ for all j. We first show that $N(\lambda_i - A) = R\left(E_{\lambda_i+\varepsilon} - E_{\lambda_i-\varepsilon}\right)$. First note that

$$\begin{aligned}
\|(\lambda_i - A)x\|^2 &= \int_{-\infty}^{+\infty} (\lambda_i - \lambda)^2 \,\mathrm{d}\|E_\lambda x\|^2 \\
&\geq \varepsilon^2 \left(\int_{\lambda_i+\varepsilon}^{+\infty} + \int_{-\infty}^{\lambda_i-\varepsilon}\right) \mathrm{d}\|E_\lambda x\|^2 \\
&= \varepsilon^2 \left(\left\|(I - E_{\lambda_i+\varepsilon})x\right\|^2 + \left\|E_{\lambda_i-\varepsilon} x\right\|^2\right).
\end{aligned}$$

So, if $x \in N(\lambda_i - A)$, then $\left(I - E_{\lambda_i+\varepsilon}\right)x = 0$ and $E_{\lambda_i-\varepsilon}x = 0$, that is, $x \in R\big(E_{\lambda_i+\varepsilon} - E_{\lambda_i-\varepsilon}\big)$. Now, suppose that $x \in R\left(E_{\lambda_i+\varepsilon} - E_{\lambda_i-\varepsilon}\right)$. Consider any ε' such that $0 < \varepsilon' < \varepsilon$. From the assumptions, we can find a constant $\delta > 0$ such that $||\lambda I - A|| > \delta$ for all $\lambda \in [\lambda_i + \varepsilon', \lambda_i + \varepsilon]$. Now, $(\lambda I - A)^{-1} = ||\lambda I - A||^{-1} B$ for some matrix B whose elements are polynomials of λ. It is easy to see that there is a constant M such that $\left\|(\lambda I - A)^{-1}\right\| < M$ for all $\lambda \in [\lambda_i + \varepsilon', \lambda_i + \varepsilon]$. By Lemma 7.63, $E_{\lambda_i+\varepsilon'} = E_{\lambda_i+\varepsilon}$. Similarly, $E_{\lambda_i-\varepsilon'} = E_{\lambda_i-\varepsilon}$. Therefore, from $x \in R\left(E_{\lambda_i+\varepsilon} - E_{\lambda_i-\varepsilon}\right)$ we have $x \in R\left(E_{\lambda_i+\varepsilon'} - E_{\lambda_i-\varepsilon'}\right)$. Then,

$$\|(\lambda_i - A)x\|^2 = \int_{\lambda_i-\varepsilon'}^{\lambda_i+\varepsilon'} (\lambda_i - \lambda)^2 \,\mathrm{d}\|E_\lambda x\|^2 \leq \varepsilon'^2 \|x\|^2.$$

Since ε' can be arbitrarily small, we have $x \in N(\lambda_i - A)$. That is, $N(\lambda_i - A) = R\left(E_{\lambda_i+\varepsilon} - E_{\lambda_i-\varepsilon}\right)$.

This means that $N(\lambda_i - A)$ is a subspace of H and consequently it is also finite-dimensional (Corollary 7.22). If the dimension of $N(\lambda_i - A)$ is 0, we would have $E_{\lambda_i+\varepsilon} = E_{\lambda_i-\varepsilon}$, and then $\lambda_i \in \rho(A)$ by Lemma 7.62, contradicting Theorem 7.64. So $N(\lambda_i - A)$ has a positive dimension.

(b) If the condition holds, we can find $\varepsilon_0 > 0$ such that $\lambda_j = \lambda_i$ or $|\lambda_j - \lambda_i| > \varepsilon_0$ for all i, j. Now, suppose that $\lambda \in \sigma(A)$. Find, by Theorem 7.65, an i such that $|\lambda - \lambda_i| < \frac{\varepsilon_0}{2}$. Then, for any positive $\varepsilon < \frac{\varepsilon_0}{2}$, by Theorem 7.65, there exists a λ_j such that $|\lambda - \lambda_j| < \varepsilon$. We must have $\lambda_j = \lambda_i$. Since ε can be arbitrarily small, $\lambda = \lambda_i$. So $\sigma(A) = \{\lambda_1, ..., \lambda_n\}$. The rest follows from (a) and the Spectral Theorem. □

The representation in (b) above is exactly the same as the classical case. It seems that indistinguishable λ_is arise only in artificial constructions. In other words, we expect that the operators on finite-dimensional spaces appearing in natural physics contexts all satisfy the conditions in Theorem 7.66(b). Because, in quantum mechanics, the spectrum of an operator A is supposed to consist of the possible values of the observable corresponding to A. Any realistic observation can be performed only up to a finite precision. So, in a finite dimensional case, we can expect that the eigen-values are all mutually distinguishable.

7.7 Stone's Theorem

We follow the classical proof in Weidmann [37], pp. 220–223. Let $\{U(t) : t \in \mathbb{R}\}$ be a family of unitary operators on H. We say that the family is a (one-parameter) unitary group, if $U(0) = I$ and $U(s)U(t) = U(s+t)$ for $s, t \in \mathbb{R}$. The group is strongly continuous, if for each $x \in H$, $t \mapsto U(t)x$ is a continuous function from $\mathbb{R}$ into H. Note that $U(t)^n = U(nt)$. Therefore, unitary operators in a unitary group are multiplicable. It is obvious that $\{U(t) : t \in \mathbb{R}\}$ is strongly continuous, if for each x, $t \mapsto U(t)x$ is continuous at $t = 0$, that is, for any $\varepsilon > 0$, there exists $\delta > 0$ such that $\|(U(t) - I)x\| < \varepsilon$ whenever $|t| < \delta$. Furthermore, in that case, $t \mapsto U(t)x$ is uniformly continuous on $\mathbb{R}$. Note that

$$\|(U(t) - I)x\|^2 = \langle (I - U(t))x, x\rangle + \langle (I - U(-t))x, x\rangle.$$

So, $\{U(t) : t \in \mathbb{R}\}$ is strongly continuous if $t \mapsto \langle U(t)x, x\rangle$ is continuous at $t = 0$.

The infinitesimal generator

$$A \equiv \lim_{t \to 0} \frac{1}{t}(U(t) - I)$$

of $\{U(t) : t \in \mathbb{R}\}$ is an operator defined as

$$D(A) \equiv_{df} \left\{ x \in H : \lim_{t \to 0} \frac{1}{t}(U(t) - I)x \text{ exists} \right\},$$

$$Ax \equiv_{df} \lim_{t \to 0} \left(\frac{1}{t}(U(t) - I)x \right) \text{ for } x \in D(A).$$

The following theorem says that every strongly self-adjoint operator is an infinitesimal generator of a strongly continuous unitary group.

Theorem 7.67. *Suppose that A is a strongly self-adjoint operator on H and E_λ, $\lambda \in \rho_s(A)$, is the spectral family of A. For $t \in \mathbb{R}$, define*

$$U(t) \equiv \mathrm{e}^{\mathrm{i}tA} \equiv_{df} \int_{-\infty}^{+\infty} \mathrm{e}^{\mathrm{i}t\lambda} \mathrm{d}E_\lambda.$$

Then, $\{U(t) : t \in \mathbb{R}\}$ is a strongly continuous unitary group with the infinitesimal generator iA, *and $U(t)x \in D(A)$ for $x \in D(A)$.*

Proof. By Corollary 7.59, $U(t)$ is bounded and defined on the whole space, and $\{U(t) : t \in \mathbb{R}\}$ is a group. It is also easy to verify that $U(t)$ is unitary.

We prove that $\{U(t) : t \in \mathbb{R}\}$ is strongly continuous. Let $x \in H$. For any $\varepsilon > 0$, first choose $\lambda_1, \lambda_2 \in \rho_s(A)$, $\lambda_1 < \lambda_2$, such that $\left\|(I - E_{\lambda_2})x\right\|^2 + \left\|E_{\lambda_1}x\right\|^2 < \frac{\varepsilon^2}{8}$, and then choose $\delta > 0$ such that $\left|\mathrm{e}^{\mathrm{i}t\lambda} - 1\right|^2 < \frac{\varepsilon^2}{2(\|x\|^2+1)}$ whenever $|t| < \delta$ and $\lambda_1 \le \lambda \le \lambda_2$. Then, as $|t| < \delta$,

$$\begin{aligned}
&\|(U(t) - I)x\|^2 \\
&= \left(\int_{-\infty}^{\lambda_1} + \int_{\lambda_2}^{+\infty}\right) \left|\mathrm{e}^{\mathrm{i}t\lambda} - 1\right|^2 \mathrm{d}\|E_\lambda x\|^2 + \int_{\lambda_1}^{\lambda_2} \left|\mathrm{e}^{\mathrm{i}t\lambda} - 1\right|^2 \mathrm{d}\|E_\lambda x\|^2 \\
&\le 4\left(\left\|(I - E_{\lambda_2})x\right\|^2 + \left\|E_{\lambda_1}x\right\|^2\right) + \frac{\varepsilon^2}{2\left(\|x\|^2 + 1\right)} \left\|(E_{\lambda_2} - E_{\lambda_1})x\right\|^2 \\
&\le \varepsilon^2.
\end{aligned}$$

So, $\{U(t) : t \in \mathbb{R}\}$ is strongly continuous.

Now we prove that i$A = \lim_{t\to 0} \frac{1}{t}(U(t) - I)$. Let $x \in D(A)$. For any $\varepsilon > 0$, choose $\lambda_1, \lambda_2 \in \rho_s(A)$, $\lambda_1 < \lambda_2$, such that

$$\left(\int_{-\infty}^{\lambda_1} + \int_{\lambda_2}^{+\infty}\right) |\lambda|^2 \mathrm{d}\|E_\lambda x\|^2 < \frac{\varepsilon^2}{16}.$$

Note that

$$\left|\frac{1}{t}\left(\mathrm{e}^{\mathrm{i}t\lambda} - 1\right) - \mathrm{i}\lambda\right|^2 = \left|\left(\frac{\cos t\lambda - 1}{t\lambda} + \mathrm{i}\frac{\sin t\lambda - t\lambda}{t\lambda}\right)\lambda\right|^2.$$

So, we can choose $\delta > 0$ such that

$$\left|\frac{1}{t}\left(\mathrm{e}^{\mathrm{i}t\lambda} - 1\right) - \mathrm{i}\lambda\right|^2 < \frac{\varepsilon^2}{2\left(\|x\|^2 + 1\right)}$$

whenever $t < \delta$, $t \ne 0$, and $\lambda \in [\lambda_1, \lambda_2]$. Furthermore, $\left|\frac{1}{t}\left(\mathrm{e}^{\mathrm{i}t\lambda} - 1\right) - \mathrm{i}\lambda\right|^2 \le 8|\lambda|^2$ for all $\lambda \in \mathbb{R}$ and $t \ne 0$. Therefore, as $t \ne 0$ and $|t| < \delta$,

$$\begin{aligned}
&\left\|\frac{1}{t}(U(t)-I)x-\mathrm{i}Ax\right\|^2\\
&=\left(\int_{-\infty}^{\lambda_1}+\int_{\lambda_2}^{+\infty}\right)\left|\frac{\mathrm{e}^{\mathrm{i}t\lambda}-1}{t}-\mathrm{i}\lambda\right|^2\mathrm{d}\|E_\lambda x\|^2+\int_{\lambda_1}^{\lambda_2}\left|\frac{\mathrm{e}^{\mathrm{i}t\lambda}-1}{t}-\mathrm{i}\lambda\right|^2\mathrm{d}\|E_\lambda x\|^2\\
&\le 8\left(\int_{-\infty}^{\lambda_1}+\int_{\lambda_2}^{+\infty}\right)|\lambda|^2\mathrm{d}\|E_\lambda x\|^2+\frac{\varepsilon^2}{2\left(\|x\|^2+1\right)}\|(E_{\lambda_2}-E_{\lambda_1})x\|^2\\
&\le\varepsilon^2.
\end{aligned}$$

So, $\lim_{t\to 0}\frac{1}{t}(U(t)-I)x=\mathrm{i}Ax$. It remains to prove $D\left(\lim_{t\to 0}\frac{1}{t}(U(t)-I)\right)\subset D(A)$. Suppose that $\lim_{t\to 0}\frac{1}{t}(U(t)-I)x$ exists. Then, for any $\varepsilon>0$, there exists $\delta>0$, such that for any $\lambda_1,\lambda_2\in\rho_s(A)$, $\lambda_1<\lambda_2$, whenever $|t|<\delta$ and $|t'|<\delta$,

$$\begin{aligned}
&\int_{\lambda_1}^{\lambda_2}\left|\frac{1}{t}\left(\mathrm{e}^{\mathrm{i}t\lambda}-1\right)-\frac{1}{t'}\left(\mathrm{e}^{\mathrm{i}t'\lambda}-1\right)\right|^2\mathrm{d}\|E_\lambda x\|^2\\
&\le\int_{-\infty}^{+\infty}\left|\frac{1}{t}\left(\mathrm{e}^{\mathrm{i}t\lambda}-1\right)-\frac{1}{t'}\left(\mathrm{e}^{\mathrm{i}t'\lambda}-1\right)\right|^2\mathrm{d}\|E_\lambda x\|^2\\
&=\left\|\frac{1}{t}(U(t)-I)x-\frac{1}{t'}\left(U(t')-I\right)x\right\|^2<\frac{\varepsilon^2}{4}
\end{aligned}$$

Let $t'\to 0$. Since $\frac{1}{t'}\left(\mathrm{e}^{\mathrm{i}t'\lambda}-1\right)\to\mathrm{i}\lambda$ uniformly for $\lambda\in[\lambda_1,\lambda_2]$, we have

$$\int_{\lambda_1}^{\lambda_2}\left|\frac{1}{t}\left(\mathrm{e}^{\mathrm{i}t\lambda}-1\right)-\mathrm{i}\lambda\right|^2\mathrm{d}\|E_\lambda x\|^2\le\frac{\varepsilon^2}{4}.$$

Fix a t such that $|t|<\delta$. $\frac{1}{t}\left(\mathrm{e}^{\mathrm{i}t\lambda}-1\right)$ is bounded for $\lambda\in\mathbb{R}$. We can find $M>0$ such that whenever $\lambda_2>\lambda_1>M$ or $\lambda_1<\lambda_2<-M$,

$$\int_{\lambda_1}^{\lambda_2}\left|\frac{1}{t}\left(\mathrm{e}^{\mathrm{i}t\lambda}-1\right)\right|^2\mathrm{d}\|E_\lambda x\|^2\le\frac{\varepsilon^2}{4}.$$

So, for any $\varepsilon>0$, we have found $M>0$ such that whenever $\lambda_1,\lambda_2\in\rho_s(A)$ and $\lambda_2>\lambda_1>M$ or $\lambda_1<\lambda_2<-M$, we have

$$\begin{aligned}
&\int_{\lambda_1}^{\lambda_2}|\lambda|^2\mathrm{d}\|E_\lambda x\|^2\\
&\le 2\left(\int_{\lambda_1}^{\lambda_2}\left|\frac{1}{t}\left(\mathrm{e}^{\mathrm{i}t\lambda}-1\right)-\mathrm{i}\lambda\right|^2\mathrm{d}\|E_\lambda x\|^2+\int_{\lambda_1}^{\lambda_2}\left|\frac{1}{t}\left(\mathrm{e}^{\mathrm{i}t\lambda}-1\right)\right|^2\mathrm{d}\|E_\lambda x\|^2\right)\\
&\le\varepsilon^2.
\end{aligned}$$

That means $x\in D(A)$.

Finally, for any $\lambda_1, \lambda_2 \in \rho_s(A)$, $\lambda_1 < \lambda_2$,

$$\int_{\lambda_1}^{\lambda_2} |\lambda|^2 \mathrm{d}\|E_\lambda U(t)x\|^2 = \int_{\lambda_1}^{\lambda_2} |\lambda|^2 \mathrm{d}\|U(t)E_\lambda x\|^2 = \int_{\lambda_1}^{\lambda_2} |\lambda|^2 \mathrm{d}\|E_\lambda x\|^2.$$

So, $x \in D(A)$ if and only if $U(t)x \in D(A)$. □

Stone's Theorem asserts that any strongly continuous unitary group can be represented as in Theorem 7.67.

Theorem 7.68. *If $\{U(t) : t \in \mathbb{R}\}$ is a strongly continuous unitary group on H, then there exists a unique strongly self-adjoint operator A on H such that*

$$U(t) = \mathrm{e}^{\mathrm{i}tA} = \int_{-\infty}^{+\infty} \mathrm{e}^{\mathrm{i}t\lambda} \mathrm{d}E_\lambda,$$

where E_λ, $\lambda \in \rho_s(A)$, is the spectral family of A.

Proof. The uniqueness follows from Theorem 7.67. We prove the existence. Define

$$A \equiv -\mathrm{i}\lim_{t\to 0} \frac{1}{t}(U(t) - I).$$

First we want to prove that A is self-adjoint. For any real number $r > 0$, $x \in H$, define

$$T_r x = \int_0^\infty \mathrm{e}^{-rs} U(s)x\mathrm{d}s, \; S_r x = \int_0^\infty \mathrm{e}^{-rs} U(-s)x\mathrm{d}s,$$

where

$$\int_0^\infty \mathrm{e}^{-rs} U(\pm s)x\mathrm{d}s \equiv \lim_{M\to\infty} \int_0^M \mathrm{e}^{-rs} U(\pm s)x\mathrm{d}s,$$

and the integration on $[0, M]$ is understood as the limit of relevant partial Riemann sums. All the limits clearly exist because $\|U(s)x\|$ is bounded by $\|x\|$. T_r and S_r are linear operators defined on the whole space H and they are bounded by r^{-1}. Moreover, T_r and S_r are positively multiplicable. For instance,

$$(T_r)^n x = \int_0^\infty \ldots \int_0^\infty \mathrm{e}^{-r(s_1+\ldots+s_n)} U(s_1 + \ldots + s_n) x\mathrm{d}s_1 \ldots \mathrm{d}s_n.$$

Note that

$$\begin{aligned}
&\frac{1}{t}\int_0^\infty \mathrm{e}^{-rs}(U(s+t) - U(s))x\mathrm{d}s \\
&= \frac{1}{t}\int_t^\infty \mathrm{e}^{-r(s-t)} U(s)x\mathrm{d}s - \frac{1}{t}\int_0^\infty \mathrm{e}^{-rs} U(s)x\mathrm{d}s \\
&= \frac{\mathrm{e}^{rt}-1}{t}\int_t^\infty \mathrm{e}^{-rs} U(s)x\mathrm{d}s - \frac{1}{t}\int_0^t \mathrm{e}^{-rs} U(s)x\mathrm{d}s.
\end{aligned}$$

As $t \to 0$, we have

$$\frac{(\mathrm{e}^{rt}-1)}{t}\to r,\ \int_t^{\infty}\mathrm{e}^{-rs}U(s)x\mathrm{d}s\to T_rx,\ \frac{1}{t}\int_0^t\mathrm{e}^{-rs}U(s)x\mathrm{d}s\to x.$$

So, we have $AT_rx=-\mathrm{i}rT_rx+\mathrm{i}x$. Similarly, $AS_rx=\mathrm{i}rS_rx-\mathrm{i}x$. As a consequence we have $(A+\mathrm{i}r)T_rx=\mathrm{i}x$ and $(A-\mathrm{i}r)S_rx=-\mathrm{i}x$. So, $R(A\pm\mathrm{i}r)=H$.

Next, we prove that $D(A)$ is dense. Since $D(A)\supset\{T_rx:x\in H,r>0\}$, we only need to prove that $T_rrx\to x$ as $r\to\infty$ for every $x\in H$. As $\{U(t):t\in\mathbb{R}\}$ is strongly continuous, for any $\varepsilon>0$, there exists $\delta>0$ such that $\|(U(s)-I)x\|<\frac{\varepsilon}{2}$ when $|s|<\delta$. Note that $\int_0^{\infty}r\mathrm{e}^{-rs}\mathrm{d}s=1$. So,

$$\begin{aligned}\|T_rrx-x\|&=\left\|\int_0^{\infty}r\mathrm{e}^{-rs}(U(s)-I)x\mathrm{d}s\right\|\\&\le\frac{\varepsilon}{2}\int_0^{\delta}r\mathrm{e}^{-rs}\mathrm{d}s+2\|x\|\int_{\delta}^{\infty}r\mathrm{e}^{-rs}\mathrm{d}s\\&\le\frac{\varepsilon}{2}+2\|x\|\mathrm{e}^{-r\delta}.\end{aligned}$$

Clearly, $\|T_rrx-x\|\le\varepsilon$ when r is sufficiently large. So $T_rrx\to x$.

For $x,y\in D(A)$,

$$\left(-\frac{\mathrm{i}}{t}(U(t)-I)x,y\right)=\left(x,-\frac{\mathrm{i}}{-t}(U(-t)-I)y\right).$$

Let $t\to 0$ we have $(Ax,y)=(x,Ay)$. So $A\subset A^*$. Put these together, we conclude that A is self-adjoint.

To see that A is strongly self-adjoint, note that $(A+\mathrm{i}r)^{-1}x=-\mathrm{i}T_rx$. Therefore, $(A+\mathrm{i}r)^{-1}$ is positively multiplicable, since T_r is. Similarly, $(A-\mathrm{i}r)^{-1}$ is positively multiplicable.

Finally, it remains to prove that $U(t)=\mathrm{e}^{\mathrm{i}tA}$. Let $V(t)=\int_{-\infty}^{+\infty}\mathrm{e}^{\mathrm{i}t\lambda}\mathrm{d}E_\lambda$, where E_λ is the spectral family of A. Fix an $x\in D(A)$ and let

$$f(t)=\|U(t)x-V(t)x\|^2=2\|x\|^2-2\mathrm{Re}(U(t)x,V(t)x).$$

We want to prove that $f'(t)=0$ for $t\in\mathbb{R}$. We have

$$\begin{aligned}&\frac{f(s)-f(t)}{s-t}\\&=-2\mathrm{Re}\left(\left(U(t)\frac{U(s-t)-I}{s-t}x,V(s)x\right)+\left(U(t)x,V(t)\frac{V(s-t)-I}{s-t}x\right)\right).\end{aligned}$$

Let $s\to t$. By the definition of A, $\frac{U(s-t)-I}{s-t}x\to\mathrm{i}Ax$. Since $U(t)$ is uniformly bounded for $t\in\mathbb{R}$, we see that $U(t)\frac{U(s-t)-I}{s-t}x\to\mathrm{i}U(t)Ax$ uniformly for $t\in\mathbb{R}$. Similarly, by Theorem 7.67, $V(t)\frac{V(s-t)-I}{s-t}x\to\mathrm{i}V(t)Ax$ uniformly for $t\in\mathbb{R}$. Then, because of the continuity of inner product, we can see that

$$\frac{f(s)-f(t)}{s-t} \to -2\,\mathrm{Re}\left(\left(\mathrm{i}U(t)Ax, V(t)x\right)+\left(U(t)x, \mathrm{i}V(t)Ax\right)\right)$$

uniformly for $t \in \mathbb{R}$. By the definition of A and Theorem 7.67, we can easily verify that $U(t)A = AU(t)$ and $V(t)A = AV(t)$. Then, since A is self-adjoint,

$$\left(\mathrm{i}U(t)Ax, V(t)x\right)+\left(U(t)x, \mathrm{i}V(t)Ax\right) = 0.$$

So, we have $\frac{f(s)-f(t)}{s-t} \to 0$ uniformly for $t \in \mathbb{R}$. That is $f'(t) = 0$. Since $f(0) = 0$, we must have $f(t) = 0$ for all $t \in \mathbb{R}$. Therefore $U(t) = V(t) = \mathrm{e}^{\mathrm{i}tA}$ on $D(A)$. $D(A)$ is dense, so $U(t) = V(t) = \mathrm{e}^{\mathrm{i}tA}$. □

Chapter 8
Semi-Riemannian Geometry

This chapter develops the basics of differentiable manifolds and semi-Riemannian geometry for the applications in general relativity. It will introduce finitistic substitutes for basic topological notions. We will see that after basic topological notions are available, the basic notions of semi-Riemannian geometry, i.e., vector, tensor, covariant derivative, parallel transportation, geodesic and Riemann curvature, are all essentially finitistic already. Theorems on the existence of spacetime singularities are good examples for analyzing the applicability of infinite and continuous mathematical models to finite physical things. The last section of this chapter will analyze one of Hawking's singularity theorems, whose common classical proof is non-constructive. The section will show that the proof can be transformed into valid logical deductions on statements about real spacetime from literally true premises about real spacetime, even if real spacetime is discrete at the microscopic scale. Therefore, the conclusion of the theorem is physically reliable for real spacetime, in spite of the fact that the common proof of the theorem appears to assume that spacetime is literally isomorphic with a classical differentiable manifold (and hence absolutely non-discrete).

We will follow the classical presentations in Wald [36], O'Neill [27], and Naber [25] for developing semi-Riemannian geometry and for the proof of Hawking's singularity theorem. We will focus on the mathematical aspect and ignore all physical details.

8.1 Differentiable Manifolds

To define manifolds in strict finitism, we must replace non-finitistic notions in classical topology by finitistic notions. A *rational open ball* $S(x,r) \subset \mathbb{R}^n$ is an open ball such that all coordinates of its center x are rational numbers and its radius r is also a rational number. We call such x a rational point. We will include open balls of the radius 0. A rational open ball in $\mathbb{R}^n$ is thus uniquely determined by an $(n+1)$-tuple of rational numbers. A *regular open subset* of $\mathbb{R}^n$ is a sequence of rational open

F. Ye, *Strict Finitism and the Logic of Mathematical Applications*, Synthese Library 355, DOI 10.1007/978-94-007-1347-5_8, © Springer Science+Business Media B.V. 2011

balls $(S(x_i, r_i))_i$. Such a sequence corresponds to a subset $O = \cup_{i=0}^{\infty} S(x_i, r_i)$ of $\mathbb{R}^n$. We also say that O is a regular open set, but remember that a sequence of rational open balls is implied. We say that $(S(x_i, r_i))_i$ is a representation of O. Therefore, we can quantify over all regular open subsets of $\mathbb{R}^n$, by which we mean a quantification over all sequences of $(n+1)$-tuples of rational numbers. Note that if all $r_i = 0$, then $O = \cup_{i=0}^{\infty} S(x_i, r_i)$ is an empty set. Therefore, we consider empty sets as regular open sets.

Regular open sets are open sets in the sense defined in Chap. 4, but we cannot represent an arbitrary open subset of $\mathbb{R}^n$ as the union of a sequence of rational open balls. In this chapter we will consider only regular open subsets of $\mathbb{R}^n$, instead of general open subsets, because we want to approximate open subsets by balls in some uniform manner and we want to quantify over open sets. It is easy to see that common open subsets of $\mathbb{R}^n$ of ordinary regular shapes, for instance, all open balls (not limited to the rational ones), ordinary open polyhedrons, ellipsoids, etc., are all regular open sets. Any union of a sequence of regular open sets is also a regular open set. Moreover, since we can also represent an open ball as a union of open cubes with rational centers and sides (or open sets of some other regular shapes), apparently we can also use open cubes (or open sets of some other regular shapes) to define regular open sets. It will give us the same regular open sets. Using open cubes to represent regular open sets sometimes makes a proof easier. For instance, the intersection of any finite sequence of (rational) open cubes is obviously a regular open set. Then, it easily follows that the intersection of finitely many regular open sets is still a regular open set.

In classical mathematics, any open subset of $\mathbb{R}^n$ *is* the union of a sequence of rational open balls, although the sequence can be non-recursive and hence quite beyond elementary recursive mathematics. In strict finitism, we deal with only open subsets that can be represented as the unions of elementary recursive sequences of rational open balls. Since general relativity is merely an approximation to spacetime structure above the Planck scale, regular open subsets should be sufficient for representing physically meaningful open spacetime areas.

A ball $S(x,r)$ (or $Sc(x,r)$) is well-contained in a set O, denoted as $S(x,r) \Subset O$, if $S(x, r+\varepsilon) \subseteq O$ for some $\varepsilon > 0$. Let $O = \cup_{i=0}^{\infty} S(x_i, r_i)$ be a regular open subset of $\mathbb{R}^n$. Each $S(x_i, r_i)$, $r_i > 0$, can be represented again as a union

$$S(x_i, r_i) = \cup \left\{ S\left(x_i, r_i - \frac{1}{k}\right) ; k > \frac{1}{r_i} \right\}.$$

Therefore, we can always construct a representation $O = \cup_{i=0}^{\infty} S(x_i, r_i)$ such that each $S(x_i, r_i)$ is well-contained in O.

Recall that a function $f : O \to \mathbb{R}^m$ on an open set O is continuous if it is uniformly continuous on any closed ball well-contained in O.

Lemma 8.1. *If $f : O \to \mathbb{R}^m$ is a continuous function on a regular open set $O \subseteq \mathbb{R}^n$, then for any regular open set $O' \subset \mathbb{R}^m$, $O \cap f^{-1}(O')$ is a regular open set in $\mathbb{R}^n$.*

Proof. It suffices to prove that, for any rational open ball $S(x,r) \subset \mathbb{R}^m$, $O \cap f^{-1}(S(x,r))$ is a regular open set. Let $O = \cup_{i=0}^{\infty} S(x_i, r_i)$ be a representation of O

such that each $S(x_i,r_i)$ is well-contained in O. Then, f is uniformly continuous on $Sc(x_i,r_i)$. It again suffices to prove that each $S(x_i,r_i)\cap f^{-1}(S(x,r))$ is a regular open set. To represent this as the union of a sequence of rational open balls, we first construct a sequence (y_j) of rational points such that each $y_j\in S(x_i,r_i)$ and the sequence is dense in $S(x_i,r_i)$. For each j,k, we approximate $|f(y_j)-x|$ sufficiently to decide whether $|f(y_j)-x|<r-\frac{1}{k}$ or $|f(y_j)-x|>r-\frac{2}{k}$. In the former case, $S\left(f(y_j),\frac{1}{k}\right)\subseteq S(x,r)$. Let ω be a modulus of continuity for f on $Sc(x_i,r_i)$, and let $s_{j,k}=\min\left(\omega(k),r_i-|y_j-x_i|\right)$. Then, $S\left(y_j,s_{j,k}\right)\subseteq S(x_i,r_i)$ and $f\left(S\left(y_j,s_{j,k}\right)\right)\subseteq S(x,r)$. Therefore, $S\left(y_j,s_{j,k}\right)\subseteq S(x_i,r_i)\cap f^{-1}(S(x,r))$. We collect all such $S\left(y_j,s_{j,k}\right)$ into a sequence and claim that its union is $S(x_i,r_i)\cap f^{-1}(S(x,r))$. To see this, let $y\in S(x_i,r_i)\cap f^{-1}(S(x,r))$. We must find $S\left(y_j,s_{j,k}\right)$ such that $y\in S\left(y_j,s_{j,k}\right)$. Choose k such that $\frac{3}{k}<r-|f(y)-x|$. Therefore, $|f(y)-x|<r-\frac{3}{k}$. Since the sequence (y_j) is dense in $S(x_i,r_i)$, we can choose y_j such that $|y_j-y|<\min\left(\omega(k),r_i-|y-x_i|\right)$ and $|y_j-x_i|<|y-x_i|$. Then, $|f(y_j)-f(y)|<\frac{1}{k}$. Hence, $|f(y_j)-x|<r-\frac{2}{k}$. This means that in the construction above, we must have included $S\left(y_j,s_{j,k}\right)$ in the final sequence. Note that since $|y_j-x_i|<|y-x_i|$, we have $|y_j-y|<\min\left(\omega(k),r_i-|y_j-x_i|\right)=s_{j,k}$. Therefore, $y\in S\left(y_j,s_{j,k}\right)$. □

A function $f:O\to\mathbb{R}^m$ is infinitely differentiable if it is infinitely differentiable in the sense defined in Chap. 3 on any closed ball $Sc(x,r)$ well-contained in O. We use $C^\infty(O,\mathbb{R}^m)$ to denote the set of all infinitely differentiable functions from O into $\mathbb{R}^m$. $f\in C^\infty(O,\mathbb{R}^m)$ implies that any order of partial derivative g of f is continuous on O, and therefore, by the lemma above, for any regular open set O' in $\mathbb{R}^m$, $g^{-1}(O')\cap O$ is a regular open set in $\mathbb{R}^n$.

Let O,O' be two regular open subsets of $\mathbb{R}^n$. A function $f:O\to O'$ is a diffeomorphism between O and O', if f is a one-one function from O onto O', and $f\in C^\infty(O,\mathbb{R}^n)$ and $f^{-1}\in C^\infty(O',\mathbb{R}^n)$. We use $DMor(O,O')$ to denote the set of diffeomorphisms between O and O'. Therefore, $f\in DMor(O,O')$ implies that both f and f^{-1} preserve regular open sets. In defining manifolds, diffeomorphisms are treated as coordinate transformations. Most common coordinate transformations on $\mathbb{R}^n$ (restricted to their well-defined regions) are diffeomorphisms.

Then, a C^∞ *manifold* can be defined similarly as in the classical theory of manifolds.

Definition 8.2. An n-dimensional C^∞ (differentiable) manifold M consists of a set (also denoted by M) with an inequality relation $\neq$, a family $\{U_i\}_{i\in I}$ of subsets of M indexed by a finite or countable set I, and a corresponding indexed family of functions $\{\mu_i\}_{i\in I}$, $\mu_i:U_i\to\mathbb{R}^n$, such that each $U_i\subseteq M$, and

(1) $\{U_i\}_{i\in I}$ covers M, that is, $\cup_{i\in I}U_i=M$, and
(2) each μ_i maps U_i 1-1 onto a regular open subset $\mu_i(U_i)$ of $\mathbb{R}^n$, and μ_i respects inequalities, that is, for $p,q\in U_i$, $p\neq q$ if and only if $\mu_i(p)\neq\mu_i(q)$ in $\mathbb{R}^m$, and
(3) any two μ_i and μ_j are compatible in the following sense: $\mu_i(U_i\cap U_j)$ and $\mu_j(U_i\cap U_j)$ are regular open subsets of $\mathbb{R}^n$ and

$$\mu_i\circ\mu_j^{-1}\in DMor\left(\mu_j(U_i\cap U_j),\mu_i(U_i\cap U_j)\right).$$

M is called the *base set* for the manifold. $\langle U_i, \mu_i \rangle$ is called a *chart* or *local coordinate system* for M, and $\langle U_i, \mu_i \rangle_{i \in I}$ is called an *atlas* for M. Note that $\mu_i(U_i \cap U_j)$ could be empty. We require I to be a finite or countably infinite set. This does not compromise generality for physics applications. For a subset $S \subseteq M$, $p \notin S$ means that for any $q \in S$, $p \neq q$.

Note that a classical definition of manifolds usually requires the atlas of a manifold to include all possible charts that are compatible with the charts in the atlas. This requires a quantification over all subsets of the base set of the manifold. We drop this requirement in our definition here. Instead, we say that a function $\mu : U \to \mathbb{R}^n$ from a subset U of M into $\mathbb{R}^n$ is an *admissible chart* for the manifold M, if μ maps U 1-1 onto a regular open subset of $\mathbb{R}^n$ and μ is compatible with all the charts in the given atlas of M. Suppose that $\left\langle U'_j, \mu'_j \right\rangle_{j \in J}$ is another indexed family of charts such that each $\left\langle U'_j, \mu'_j \right\rangle$ in the family is admissible for M, and any two $\left\langle U'_j, \mu'_j \right\rangle$, $\langle U'_k, \mu'_k \rangle$ in the family are compatible, and $\left\{ U'_j \right\}_{j \in J}$ also covers M. Then we say that $\left\langle U'_j, \mu'_j \right\rangle_{j \in J}$ is an (admissible) *alternative atlas* for M.

Each $\mu_i(U_i)$ has a representation $\mu_i(U_i) = \cup_{k=0}^{\infty} S(x_{i,k}, r_{i,k})$ as a regular open set in $\mathbb{R}^n$, where the sequence $(x_{i,k}, r_{i,k})_k$ (of tuples of rational numbers) can be constructed from i. For each i, k, we can construct the sequence $(S(x_{i,k,l}, r_{i,k,l}))_l$ of *all* rational open balls such that $S(x_{i,k,l}, r_{i,k,l}) \subseteq S(x_{i,k}, r_{i,k})$. We will call the sequence $(\mu_i^{-1}(S(x_{i,k,l}, r_{i,k,l})))_{i,k,l}$ the *open basis* of M. (Apparently, this is a topological basis for the topological space M in the classical theory.) Each $\mu_i^{-1}(S(x_{i,k,l}, r_{i,k,l}))$ is called a *basic open subset* of M. Then, a *regular open subset* of M is any union of a sequence of basic open subsets of M. Therefore, we can quantify over all regular open subsets of M, by which we mean a quantification over all sequences of basic open subsets, that is, all sequences whose members are from $(x_{i,k,l}, r_{i,k,l})_{i,k,l}$.

We can show that this definition of regular open subsets of M is invariant for all alternative atlases for M.

Lemma 8.3. *Suppose that M is an n-dimensional C^∞ manifold, and $\mu : U \to \mathbb{R}^n$ is an admissible chart for M. Then, for any regular open set $O \subseteq \mu(U)$, $\mu^{-1}(O)$ is a regular open subset of M. In particular, U is a regular open subset of M.*

Proof. Let $\langle U_i, \mu_i \rangle_{i \in I}$ be the atlas for M. For each i, since μ and μ_i are compatible charts, $\mu(U \cap U_i)$ is a regular open set in $\mathbb{R}^n$. Therefore, $\mu(U \cap U_i) \cap O$ is a regular open set contained in $\mu(U \cap U_i)$. By compatibility again, $\mu_i \circ \mu^{-1}$ is a diffeomorphism from $\mu(U \cap U_i)$ to $\mu_i(U \cap U_i)$. Therefore, $\mu_i \circ \mu^{-1}(\mu(U \cap U_i) \cap O)$ must be a regular open set. Since μ_i, μ are one-one,

$$\begin{aligned} \mu_i \circ \mu^{-1}(\mu(U \cap U_i) \cap O) &= \mu_i((U \cap U_i) \cap \mu^{-1}(O)) \\ &= \mu_i(U_i \cap \mu^{-1}(O)). \end{aligned}$$

Therefore, by definition, $U_i \cap \mu^{-1}(O)$ is a regular open subset of M. Then, $\mu^{-1}(O) = \cup_{i \in I}(U_i \cap \mu^{-1}(O))$ is a regular open subset of M. □

Corollary 8.4. *If $\left\langle U'_j, \mu'_j \right\rangle_{j \in J}$ is an alternative atlas for M, then regular open subsets of M defined by using the atlas $\left\langle U'_j, \mu'_j \right\rangle_{j \in J}$ are the same as those defined by using the original atlas of M.*

Corollary 8.5. *If $\langle U, \mu \rangle$ is an admissible chart for M, then a subset of U is a regular open subset of M if and only if its image under μ is a regular open subset of $\mathbb{R}^n$.*

Proof. Let $\langle U_i, \mu_i \rangle_{i \in I}$ be the atlas for M and let $U' \subseteq U$. The 'if' part is obvious by the lemma. Suppose that U' is a regular open subset of M. That is, $U' = \bigcup_{j=0}^{\infty} \mu_{i(j)}^{-1} (S(x_j, r_j))$, where $S(x_j, r_j) \subseteq \mu_{i(j)} \left(U_{i(j)}\right)$ for each j. Each $\mu_{i(j)}$ is compatible with μ. Therefore, $\mu \circ \mu_{i(j)}^{-1}$ maps regular open sets in $\mu_{i(j)} \left(U_{i(j)} \cap U\right)$ to regular open sets in $\mu \left(U_{i(j)} \cap U\right)$. Since $\mu_{i(j)}^{-1} (S(x_j, r_j)) \subseteq U' \subseteq U$, $S(x_j, r_j) \subseteq \mu_{i(j)} \left(U_{i(j)} \cap U\right)$. Therefore, $\mu \left(\mu_{i(j)}^{-1} (S(x_j, r_j))\right)$ is a regular open set. Then, $\mu(U') = \bigcup_{j=0}^{\infty} \mu \left(\mu_{i(j)}^{-1} (S(x_j, r_j))\right)$ is also a regular open set. □

The lemma and the corollaries imply that after giving an atlas for a manifold, we can actually quantify over all admissible charts for the manifold and all alternative atlases for the manifold. A quantification over all admissible charts for M actually quantifies over every regular open subset U of M and every function $\mu : U \to \mathbb{R}^n$, and a quantification over all alternative atlases of M actually quantifies over all sequences of admissible charts.

Lemma 8.6. *An intersection of finitely many regular open subsets of a manifold M is still a regular open subset of M.*

Proof. It suffices to prove that an intersection of finitely many basic open sets is a regular open set. We first prove this for two basic open sets. So, let $\langle U_1, \mu_1 \rangle, \langle U_2, \mu_2 \rangle$ be two admissible charts for M and let $O_1 \subseteq \mu_1 (U_1), O_2 \subseteq \mu_2 (U_2)$ be two open balls in $\mathbb{R}^m$. We need to show that $\mu_1^{-1} (O_1) \cap \mu_2^{-1} (O_2)$ is a regular open set. Since the two charts are compatible, $\mu_1 (U_1 \cap U_2)$ is a regular open set in $\mathbb{R}^m$ and $\mu_2 \circ \mu_1^{-1} \in C^\infty (\mu_1 (U_1 \cap U_2), \mu_2 (U_1 \cap U_2))$. Therefore, $O_1 \cap \mu_1 (U_1 \cap U_2) \subseteq \mu_1 (U_1)$ is a regular open set in $\mathbb{R}^m$. Then,

$$\mu_2 \circ \mu_1^{-1} (O_1 \cap \mu_1 (U_1 \cap U_2)) = \mu_2 \left(\mu_1^{-1} (O_1)\right) \cap \mu_2 (U_1 \cap U_2)$$

is a regular open set in $\mathbb{R}^m$. This implies that $\mu_2 \left(\mu_1^{-1} (O_1)\right) \cap \mu_2 (U_1 \cap U_2) \cap O_2$ is a regular open set in $\mathbb{R}^m$. Applying μ_2^{-1} we see that

$$\mu_1^{-1} (O_1) \cap (U_1 \cap U_2) \cap \mu_2^{-1} (O_2) = \mu_1^{-1} (O_1) \cap \mu_2^{-1} (O_2)$$

is a regular open subset of M.

Then, consider a finite sequence $\langle U_1, \mu_1 \rangle$, ..., $\langle U_k, \mu_k \rangle$ of admissible charts for M and consider open balls $O_1 \subseteq \mu_1 (U_1)$, ..., $O_k \subseteq \mu_k (U_k)$. For each $i = 2, ..., k$, $\mu_1^{-1} (O_1) \cap \mu_i^{-1} (O_i)$ is a regular open subset of M. Applying μ_1 we see that $O_1 \cap \mu_1 \circ \mu_i^{-1} (O_i)$ is a regular open set in $\mathbb{R}^m$. Therefore, $O_1 \cap \bigcap_{i=2}^{k} \mu_1 \circ \mu_i^{-1} (O_i)$ is a regular open set in $\mathbb{R}^m$. Applying μ_1^{-1} we see that $\cap_{i=1}^k \mu_i^{-1}(O_i)$ is a regular open subset of M. □

Given an admissible chart $\langle U,\mu\rangle$ of M and a closed ball $C \subset \mu(U)$, we call $\mu^{-1}(C)$ a basic compact subset of M. A *regular compact subset* of M is a union of finitely many basic compact subsets of M.

Let M, M' be m and n dimensional manifolds with the atlases $\langle U_i,\mu_i\rangle_{i\in I}$ and $\left\langle U_j',\mu_j'\right\rangle_{j\in J}$ respectively. We can define the product $M \times M'$ as an $(m+n)$ dimensional manifold with the base set $M \times M'$ and the atlas $\left\langle U_i \times U_j', \mu_i \times \mu_j'\right\rangle_{(i,j)\in I\times J}$, where $\mu_i \times \mu_j'$ is defined naturally. Note that open cubes in $\mathbb{R}^{m+n}$ are the products of open cubes in $\mathbb{R}^m$ and $\mathbb{R}^n$ respectively. Therefore, using open cubes in place of open balls in the representations of regular open sets, we see that the products of regular open sets in $\mathbb{R}^m$ and $\mathbb{R}^n$ are regular open sets in $\mathbb{R}^{m+n}$, and on the other side, any regular open set in $\mathbb{R}^{m+n}$ can be represented as a union of the products of open cubes in $\mathbb{R}^m$ and $\mathbb{R}^n$. Then, it is easy to verify that $\left\langle U_i \times U_j', \mu_i \times \mu_j'\right\rangle_{(i,j)\in I\times J}$ is an atlas, that is, each $\left\langle U_i \times U_j', \mu_i \times \mu_j'\right\rangle$ is a chart and the charts are mutually compatible. Moreover, regular open sets of $M \times M'$ are the unions of products of regular open sets.

Let M, M' be manifolds as above again and let $U \subseteq M$ be a regular open subset of M. A function $f : U \to M'$ is a C^∞ function if for any charts $\langle U_i,\mu_i\rangle$ and $\left\langle U_j',\mu_j'\right\rangle$, $\mu_i\left(U \cap U_i \cap f^{-1}\left(U_j'\right)\right) = O$ is a regular open set in $\mathbb{R}^m$, and $\mu_j' \circ f \circ \mu_i^{-1} \in C^\infty(O,\mathbb{R}^n)$. In particular, a function $f : U \to \mathbb{R}$ is C^∞ if for any chart $\langle U_i,\mu_i\rangle$, $f \circ \mu_i^{-1} \in C^\infty(\mu_i(U \cap U_i),\mathbb{R})$. $C^\infty(U,M')$ denotes the set of all C^∞ function from U to M'. When $M = \mathbb{R}^m$ and U is a regular open set O in $\mathbb{R}^m$ and $M' = \mathbb{R}^n$, this definition coincides with the definition of $C^\infty(O,\mathbb{R}^n)$ above. $f : M \to M'$ is a diffeomorphism if f is one-one, onto, and $f \in C^\infty(M,M')$, and $f^{-1} \in C^\infty(M',M)$. When both M and M' are regular open sets in $\mathbb{R}^n$, this similarly coincides with the definition of diffeomorphism between regular open sets in $\mathbb{R}^n$ above.

A curve (also called differentiable or smooth curve) on an open interval (a,b) (where a,b can be $\mp\infty$ respectively) in M is a function $\gamma \in C^\infty((a,b),M)$ that satisfies a further condition: For any compact interval $[c,d] \subset (a,b)$, we can divide $[c,d]$ into finitely many subintervals such that the image under γ of each subinterval is contained in the base set U of a chart $\langle U,\mu\rangle$ of M. In the classical theory, this follows from the fact that the image of $[c,d]$ must be compact in M. However, we have to state this condition explicitly in the definition. This allows us to do constructions on a curve uniformly. We also consider a curve γ from a closed interval $[a,b] \subseteq \mathbb{R}$ to M, called a curve from $\gamma(a)$ to $\gamma(b)$, by which we mean that for some $\varepsilon > 0$, γ is a curve on $(a-\varepsilon, b+\varepsilon)$ in M. Note that we always assume that a curve is differentiable.

Let M, M' be m and n dimensional manifolds and suppose that $m \leq n$. A function $f : M \to M'$ is an embedding of M into M', if f is one-one, and there are alternative atlases $\langle U_i,\mu_i\rangle_{i\in I}$ and $\left\langle U_j',\mu_j\right\rangle_{j\in J}$ for M and M' respectively, such that for each

$\langle U_i, \mu_i \rangle$, there exists $\left\langle U'_j, \mu'_j \right\rangle$, such that $f(U_i) \subseteq U'_j$, and the function $\left(\mu'_j\right)^{\leq m} \circ f :$ $U_i \to \mathbb{R}^m$, $\left(\mu'_j\right)^{\leq m} \circ f(x) = \left(\left(\mu'_j\right)^1 (f(x)), \ldots, \left(\mu'_j\right)^m (f(x))\right)$, is an admissible chart for M, where $\left(\mu'_j\right)^k$ is the kth component of the function μ'_j. It is easy to show that $\left(\mu'_j\right)^{\leq m} \circ f : U_i \to \mathbb{R}^m$ is an admissible chart for M if and only if $\left(\mu'_j\right)^{\leq m} \circ f \circ$ $\mu_i^{-1} \in DMor\left(\mu_i(U_i), \left(\mu'_j\right)^{\leq m} \circ f(U_i)\right)$. Therefore, to embed M into M', we must construct alternative atlases for M and M' respectively, such that each chart in the former atlas is mapped by the embedding into an $\mathbb{R}^m$ slice of one of the chart in the latter atlas. An m-dimensional *submanifold* of M' is a manifold M'', such that there is an embedding $f : M \to M'$ of an m-dimensional manifold M into M', and the base set of M'' is the image $f(M)$, and, using the notations above, the atlas of M'' is $\left\langle f(U_i), \left(\mu'_{j(i)}\right)^{\leq m} \right\rangle_{i \in I}$.

A common way of defining manifolds in classical mathematics is to take subsets of $\mathbb{R}^n$ with some regular shape and glue the edges or sides. For instance, in classical mathematics, to construct a Klein bottle, we take the square $[0,1] \times [0,1]$, and identify the line segment $[0,1] \times \{0\}$ with the line segment $[0,1] \times \{1\}$, and identify the line segment $\{0\} \times [0,1]$ with the line segment $\{1\} \times [0,1]$ with the direction reversed. An atlas on the resulted set can be easily constructed. However, this does not work directly for strict finitism (neither for constructive mathematics). The problem is that gluing the edges of a square this way actually leaves gaps. We can illustrate this by a simple example. Consider a closed interval $[0,1]$. In classical mathematics, we can identify the end points 0 and 1 and get the manifold S^1. A chart of the manifold can have a function f mapping $[0,\varepsilon) \cup (1-\varepsilon, 1]$ 1-1 onto to the open interval $(-\varepsilon, \varepsilon)$. For instance, we can let $f(x) = x$ for $x \in [0,\varepsilon)$, and $f(x) = x - 1$ for $x \in (1-\varepsilon, 1]$. However, this function is not a surjection in constructive mathematics. There are real numbers r in $(-\varepsilon, \varepsilon)$ for which we cannot decide whether $r \geq 0$ or $r \leq 0$. If we could construct $f^{-1}(r)$, we would be able to decide whether $f^{-1}(r) \in [0,\varepsilon)$ or $f^{-1}(r) \in (1-\varepsilon, 1]$, and then we would be able to decide whether $r \geq 0$ or $r \leq 0$. Therefore, such an r is not in the range of f. Apparently, the problem is that when we glue the end points 0 and 1 of the interval $[0,1]$ in order to get S^1, we still leave a gap and do not really get S^1. This is similar to the fact that we cannot equate $[-1,0] \cup [0,1]$ with $[-1,1]$. There is a gap around 0 in the former.

To overcome the problem, we have to fill the gap. We can first define a new metric $d(p,q) = \min(|p-q|, 1-|p-q|)$ for rational numbers p,q in $[0,1]$ (with 0 and 1 being identified). Then, we can define Cauchy sequences (p_n) of rational numbers in $[0,1]$ using the metric $d(p,q)$ and construct the completion of the metric space of rational numbers in $[0,1]$. These will include Cauchy sequences for which we cannot decide if they converge to a real number in $[0,\varepsilon)$ or to a real number in $(1-\varepsilon, 1]$, that is, Cauchy sequences that fill the gap when we glue the end points of $[0,1]$. It is then easy to construct a 1-1 function from the open ball $S(0,\varepsilon)$ in this complete metric space to the open interval $(-\varepsilon, \varepsilon)$: A Cauchy sequence (p_n) of rational numbers in $[0,1]$ with respect to the metric $d(p,q)$ can be mapped to (p'_n), where $p'_n = p_n$ if $p_n \leq \frac{1}{2}$, and $p'_n = p_n - 1$ if $p_n > \frac{1}{2}$. It is easy to verify that this is a 1-1 correspondence between $S(0,\varepsilon)$ and $(-\varepsilon,\varepsilon)$ when $\varepsilon < \frac{1}{2}$. This gives a chart around the glue point.

We can similarly construct a Klein bottle by gluing the edges of the square $[0,1] \times [0,1]$ and filling the gaps.

8.2 Vectors, Dual Vectors and Tensors

As in the classical theory, tangent vectors at a point in a manifold can be defined as directional derivative operators on the point. More specifically, suppose that M is an m-dimensional manifold and $p \in M$. Let $C^\infty_M(p)$ be the set of functions f such that for some regular open subset U of M with $p \in U$, $f \in C^\infty(U,\mathbb{R})$. Note that this definition quantifies over all regular open subsets of M and all such f have the same signature (which is necessary for defining the set $C^\infty_M(p)$). Also note that every $f \in C^\infty_M(p)$ comes with a regular open subset U of M as its witness. Let $f,g \in C^\infty_M(p)$ with regular open subsets U_1, U_2 of M such that $p \in U_1 \cap U_2$, $f \in C^\infty(U_1,\mathbb{R})$, $g \in C^\infty(U_2,\mathbb{R})$. We define $f = g$ if and only if there exists a regular open subset U such that $p \in U \subseteq U_1 \cap U_2$ and $f(x) = g(x)$ for all $x \in U$. Moreover, since the intersection of any finite sequence of regular open subsets of M is still a regular open subset of M, it is easy to show that if $f,g \in C^\infty_M(p)$ then $fg \in C^\infty_M(p)$, and if $f_1,\dots,f_k \in C^\infty_M(p)$, $a^1,\dots,a^k \in \mathbb{R}$, then $\sum_{i=1}^k a^i f_i \in C^\infty_M(p)$. Furthermore, this definition is invariant for all admissible atlases of M.

Then, we can define tangent vectors:

Definition 8.7. A tangent vector v of a manifold M at $p \in M$ is a function $v : C^\infty_M(p) \to \mathbb{R}$ such that

(1) v is linear: for any finite sequence $f_1,\dots,f_k$ of functions in $C^\infty_M(p)$ and any finite sequence $a^1,\dots,a^k$ of real numbers, $v\left(\sum_{i=1}^k a^i f_i\right) = \sum_{i=1}^k a^i v(f_i)$, and
(2) v satisfies the Leibniz rule for derivation: for $f,g \in C^\infty_M(p)$, $v(fg) = v(f)g(p) + f(p)v(g)$.

The set of all tangent vectors at p is denoted as V_p. $V_M = \cup\{V_p : p \in M\}$ is called the tangent bundle of M. The equality for members of V_p is the standard one. That is, $v = v'$ if and only if $v(f) = v'(f)$ for all $f \in C^\infty_M(p)$. We can define linear combinations on V_p straightforwardly: for $a^1,\dots,a^k \in \mathbb{R}$, $v_1,\dots,v_k \in V_p$,

$$\left(\sum_{i=1}^k a^i v_i\right)(f) = \sum_{i=1}^k a^i v_i(f).$$

Then it is easy to see that V_p becomes a linear space in the sense defined in Chap. 7.

Given any admissible chart $\langle U,\mu\rangle$ for M such that $p \in U$, we can construct coordinate tangent vectors $\partial_{\mu,i} = \partial_{\mu,i}(p)$, $i = 1,\dots,m$, as follows: Let $f \in C^\infty_M(p)$. Then, $f \circ \mu^{-1} \in C^\infty(\mu(U),\mathbb{R})$. Let $\mathbf{x} = (x_1,\dots,x_m)$. We define

$$\partial_{\mu,i}(f) = \partial_i\left(f\circ\mu^{-1}\right)(\mu(p)) = \frac{\partial\left(f\circ\mu^{-1}\right)(\mathbf{x})}{\partial x_i}|_{\mathbf{x}=\mu(p)}.$$

Therefore, $\partial_{\mu,i}$, which is also denoted as $\frac{\partial}{\partial x_i}$, is the partial derivative operator for the i-th argument with respect to the coordinate system $\langle U,\mu\rangle$. Then, for any vector $\mathbf{a} = \left(a^1,...,a^m\right)$ of real numbers, $\partial_{\mu,\mathbf{a}} = \sum_{i=1}^m a^i\partial_{\mu,i}$ is the directional partial derivative at the direction $\mathbf{a}$ with respect to the coordinate system $\langle U,\mu\rangle$.

Note that if $\langle U',\mu'\rangle$ is another admissible chart for M such that $p\in U'$, then $\partial_{\mu',i}(p)$ is generally not equal to $\partial_{\mu,i}(p)$. However, since $\mu\circ\mu'^{-1}$ is infinitely differentiable at $\mu'(p)\in\mu'(U')$, we can express $\partial_{\mu',i}$ as a linear combination of $\partial_{\mu,i}$, $i=1,...,m$. To see this, let us write the function $\mu\circ\mu'^{-1}:\mu'(U')\to\mathbb{R}^m$ as $\mathbf{x}=\mu\circ\mu'^{-1}(\mathbf{x}')$, where $\mathbf{x}=(x_1,...,x_m)$, $\mathbf{x}'=(x'_1,...,x'_m)$. That is, the i-th component of the function $\mu\circ\mu'^{-1}$ is $\left(\mu\circ\mu'^{-1}\right)_i = x_i = x_i(\mathbf{x}')$. Then, $\mu\circ\mu'^{-1}(\mu'(p)) = \mu(p)$, and by the chain rule for partial derivatives,

$$\begin{aligned}\partial_{\mu',i}(f) &= \partial_i\left(f\circ\mu'^{-1}\right)(\mu'(p))\\ &= \partial_i\left(f\circ\mu^{-1}\circ\mu\circ\mu'^{-1}\right)(\mu'(p))\\ &= \sum_{j=1}^m \partial_j\left(f\circ\mu^{-1}\right)(\mu(p))\,\partial_i\left(\mu\circ\mu'^{-1}\right)_j(\mu'(p))\\ &= \sum_{j=1}^m \frac{\partial x_j}{\partial x'_i}|_{\mathbf{x}'=\mu'(p)}\partial_{\mu,j}(f).\end{aligned}$$

We will omit the subscript $|_{\mathbf{x}'=\mu'(p)}$ in the following. Therefore,

$$\partial_{\mu',i} = \sum_{j=1}^m \frac{\partial x_j}{\partial x'_i}\partial_{\mu,j},\text{ or }\frac{\partial}{\partial x'_i} = \sum_{j=1}^m \frac{\partial x_j}{\partial x'_i}\frac{\partial}{\partial x_j}.$$

Then, for a tangent vector $v=\sum_{i=1}^m a^i\partial_{\mu',i}$, we have

$$v = \sum_{i=1}^m a^i\sum_{j=1}^m \frac{\partial x_j}{\partial x'_i}\partial_{\mu,j} = \sum_{j=1}^m\left(\sum_{i=1}^m \frac{\partial x_j}{\partial x'_i}a^i\right)\partial_{\mu,j}.$$

In other words, $v=\sum_{j=1}^m b^j\partial_{\mu,i}$, with $b^j=\sum_{i=1}^m \frac{\partial x_j}{\partial x'_i}a^i$. This is called the vector transformation law.

As in the classical theory, we can also prove that every tangent vector can be expressed as a linear combination of partial derivative operators. Let $\langle U,\mu\rangle$ be an admissible chart for M such that $p\in U$. For $i=1,...,m$, let x_i^* be the ith-component function in the coordinate system μ, that is, $x_i^*(q) = (\mu(q))_i$ for $q\in U$. We need a lemma:

Lemma 8.8. *For any $f\in C_M^\infty(p)$, there exist $g_i^*\in C_M^\infty(p)$, $i=1,...,m$, and a regular open subset U', such that $p\in U'$, and*

$$f(q) = f(p) + \sum_{i=1}^{m} (x_i^*(q) - x_i^*(p)) g_i^*(q)$$

for all $q \in U'$.

Proof. Since $f \in C_M^\infty(p)$, for some regular open subset $U_1 \subseteq U$, $p \in U_1$, we have $f \circ \mu^{-1} \in C^\infty(\mu(U_1), \mathbb{R})$. Therefore, $f \circ \mu^{-1}$ has its first order Taylor expansion at $\mu(p)$ (see Sect. 3.5). That is, for some $r > 0$ such that $O = Sc(\mu(p), r) \subseteq \mu(U_1)$,

$$f \circ \mu^{-1}(\mathbf{x}) = f \circ \mu^{-1}(\mu(p)) + \sum_{i=1}^{m} (x_i - (\mu(p))_i) g_i(\mathbf{x}),$$

for any $\mathbf{x} \in O$, where $g_i \in C^\infty(O, \mathbb{R})$. Then, for any $q \in U' = \mu^{-1}(O)$, let $\mathbf{x} = \mu(q)$ in the above equation, we have

$$\begin{aligned} f(q) &= f(p) + \sum_{i=1}^{m} ((\mu(q))_i - (\mu(p))_i) g_i(\mu(q)) \\ &= f(p) + \sum_{i=1}^{m} (x_i^*(q) - x_i^*(p)) g_i^*(q), \end{aligned}$$

where $g_i^* \in C_M^\infty(p)$ is defined as $g_i^*(q) = g_i(\mu(q))$ on U'. □

Now, given $f \in C_M^\infty(p)$, by the lemma and the equality condition for the set $C_M^\infty(p)$,

$$f = f(p) + \sum_{i=1}^{m} (x_i^* - x_i^*(p)) g_i^*.$$

Note that by the Leibnitz rule,

$$v(1) = v(1 \cdot 1) = v(1) + v(1).$$

Therefore, for any $v \in V_p$, $v(1) = 0$, and hence $v(c) = 0$ for any constant function $c \in C_M^\infty(p)$, since v is linear. Applying the tangent vector $\partial_{\mu,i}$ to the above expression for f and noting that $\partial_{\mu,i}(x_i^*) = 1$ and $\partial_{\mu,i}\left(x_j^*\right) = 0$ for $j \neq i$, we have

$$\partial_{\mu,i}(f) = g_i^*(p).$$

Then, given any $v \in V_p$, applying v to the above expression for f, we have

$$v(f) = \sum_{i=1}^{m} v(x_i^*) g_i^*(p) = \sum_{i=1}^{m} v(x_i^*) \partial_{\mu,i}(f).$$

Therefore, $v = \sum_{i=1}^{m} v(x_i^*) \partial_{\mu,i}$. This means that $\partial_{\mu,1}, \ldots, \partial_{\mu,m}$ constitute a basis for the linear space V_p. It is called a coordinate basis.

Suppose that M' is a k-dimensional submanifold of M. For $p \in M'$, we can identify a tangent vector v of M' at p with a tangent vector of M at p. There exists a coordinate system $\langle U, \mu \rangle$ of M with $p \in U$ and a coordinate system $\langle U', \mu' \rangle$ of M'

with $p \in U'$ such that $\mu'(U')$ is a k-dimensional slice of $\mu(U)$. Then, as a tangent vector of M', v has an expansion $v = \sum_{i=1}^{k} v^i \partial_{\mu',i}$ for some v^i, $i = 1,...,k$. Apparently, v uniquely corresponds to the tangent vector $\sum_{i=1}^{k} v^i \partial_{\mu,i}$ of M. It is easy to see that this definition is independent of the chosen coordinate systems (because of the vector transformation law).

Recall that a differentiable curve γ in M with the parameter on an interval $[a,b]$ is a function $\gamma \in C^\infty((a-\varepsilon, b+\varepsilon), M)$ for some $\varepsilon > 0$. The tangent vector of γ at a point $p = \gamma(t)$ is a vector $v \in V_p$ such that for any $f \in C_M^\infty(p)$, $v(f) = \frac{\mathrm{d}}{\mathrm{d}t}(f \circ \gamma)(t)$. Write the function $\mu \circ \gamma \in C^\infty((a-\varepsilon, b+\varepsilon), \mathbb{R}^m)$ as $(\gamma^{\mu,1}, ..., \gamma^{\mu,m})$, we have

$$\frac{\mathrm{d}}{\mathrm{d}t}(f \circ \gamma)(t) = \frac{\mathrm{d}}{\mathrm{d}t}\left(\left(f \circ \mu^{-1}\right) \circ (\mu \circ \gamma)\right)(t) = \sum_{i=1}^{m} \frac{\mathrm{d}\gamma^{\mu,i}}{\mathrm{d}t} \partial_{\mu,i} f.$$

Therefore, the components of v in the coordinate basis of μ are exactly $\frac{\mathrm{d}}{\mathrm{d}t}\gamma^{\mu,1}, ..., \frac{\mathrm{d}}{\mathrm{d}t}\gamma^{\mu,m}$. Sometimes we denote this vector as $\gamma'(t)$, when no ambiguity will arise.

A dual vector (or cotangent vector) at p is a linear function from V_p to $\mathbb{R}$. V_p^* denotes the set of dual vectors at p, and $V_M^* = \cup\{V_p^* : p \in M\}$ is the cotangent bundle of M. Obviously, V_p^* is a linear space with the naturally defined linear combination. Let $e_1, ..., e_m$ be a basis for V_p. Then, a dual vector ω is completely determined by its values for the basis, $\omega(e_i)$, $i = 1, ..., m$, that is, $\omega\left(\sum_{i=1}^{m} a^i e_i\right) = \sum_{i=1}^{m} a^i \omega(e_i)$. Let $e^j \in V_p^*$ be defined as $e^j(e_i) = \delta_i^j$, where δ_i^j is the Kronecker symbol, that is, $\delta_i^i = 1$ and $\delta_i^j = 0$ whenever $i \neq j$. Then, $\omega = \sum_{j=1}^{m} \omega(e_j) e^j$. That is, $e^1, ..., e^m$ constitute a basis for V_p^*. This is called the dual basis corresponding to $e_1, ..., e_m$. When $e_i = \partial_{\mu,i} = \frac{\partial}{\partial x_i}$, $i = 1, ..., m$, are the partial derivative operators in the coordinate system μ at p, the corresponding dual vectors are denoted as $\mathrm{d}x^1, ..., \mathrm{d}x^m$. Therefore, $\mathrm{d}x^j\left(\partial_{\mu,i}\right) = \delta_i^j$.

A *tensor* of the type (k,l) at p is a multi-linear function

$$T : V_p^* \times ... \times V_p^* \times V_p \times ... \times V_p \to \mathbb{R},$$

where there are k copies of V_p^* and l copies of V_p. T is multi-linear in the sense that T is linear for each of its $k+l$ arguments. Let $e_1, ..., e_m$ be a basis for V_p and let $e^1, ..., e^m$ be the corresponding dual basis for V_p^*. A tensor T of the type (k,l) is completely determined by the values

$$T^{i_1 ... i_k}{}_{j_1 ... j_l} = T\left(e^{i_1}, ..., e^{i_k}, e_{j_1}, ..., e_{j_l}\right), i_1, ..., j_l = 1, ..., m.$$

These values are called the components of T in the basis $e_1, ..., e_m$. We use $\mathscr{T}_p^{(k,l)}$ to denote the set of tensors of the type (k,l) at p. $\mathscr{T}_p^{(k,l)}$ becomes a linear space with the naturally defined linear combination.

Suppose that $e'_1, ..., e'_m$ constitute another basis for V_p with the corresponding dual basis $e'^1, ..., e'^m$ for V_p^*, and suppose that $e'_i = a_i^j e_j$, $e'^i = b^i_j e^j$. Here we use the common summation convention that an index letter appearing both at a superscript position and a subscript position means that the index is summed over. That is, $a_i^j e_j$ is actually $\sum_{j=1}^m a_i^j e_j$. Note that

$$e'^j\left(e'_i\right) = b_k^j e^k\left(a_i^l e_l\right) = b_k^j a_i^l e^k\left(e_l\right) = b_k^j a_i^l \delta_l^k = a_i^k b_k^j.$$

Since we also have $e'^j\left(e'_i\right) = \delta_j^i$, we see that $a_i^k b_k^j = \delta_i^j$. That is, the matrix $\left(b_j^i\right)$ is the inverse of $\left(a_i^j\right)$. Then, a straightforward computation shows that the components of T in the basis $e'_1, ..., e'_m$ are

$$\begin{aligned} T'^{i_1...i_k}{}_{j_1...j_l} &= T\left(e'^{i_1}, ..., e'^{i_k}, e'_{j_1}, ..., e'_{j_l}\right) \\ &= b_{h_1}^{i_1} ... b_{h_k}^{i_k} a_{j_1}^{n_1} ... a_{j_l}^{n_l} T^{h_1...h_k}{}_{n_1...n_l}. \end{aligned}$$

This is the component transformation for a tensor. In particular, when $e_i = \partial_{\mu,i}$, $e'_i = \partial_{\mu',i}$, $i = 1, ..., m$, are the partial derivative operators in the coordinate systems μ and μ' respectively, we have $a_i^j = \frac{\partial x_j}{\partial x'_i}$. By the chain rule, $\frac{\partial x_k}{\partial x'_i}\frac{\partial x'_j}{\partial x_k} = \frac{\partial x'_j}{\partial x'_i} = \delta_i^j$. Therefore, $b_j^i = \frac{\partial x'_i}{\partial x_j}$ and the component transformation for a tensor becomes

$$T'^{i_1...i_k}{}_{j_1...j_l} = \frac{\partial x'_{i_1}}{\partial x_{h_1}} \cdots \frac{\partial x'_{i_k}}{\partial x_{h_k}} \frac{\partial x_{n_1}}{\partial x'_{j_1}} \cdots \frac{\partial x_{n_l}}{\partial x'_{j_l}} T^{h_1...h_k}{}_{n_1...n_l}.$$

This is the coordinate transformation law for tensor components.

We sometimes use the symbols $T^{i_1...i_k}{}_{j_1...j_l}$ for tensor components in a particular basis to denote a tensor T. When doing this we should be aware that the components of the tensor in another basis may have a different format. Then, given any vectors $v_j = a_j^n e_n$, $j = 1, ..., l$ and dual vectors $\omega^i = b_h^i e^h$, $i = 1, ..., k$,

$$\begin{aligned} &T\left(\omega^1, ..., \omega^k, v_1, ..., v_l\right) \\ &= b_{h_1}^1 ... b_{h_k}^k a_1^{n_1} ... a_l^{n_l} T\left(e^{h_1}, ..., e^{h_k}, e_{n_1}, ..., e_{n_l}\right) \\ &= b_{h_1}^1 ... b_{h_k}^k a_1^{n_1} ... a_l^{n_l} T^{h_1...h_k}{}_{n_1...n_l}. \end{aligned}$$

A tensor T of the type (k,l) is symmetric in its p-th and q-th vector arguments, if for any dual vectors and vectors $\omega^1, ..., \omega^k, v_1, ..., v_l$,

$$T\left(\omega^1, ..., \omega^k, ..., v_p, ..., v_q, ...\right) = T\left(\omega^1, ..., \omega^k, ..., v_q, ..., v_p, ...\right).$$

Symmetry in a pair of dual vector arguments can be defined similarly. T is symmetric if it is symmetric in all pairs of its vector arguments and dual vector arguments. It is easy to verify that if the components $T^{i_1...i_k}{}_{j_1...j_l}$ for T in a basis are such that

$$T^{i_1...i_k}{}_{j_1...j_p...j_q...j_l} = T^{i_1...i_k}{}_{j_1...j_q...j_p...j_l}$$

for all possible indices, then T is symmetric in its p-th and q-th vector arguments. For a tensor T with the components $T^{i_1...i_k}{}_{j_1...j_l}$ in a basis, we define a new tensor $T_{(1...p)}$ of the same type with the components

$$T^{i_1...i_k}{}_{(j_1...j_p)...j_l} = \frac{1}{p!} \sum_{\pi(j_1...j_p)} T^{i_1...i_k}{}_{\pi(j_1...j_p)...j_l}$$

in the same basis, where the sum ranges over all $p!$ permutations $\pi(j_1...j_p)$ of the given sequence $(j_1...j_p)$ of numbers. Apparently, $T^{i_1...i_k}{}_{(j_1...j_p)...j_l}$ are the components of the tensor $T_{(1...p)}$ such that

$$T_{(1...p)}(...,v_1...,v_p,...) = \frac{1}{p!} \sum_{(j_1...j_p)} T(...,v_{j_1}...,v_{j_p},...),$$

where $(j_1...j_p)$ ranges over all $p!$ permutations of $(1...p)$. $T_{(1...p)}$ is called a symmetrization of T with respect to the vector arguments at $(1...p)$. The definition is invariant for all bases. That is, if the tensor T has components $T'^{i_1...i_k}{}_{j_1...j_l}$ in another basis, then the tensor $T_{(1...p)}$ has the corresponding components $T'^{i_1...i_k}{}_{(j_1...j_p)...j_l}$ in the new basis. Therefore, we will use $T^{i_1...i_k}{}_{(j_1...j_p)...j_l}$ to denote the symmetrization. In particular, we have

$$T_{(j_1 j_2)} = \frac{1}{2}(T_{j_1 j_2} + T_{j_2 j_1}).$$

Other types of symmetrization can be defined similarly, for instance, $T^{i}{}_{j_1(j_2 j_3 j_4) j_5}$, $T^{i_1(i_2 i_3)}{}_{j_1 j_2}$ etc. Similarly, we can define anti-symmetrization. For instance,

$$T^{i_1...i_k}{}_{[j_1...j_p]...j_l} = \frac{1}{p!} \sum_{\pi(j_1...j_p)} \sigma_\pi T^{i_1...i_k}{}_{\pi(j_1...j_p)...j_l},$$

where σ_π is the sign of the permutation π. That is, it is $+1$ when π is an even permutation, and it is -1 when π is an odd permutation. In particular,

$$T_{[j_1 j_2]} = \frac{1}{2}(T_{j_1 j_2} - T_{j_2 j_1}).$$

Note that a tangent vector v induces a linear function from cotangent vectors ω to real numbers: $\omega \mapsto \omega(v)$. Therefore, a tangent vector can be seen as a tensor of the type $(1,0)$. Similarly, a cotangent vector is a tensor of the type $(0,1)$. Let v be a vector. When we treat v as a type $(1,0)$ tensor, its components in a basis $e_1,...,e_m$ are $v^i = e^i(v)$. On the other hand, the expansion of v in the basis is exactly $v = v^i e_i$. Similarly, for a dual vector ω, its components are $\omega_i = \omega(e_i)$ and $\omega = \omega_i e^i$.

Let T, T' be tensors of the types (k,l) and (k',l') respectively. The outer product $T \otimes T'$ is a tensor of the type $(k+k', l+l')$ defined as

$$T \otimes T'\left(\omega^1, ..., \omega^k, \omega'^1, ..., \omega'^{k'}, v_1, ..., v_l, v_1', ..., v_{l'}'\right)$$
$$= T\left(\omega^1, ..., \omega^k, v_1, ..., v_l\right) T'\left(\omega'^1, ..., \omega'^{k'}, v_1', ..., v_{l'}'\right).$$

Therefore, given a basis $e_1, ..., e_m$ and the corresponding dual basis $e^1, ..., e^m$,

$$e_{i_1} \otimes ... \otimes e_{i_k} \otimes e^{j_1} \otimes ... \otimes e^{j_l}$$

is a tensor of the type (k,l), and it is easy to see that these tensors form a basis for the linear space $\mathscr{T}_p^{(k,l)}$. That is, for any tensor $T \in \mathscr{T}_p^{(k,l)}$,

$$T = T^{i_1 ... i_k}{}_{j_1 ... j_l} e_{i_1} \otimes ... \otimes e_{i_k} \otimes e^{j_1} \otimes ... \otimes e^{j_l},$$

where $T^{i_1 ... i_k}{}_{j_1 ... j_l}$ are exactly the components of T at the basis $e_1, ..., e_m$.

A contraction (or trace) of a tensor $T^{i_1 ... h ... i_k}{}_{j_1 ... n ... j_l}$ of the type $(k+1, l+1)$ at the position h, n is a tensor of the type (k,l), whose components in the same basis are $T^{i_1 ... h ... i_k}{}_{j_1 ... h ... j_l}$. Using the component transformation above, a straightforward computation can verify that if the original tensor $T^{i_1 ... h ... i_k}{}_{j_1 ... n ... j_l}$ has components $T'^{i_1 ... h ... i_k}{}_{j_1 ... n ... j_l}$ in another basis, then the components of the contraction in the new basis are also exactly $T'^{i_1 ... h ... i_k}{}_{j_1 ... h ... j_l}$. That is, this definition of the contraction of a tensor, while referring to a particular basis, is actually invariant for all bases.

A tangent vector filed is a function $v : U \to V_M$ from a regular open subset U of M to the tangent bundle of M, such that for each $p \in U$, $v_p = v(p) \in V_p$. For $f \in C^\infty(U, \mathbb{R})$, $v(f)$ is a function from U to $\mathbb{R}$ defined as $v(f)(p) = v_p(f)$ for $p \in U$. We say that v is a C^∞ (or smooth) vector field on U if $v(f) \in C^\infty(U, \mathbb{R})$ for any $f \in C^\infty(U, \mathbb{R})$. Let $\langle U', \mu \rangle$ be any chart for M. For any $p \in U \cap U'$, there exist $a^1(p), ..., a^m(p)$ such that $v_p = a^i(p)\, \partial_{\mu,i}(p)$, where $\partial_{\mu,i}(p)$ is the partial derivative operator at p in the coordinate system μ. Therefore, for each $i = 1, ..., m$, $a^i : U \cap U' \to \mathbb{R}$. Let $x^i \in C^\infty(U, \mathbb{R})$ be the i-th coordinate function in the coordinate system μ. Apparently, $v\left(x^i\right) = a^i$ on $U \cap U'$. Therefore, if v is a C^∞ vector field, then $a^i \in C^\infty(U, \mathbb{R})$. On the other hand, let $f \in C^\infty(U, \mathbb{R})$. Then $v(f)(p) = a^i(p) \frac{\partial(f \circ \mu^{-1})}{\partial x_i}(\mu(p))$. Obviously, $v(f) \in C^\infty(U \cap U', \mathbb{R})$ if all $a^i \in C^\infty(U, \mathbb{R})$. That is, v is a C^∞ vector field, if and only if all its component functions in any coordinate derivative basis are C^∞ functions.

Cotangent vector fields and tensor fields of a given type can be defined similarly. For a cotangent vector field ω on U and a tangent vector field v on U, $\omega(v)$ is a function from U to $\mathbb{R}$: $\omega(v)(p) = \omega_p(v_p)$. ω is a C^∞ cotangent vector field if $\omega(v) \in C^\infty(U, \mathbb{R})$ for any C^∞ vector field v on U. Similarly, a tensor field on U is C^∞ if it produces C^∞ functions on U when applied to C^∞ tangent vector fields and cotangent vector fields. It similarly follows that a cotangent vector field or tensor field on U is C^∞ if its components in a coordinate derivative basis are C^∞ functions.

8.3 Metric

The components of a tensor T of the type $(0,2)$ in a basis $e_1,...,e_m$ make up an $m\times m$ matrix (T_{ij}). Suppose that another basis $e'_1,...,e'_m$ is given by $e'_i = a_i^j e_j$. Then, the components of T in the new basis are $T'_{ij} = a_i^k a_j^l T_{kl}$. Therefore,

$$\left(T'_{ij}\right) = \left(a_i^k\right)\left(T_{kl}\right)\left(a_j^l\right)',$$

where $\left(a_i^j\right)'$ is the matrix transposition of $\left(a_i^j\right)$ and the right hand side is a matrix product. We say that T is non-degenerated if its component matrix in a basis has a non-zero determinant. Simple computations on real numbers can show that a matrix is invertible if and only if it has a non-zero determinant. The basis transformation matrix $\left(a_i^j\right)$ is invertible (for we can also express $e_1,...,e_m$ as linear combinations of $e'_1,...,e'_m$). Therefore, its determinant $\left|a_i^j\right| \neq 0$. Then, the determinants of $\left(T'_{ij}\right)$ and (T_{ij}) are related by $\left|T'_{ij}\right| = \left|a_i^j\right|^2 \left|T_{ij}\right|$. That is, $\left|T'_{ij}\right|$ is non-zero if and only if $\left|T_{ij}\right|$ is non-zero. The definition of non-degeneracy of T is thus independent of the chosen basis. Note that this is sightly different from the common classical definition of non-degeneracy.

Obviously, T is symmetric if and only if its components on any basis make up a symmetric matrix. If T is non-degenerated and symmetric, then we can construct a new basis $e_1^*,...,e_m^*$ such that the component matrix of T in the new basis is a diagonal matrix, which means that $T\left(e_i^*, e_j^*\right) = 0$ for $i \neq j$. There is an obstacle when we try to carry out the diagonalization process for a symmetric matrix in the classical theory of matrix, because we cannot generally decide whether an entry in the matrix is non-zero. However, we can overcome this difficulty when the symmetric matrix (T_{ij}) has a non-zero determinant.

To diagonalize the matrix (T_{ij}) we must construct invertible matrices $A_1,...,A_k$ such that

$$\left(T_{ij}^*\right) = A_k...A_1\left(T_{ij}\right)A'_1...A'_k$$

becomes a diagonal matrix. Then, using $A_1,...,A_k$ to transform the basis successively, we will finally get the new basis in which T has the component matrix $\left(T_{ij}^*\right)$. First, from the assumption that $\left|T_{ij}\right| \neq 0$, we can find a matrix entry $T_{ij} \neq 0$. This means that $T\left(e_i, e_j\right) \neq 0$. Now,

$$T\left(e_i + e_j, e_i + e_j\right) = T\left(e_i, e_i\right) + 2T\left(e_i, e_j\right) + T\left(e_j, e_j\right).$$

By approximating these real numbers sufficiently, we see that either $T\left(e_i, e_i\right) \neq 0$, or $T\left(e_j, e_j\right) \neq 0$, or $T\left(e_i + e_j, e_i + e_j\right) \neq 0$. In the first or the second case, we apply the basis transformation of switching e_1 with e_i or e_j, respectively. In the third case, we apply the basis transformation of replacing e_i and e_j by $\frac{1}{\sqrt{2}}\left(e_i + e_j\right)$

and $\frac{1}{\sqrt{2}}(e_i - e_j)$ and then switching e_1 with $\frac{1}{\sqrt{2}}(e_i + e_j)$. In the new basis $e'_1, ..., e'_m$, we must have $T(e'_1, e'_1) \neq 0$. That is, the corresponding basis transformation A_1 is such that, in the new basis, the component matrix for T is $\left(T'_{ij}\right) = A_1 (T_{ij}) A'_1$ with $T'_{11} \neq 0$. Then we can perform the diagonalization as in the classical case. That is, for all $i > 1$, we multiply the first row of $\left(T'_{ij}\right)$ by $-T'_{i1}/T'_{11}$ and add it to the i-th row and multiply the first column of the resulted matrix by $-T'_{1i}/T'_{11} = -T'_{i1}/T'_{11}$ and add it to the i-th column. This is equivalent to transforming $\left(T'_{ij}\right)$ into $A_2 \left(T'_{ij}\right) A'_2$ for some appropriate basis transformation matrix A_2, and the resulted matrix $\left(T''_{ij}\right)$ is such that $T''_{11} \neq 0$ while $T''_{1i} = T''_{i1} = 0$ for all $i > 1$. We can repeat the process for the rest $(m-1) \times (m-1)$ submatrix. Note that this submatrix must have a non-zero determinant as well. Moreover, note that the process involves only summation and multiplication of real numbers. It is iteratable within strict finitism. Obviously, we can normalize the final new basis $e^*_1, ..., e^*_m$ so that $T(e^*_i, e^*_i) = \pm 1$. That is, the final diagonal component matrix for T has ± 1 as the diagonal entries. Therefore, we have

Lemma 8.9. *For any non-degenerated, symmetric type* $(0,2)$ *tensor* T*, there exists a basis* $e_1, ..., e_m$ *such that* $T(e_i, e_i) = \pm 1$ *for all* $i = 1, ..., m$ *and* $T(e_i, e_j) = 0$ *for* $i \neq j$*.*

The basis in the lemma is called an orthonormal basis for T. The number of e_i such that $T(e_i, e_i) = -1$ is called the index of T, and the sequence of ± 1 at the diagonal of the diagonal component matrix for T is called the signature of T in the orthonormal basis. We can show that this definition of index is invariant for all orthonormal bases for T. To see this, let $e'_1, ..., e'_m$ be another basis such that $T(e'_i, e'_i) = \pm 1$ for all i. We may assume that for some k, k', $T(e_i, e_i)$ is $+1$ for $i \leq k$, and it is -1 for $i > k$, and $T(e'_i, e'_i)$ is $+1$ for $i \leq k'$, and it is -1 for $i > k'$. We need to show that $k = k'$. Since $k = k'$ is decidable, we can assume that $k > k'$ and try to deduce a contradiction. Each e_i, $i \leq k$, can be expressed as $e_i = a^j_i e'_j$. We decompose this into

$$e_i = v_i + v^*_i, \text{ where } v_i = \sum_{j=1}^{k'} a^j_i e'_j, \; v^*_i = \sum_{j=k'+1}^{m} a^j_i e'_j.$$

We use explicit summation symbols here since they are not sums over all values of the index j. Note that for any linear combination v of $e'_1, ..., e'_{k'}$ and any linear combination v^* of $e'_{k'+1}, ..., e'_m$, we have $T(v,v) \geq 0$, $T(v^*, v^*) \leq 0$, and $T(v, v^*) = 0$. We want to find real numbers $r^1, ..., r^k$, such that $\max\left(\left|r^1\right|, ..., \left|r^k\right|\right) \geq 1$, but for $v = \sum_{i=1}^k r^i v_i$ we have $T(v,v) < 1$. This will lead to a contradiction. Because, let $e = \sum_{i=1}^k r^i e_i$, $v^* = \sum_{i=1}^k r^i v^*_i$. Then, $e = v + v^*$. v is a linear combination of $e'_1, ..., e'_{k'}$ and v^* is a linear combination of $e'_{k'+1}, ..., e'_m$. Therefore,

$$T(e,e) = T(v,v) + 2T(v,v^*) + T(v^*,v^*) = T(v,v) + T(v^*,v^*).$$

But $T(e,e) = \sum_{i=1}^k \left(r^i\right)^2 \geq 1$ and $T(v^*, v^*) \leq 0$. This is a contradiction.

To find such real numbers $r^1, ..., r^k$, note that

$$v = \sum_{i=1}^{k} r^i v_i = \sum_{i=1}^{k} \sum_{j=1}^{k'} r^i a_i^j e'_j = \sum_{j=1}^{k'} \left(\sum_{i=1}^{k} r^i a_i^j \right) e'_j.$$

Therefore,

$$T(v,v) = \sum_{j=1}^{k'} \left(\sum_{i=1}^{k} r^i a_i^j \right)^2.$$

We need a lemma, which is an approximate version of the existential theorem on the solutions of a system of homogeneous linear equations.

Lemma 8.10. *Suppose that $k > k'$. Then, given any real numbers a_i^j, $i = 1,...,k$, $j = 1,...,k'$, for any $\varepsilon > 0$, there exist real numbers $x^1,...,x^k$, such that* $\max\left(\left|x^1\right|,...,\left|x^k\right|\right) \geq 1$, *and* $\left|\sum_{i=1}^{k} a_i^j x^i\right| < \varepsilon$, *for* $j = 1,...,k'$.

Proof. In the classical theory of homogeneous linear equations, we can solve the equations $\sum_{i=1}^{k} a_i^j x^i = 0$, $j = 1,...,k'$, with arbitrary large x^i's since $k > k'$. However, we cannot do this in strict finitism, because we cannot decide whether $a_i^j = 0$. We can only get approximate solutions, as the lemma states. We can proceed as follows. First, for each j decide whether $\left|a_1^j\right| < \varepsilon$ or $\left|a_1^j\right| > 0$. In case $\left|a_1^j\right| < \varepsilon$ for all $j = 1,...,k'$, we let $x^1 = 1$ and $x^i = 0$ for $i > 1$. Otherwise, we can find a j such that $\left|a_1^j\right| > 0$. We may assume that $\left|a_1^1\right| > 0$. Then, we can reduce the inequalities $\left|\sum_{i=1}^{k} a_i^j x^i\right| < \varepsilon$ to the following inequalities, by dividing the first inequality by $\left|a_1^1\right|$, multiplying the resulted first inequality by $-a_1^j$ and adding to the j-th inequality, and adjusting the right hand side of the inequalities appropriately:

$$\left| x^1 + \frac{a_2^1}{a_1^1} x^2 + \frac{a_3^1}{a_1^1} x^3 ... \right| < \frac{\varepsilon}{2\left(1 + \max\left(\left|a_1^1\right|,...,\left|a_1^{k'}\right|\right)\right)},$$

$$\left| \left(a_2^2 - a_1^2 \frac{a_2^1}{a_1^1} \right) x^2 + \left(a_3^2 - a_1^2 \frac{a_3^1}{a_1^1} \right) x^3 + ... \right| < \frac{\varepsilon}{2\left(1 + \max\left(\left|a_1^1\right|,...,\left|a_1^{k'}\right|\right)\right)},$$

$$...$$

If $x^1,...,x^k$ satisfy these inequalities, then they will also satisfy the original inequalities $\left|\sum_{i=1}^{k} a_i^j x^i\right| < \varepsilon$, $j = 1,...,k'$. Then, we can repeat the process by ignoring the first inequality and considering the new coefficients $a_2^{\prime j}$, $j = 2,...,k'$, for the variable x^2 in the above inequalities. If all $\left|a_2^{\prime j}\right| < \varepsilon' = \frac{\varepsilon}{2\left(1 + \max\left(\left|a_1^1\right|,...,\left|a_1^{k'}\right|\right)\right)}$, then we can again let $x^2 = 1$, $x^i = 0$ for $i = 3,...,k$, and $x^1 = -\frac{a_2^1}{a_1^1} x^2$. All the inequalities will be satisfied. Otherwise, we will find a $\left|a_2^{\prime j}\right| > 0$ and proceed as before. In the end we will get inequalities

$$\left|x^1 + b_2^1 x^2 + \dots\dots\dots\dots\dots\dots\dots\right| < \delta,$$
$$\left|x^2 + b_3^2 x^3 + \dots\dots\dots\dots\dots\right| < \delta,$$
$$\dots$$
$$\left|x^{k'} + b_{k'+1}^{k'} x^{k'+1} + \dots\right| < \delta.$$

Then, we can let $x^{k'+1} = 1$ and set values for other x^i to satisfy the inequalities. The construction involves only addition and multiplication of real numbers. It is repeatable within strict finitism. □

From the lemma it easily follows that we can find real numbers $r^1, \dots, r^k$ to make each $\left(\sum_{i=1}^{k} r^i a_i^j\right)^2$ arbitrarily small and thus make $T(v,v) < 1$. Therefore, all orthonormal bases for T have the same number of vectors e such that $T(e,e) = -1$. That is, the definition of index of T is invariant for all orthonormal bases for T.

Note that non-degeneracy is a point-wise condition. It is defined for each point $p \in M$. Similar to the case of continuity vs. uniform continuity, we need a uniform notion. We say that T is uniformly non-degenerated, if for any basic regular compact subset B of M and the corresponding chart that is a witness for B to be a basic regular compact subset, there exists a constant $c > 0$, such that $\left\|T_{ij}(p)\right\| \geq c$ for $p \in B$, where $\left\|T_{ij}(p)\right\|$ is the absolute value of the component matrix of T at the point p in the chart. Suppose that in another chart $T'_{ij} = \frac{\partial x_k}{\partial x'_i}\frac{\partial x_l}{\partial x'_j} T_{kl}$. In the matrix notation we have $\left(T'_{ij}\right) = \left(\frac{\partial x_k}{\partial x'_i}\right)(T_{kl})\left(\frac{\partial x_l}{\partial x'_j}\right)'$. Now, for a coordinate transformation, the inverse matrix $\left(\frac{\partial x_k}{\partial x'_i}\right)^{-1} = \left(\frac{\partial x'_l}{\partial x_k}\right)$ has uniformly continuous (actually C^∞) functions as entries. Therefore, the absolute value of its determinant, $\left\|\left(\frac{\partial x_k}{\partial x'_i}\right)^{-1}\right\| = \left\|\frac{\partial x_k}{\partial x'_i}\right\|^{-1}$, is bounded on B. Then, $\left\|\frac{\partial x_k}{\partial x'_i}\right\| \geq c$ on B for some constant $c > 0$. This implies that $\left\|T'_{ij}\right\| \geq c$ on B for some constant $c > 0$. Therefore, the definition of uniform non-degeneracy is independent of the chosen coordinate system. Note that in the classical theory, uniform non-degeneracy follows from non-degeneracy.

Now we can define metrics.

Definition 8.11. A semi-Riemann metric g of the index k on an m-dimensional manifold M is a C^∞ uniformly non-degenerated tensor field on M of the type $(0,2)$, such that for every $p \in M$, g_p is symmetric and has the index k. When $k = 0$, g is called a Riemann metric. When $k = 1$ or $m-1$, g is called a Lorentz metric. A semi-Riemann (or Lorentz) manifold is a manifold with a semi-Riemann (or Lorentz) metric.

A Riemann metric g is positively definite. That is, $g(v,v) \geq 0$, and $g(v,v) = 0$ implies that $v = 0$, for all $v \in V_M$. However, for a semi-Riemann metric g, there can be non-zero vector v such that $g(v,v) = 0$. In the rest of the chapter we assume that M is an m-dimensional semi-Riemann manifold with a semi-Riemann metric g.

Since g is non-degenerated (at every point of M), its component matrix (g_{ij}) in a basis has an inverse matrix, which we denote as $\left(g^{ij}\right)$. That is, $g^{ij} g_{jk} = \delta_k^i$. Recall that from matrix algebra, g^{ij} can be computed from the determinants of (g_{ij}) and its submatrices. From there it is easy to see that since (g_{ij}) is symmetric, (g^{ij}) must also be symmetric. We define g^{-1} as a tensor of the type $(2,0)$ such that its components in the same basis are g^{ij}. Suppose that in another basis the components of g are $g'_{ij} = a^k_i a^h_j g_{kh}$. Then, the components of g^{-1} are $g'^{ij} = b^i_k b^j_h g^{kh}$, where $\left(b^j_i\right)$ is the inverse of $\left(a^j_i\right)$. A straightforward computation shows that $\left(g'^{ij}\right)$ is the inverse of $\left(g'_{ij}\right)$. Therefore, the definition of g^{-1} is independent of the basis. We will simply denote g^{-1} as g^{ij}.

Using the tensors g_{ij} and g^{ij}, we can transform vectors to dual vectors and reversely. Given a vector v with the components v^i, let $v_i = g_{ij}v^j$. v_i are the components of a dual vector v^* in the same basis. v^* is actually the linear function $g(\cdot, v)$. Therefore, this definition is actually independent of the chosen basis. We will simply write the vector and the corresponding dual vector as v^i and v_i and say that v_i is obtained from v^i by lowering the index i. Similarly, given a dual vector ω with the components ω_i, $\omega^i = g^{ij}\omega_j$ becomes the components of a vector ω_*. It is obtained by raising the index. Since $\left(g^{ij}\right)$ is the inverse matrix of (g_{ij}), a straightforward computation shows that raising and then lowering the same index again (or reversely) will go back to the original dual vector (or vector). Note that $v \mapsto v^*$ and $\omega \mapsto \omega_*$ are linear functions. Raising and lowering index can be defined for tensors of other types similarly. For instance, let $T^{ij}_{\ \ k}$ be (the components of) a tensor. We define $T^{i\ k}_{\ j}$ to be $g_{jh}g^{kl}T^{ih}_{\ \ l}$. $T^{i\ k}_{\ j}$ are actually the components of the tensor that maps (ω_1, v, ω_2) into $T(\omega_1, v^*, (\omega_2)_*)$. Note that $g^{ik}g^{jl}g_{kl} = g^{ij}$. That is, raising the indices of g_{ij} will get g^{ij}.

8.4 Covariant Derivative

Given an admissible chart $\langle U, \mu \rangle$ for M, for $p \in U$, $i, j, k = 1, \ldots, m$, we define the Christoffel symbols $\Gamma^i_{jk;\mu}(p)$ at p with respect to μ as

$$\Gamma^i_{jk;\mu}(p) = \frac{1}{2} g^{ih}\left(\partial_{\mu,j}g_{kh} + \partial_{\mu,k}g_{jh} - \partial_{\mu,h}g_{jk}\right),$$

where items at the right hand side all take their values at p, and g_{ij} are the components of g in the coordinate basis $\partial_{\mu,1}, \ldots, \partial_{\mu,m}$ for the chart. Moreover, recall that $\partial_{\mu,j}g_{kh} = \partial_j\left(g_{kh} \circ \mu^{-1}\right) = \frac{\partial\left(g_{kh}\circ\mu^{-1}\right)(\mathbf{x})}{\partial x_j}\big|_{\mathbf{x}=\mu(p)}$. We will write $\Gamma^i_{jk;\mu}(p)$ as $\Gamma^i_{jk;\mu}$ or Γ^i_{jk} when no ambiguity will arise, and we do not consider Γ^i_{jk} as components of any tensor. Note that Γ^i_{jk} is symmetric with respect to its indices j, k:

$$\Gamma^i_{jk} = \Gamma^i_{kj}.$$

Suppose that $\langle U', \mu' \rangle$ is another admissible chart for M. Then, the components of g and g^{-1} in the new coordinate basis $\partial_{\mu',1}, \ldots, \partial_{\mu',m}$ will be

$$g'_{ij} = \frac{\partial x_h}{\partial x'_i} \frac{\partial x_l}{\partial x'_j} g_{hl}, \; g'^{ij} = \frac{\partial x'_i}{\partial x_h} \frac{\partial x'_j}{\partial x_l} g^{hl},$$

where $\mathbf{x}' = \mu' \circ \mu^{-1}(\mathbf{x})$ on $\mu(U)$ are the coordinate transformation functions. The Christoffel symbols with respect to the new chart are

$$\Gamma^i_{jk;\mu'} = \frac{1}{2} g'^{ih} \left(\partial_{\mu',j} g'_{kh} + \partial_{\mu',k} g'_{jh} - \partial_{\mu',h} g'_{jk} \right).$$

By the chain rule, $\partial_{\mu',j} g'_{kh} = \frac{\partial x_n}{\partial x'_j} \partial_{\mu,n} g'_{kh}$. Substitute these into the above expression, we obtain, by some straightforward computations based on the chain rule, that

$$\Gamma^i_{jk;\mu'} = \frac{\partial x_h}{\partial x'_j} \frac{\partial x_l}{\partial x'_k} \frac{\partial x'_i}{\partial x_n} \Gamma^n_{hl;\mu} + \frac{\partial x'_i}{\partial x_n} \frac{\partial^2 x_n}{\partial x'_j \partial x'_k}.$$

Note that $\frac{\partial x'_i}{\partial x_n} \frac{\partial x_n}{\partial x'_j} = \delta^i_j$. Therefore,

$$\frac{\partial}{\partial x'_k} \left(\frac{\partial x'_i}{\partial x_n} \frac{\partial x_n}{\partial x'_j} \right) = \frac{\partial x'_i}{\partial x_n} \frac{\partial^2 x_n}{\partial x'_j \partial x'_k} + \frac{\partial^2 x'_i}{\partial x_n \partial x_l} \frac{\partial x_l}{\partial x'_k} \frac{\partial x_n}{\partial x'_j} = 0.$$

Then, the expression for $\Gamma^i_{jk;\mu'}$ can also be written as

$$\Gamma^i_{jk;\mu'} = \frac{\partial x_h}{\partial x'_j} \frac{\partial x_l}{\partial x'_k} \frac{\partial x'_i}{\partial x_n} \Gamma^n_{hl;\mu} - \frac{\partial^2 x'_i}{\partial x_n \partial x_l} \frac{\partial x_n}{\partial x'_j} \frac{\partial x_l}{\partial x'_k}.$$

This means that $\Gamma^i_{jk;\mu'}$ and $\Gamma^i_{jk;\mu}$ are not related to each other as the components of a tensor in different bases are.

Given any vector filed v on U with components v^i in the coordinate basis of the chart μ, for any $i, j = 1, \ldots, m$, we define

$$\nabla_{j;\mu} v^i = \partial_{\mu,j} v^i + \Gamma^i_{jk;\mu} v^k.$$

In a new chart μ', we will have $\nabla_{j;\mu'} v'^i = \partial_{\mu',j} v'^i + \Gamma^i_{jk;\mu'} v'^k$, where $v'^i = \frac{\partial x'_i}{\partial x_h} v^h$ are the components of v in the new coordinate basis and $\Gamma^i_{jk;\mu'}$ are given above. Some straightforward computations will show that

$$\nabla_{j;\mu'} v'^i = \frac{\partial x'_i}{\partial x_h} \frac{\partial x_l}{\partial x'_j} \nabla_{l;\mu} v^h.$$

We will treat $\nabla_{j;\mu} v^i$ as the components $(\nabla v)_{ji}$ of a $(1,1)$ tensor ∇v in the coordinate basis of μ. Then, the components of the tensor in the new basis will be exactly

$\nabla_{j;\mu'}v'^i$. That is, the definition of the tensor ∇v is actually independent of the chosen coordinate basis. ∇v is called the covariant derivative of v. We will simply denote the components of the tensor ∇v as

$$\nabla_j v^i = \partial_j v^i + \Gamma^i_{jk} v^k.$$

Note that (∇v) is a tensor field on U, and for $p \in U$, $(\nabla v)_p$ depends on v_q for $q \neq p$, because the partial derivative $\partial_j v^i(p)$ depends on $v^i(q)$ for $q \neq p$. In other words, ∇ is not a function from V_p to $\mathcal{T}_p^{(1,1)}$.

The covariant derivative $\nabla\omega$ of a dual vector ω with the components ω_i is a $(0,2)$ tensor with components $(\nabla\omega)_{ij}$ defined as

$$\nabla_i \omega_j = \partial_i \omega_j - \Gamma^k_{ij} \omega_k. \tag{8.1}$$

It is easy to verify that this is also independent of the chosen coordinate basis. More generally, the covariant derivative ∇T of a tensor T of the type (k,l) is a tensor of the type $(k,l+1)$ with the components

$$(\nabla T)^{i\dots}{}_{kj\dots} = \nabla_k T^{i\dots}{}_{j\dots} = \partial_k T^{i\dots}{}_{j\dots} + \Gamma^i_{kh} T^{h\dots}{}_{j\dots} + \dots - \Gamma^l_{kj} T^{i\dots}{}_{l\dots} - \dots. \tag{8.2}$$

It is also easy to verify that this is independent of the chosen coordinate basis. Moreover, for a C^∞ scalar function f on M, the covariant derivative ∇f of f is a dual vector with the components

$$\nabla_i f = \partial_i f.$$

The covariant derivative operator ∇ satisfies the following conditions

(1) Linearity: for any n tensors T_k and real numbers r_k, $k = 1,\dots,n$,

$$\nabla\left(\sum_{k=1}^n r_k T_k\right) = \sum_{k=1}^n r_k \nabla T_k.$$

(2) Leibnitz rule: for any tensors T, T',

$$\nabla\left(T \otimes T'\right) = (\nabla T) \otimes T' + T \otimes \nabla T'.$$

(3) Commutativity with contraction: for any tensor $T^{\dots i\dots}{}_{\dots j\dots}$,

$$\nabla_k\left(T^{\dots i\dots}{}_{\dots i\dots}\right) = \nabla_k T^{\dots i\dots}{}_{\dots i\dots}.$$

(4) Being the partial derivative operator on scalar functions: given any coordinate system, for a C^∞ scalar function f on M, the components of ∇f (as a $(0,1)$ tensor) are $(\nabla f)_i = \partial_i f$.
(5) Torsion-free: for any C^∞ scalar function f on M, $\nabla\nabla f$ is a symmetric tensor of the type $(0,2)$.
(6) Metric compatibility: $\nabla g = 0$.

The conditions (1), (2) and (4) are trivial from the definition. To see (3), let T be the tensor in (8.2). Then,

$$\nabla_k T^{i\dots}_{i\dots} = \partial_k T^{i\dots}_{i\dots} + \Gamma^i_{kh} T^{h\dots}_{i\dots} + \dots - \Gamma^l_{ki} T^{i\dots}_{l\dots} - \dots.$$

Note that $\Gamma^i_{kh} T^{h\dots}_{i\dots} = \Gamma^l_{ki} T^{i\dots}_{l\dots}$. Then it is easy to see that $\nabla_k T^{i\dots}_{i\dots} = \nabla_k \left(T^{\dots i \dots}_{\dots i \dots} \right)$. Since $\nabla_i f = \partial_i f$, by (8.1),

$$(\nabla\nabla f)_{ji} = \nabla_j \nabla_i f = \partial_j \partial_i f - \Gamma^k_{ji} \partial_k f.$$

Γ^k_{ji} and $\partial_j \partial_i$ are symmetric in the indices i, j. Therefore, $\nabla\nabla f$ is symmetric. Finally, the components of ∇g are

$$\nabla_k g_{ij} = \partial_k g_{ij} - \Gamma^h_{ki} g_{hj} - \Gamma^l_{kj} g_{il}.$$

Substituting the expression for Γ^h_{ki} and Γ^l_{kj} into the right hand side, a straightforward computation gives $\nabla_k g_{ij} = 0$.

As in the classical theory, we can show that covariant derivative ∇ is the unique operator that maps a type (k,l) tensor field into a type $(k, l+1)$ tensor field and satisfies the above conditions (1) to (6). To see this, let $\widetilde{\nabla}$ be another such operator. We fix a coordinate system. Condition (4) implies that for any C^∞ function f,

$$\widetilde{\nabla} f = (\partial_i f)\, \mathrm{d}x^i.$$

For each dual vector field $\mathrm{d}x^i$, $\widetilde{\nabla}\left(\mathrm{d}x^i\right)$ is a $(0,2)$ tensor field. We denote its components in the coordinate basis as $-C^i{}_{jk}$. That is,

$$\widetilde{\nabla}\left(\mathrm{d}x^i\right) = -C^i{}_{jk} \mathrm{d}x^j \otimes \mathrm{d}x^k.$$

Then, for any dual vector field $\omega = \omega_i \mathrm{d}x^i$, by the Leibnitz rule,

$$\begin{aligned}
\widetilde{\nabla}\omega &= \widetilde{\nabla}\left(\omega_i \mathrm{d}x^i\right) = \widetilde{\nabla}(\omega_i) \otimes \mathrm{d}x^i + \omega_i \widetilde{\nabla}\left(\mathrm{d}x^i\right) \\
&= (\partial_j \omega_i)\, \mathrm{d}x^j \otimes \mathrm{d}x^i - \omega_i C^i{}_{jk} \mathrm{d}x^j \otimes \mathrm{d}x^k \\
&= (\partial_j \omega_k)\, \mathrm{d}x^j \otimes \mathrm{d}x^k - C^i{}_{jk} \omega_i \mathrm{d}x^j \otimes \mathrm{d}x^k \\
&= \left(\partial_j \omega_k - C^i{}_{jk} \omega_i\right) \mathrm{d}x^j \otimes \mathrm{d}x^k.
\end{aligned}$$

That is, the components of $\widetilde{\nabla}\omega$ are

$$\widetilde{\nabla}_j \omega_k = \partial_j \omega_k - C^i{}_{jk} \omega_i.$$

Note that when f is the i-th coordinate function x_i^*, $x_i^*(p) = (\mu(p))_i$, f is a C^∞ function and $\widetilde{\nabla} f = \mathrm{d}x^i$. Then, $\widetilde{\nabla}\widetilde{\nabla} f = -C^i{}_{jk} \mathrm{d}x^j \otimes \mathrm{d}x^k$. By the torsion-free condition (5), $\widetilde{\nabla}\widetilde{\nabla} f$ is symmetric. Therefore, $C^i{}_{jk} = C^i{}_{kj}$.

For each vector field ∂_k, $\widetilde{\nabla}\partial_k$ is a $(1,1)$ tensor field. We denote its components as $S^i{}_{jk}$. That is,

$$\widetilde{\nabla}\partial_k = S^i{}_{jk}\partial_i \otimes \mathrm{d}x^j.$$

Consider the tensor $\partial_k \otimes \mathrm{d}x^i$. Its contraction $Tr\left(\partial_k \otimes \mathrm{d}x^i\right) = \delta^i_k$ is a constant scalar function. Therefore, by the condition (4), $\widetilde{\nabla}Tr\left(\partial_k \otimes \mathrm{d}x^i\right) = 0$. On the other side, by the condition (3), $\widetilde{\nabla}$ commutates with Tr. Therefore, we should have $Tr\widetilde{\nabla}(\partial_k \otimes \mathrm{d}x^i) = 0$. Now, by the Leibnitz rule,

$$\begin{aligned}
&\widetilde{\nabla}\left(\partial_k \otimes \mathrm{d}x^i\right)\\
&= \widetilde{\nabla}\partial_k \otimes \mathrm{d}x^i + \partial_k \otimes \widetilde{\nabla}\left(\mathrm{d}x^i\right)\\
&= S^h{}_{lk}\partial_h \otimes \mathrm{d}x^l \otimes \mathrm{d}x^i - C^i{}_{lh}\partial_k \otimes \mathrm{d}x^l \otimes \mathrm{d}x^h.
\end{aligned}$$

Note that we are using the tensor notation here, not the component notation. Each term in the sum above is a tensor. Also note that the new argument generated by the operator $\widetilde{\nabla}$ in the tensor above is the middle argument, that is, $\mathrm{d}x^l$, and contraction is performed on the other two arguments. Therefore, we have

$$Tr\left(\widetilde{\nabla}\left(\partial_k \otimes \mathrm{d}x^i\right)\right) = S^i{}_{lk}\mathrm{d}x^l - C^i{}_{lk}\mathrm{d}x^l = \left(S^i{}_{lk} - C^i{}_{lk}\right)\mathrm{d}x^l = 0.$$

Applying $\left(S^i{}_{lk} - C^i{}_{lk}\right)\mathrm{d}x^l$ to ∂_j, we have $\left(S^i{}_{jk} - C^i{}_{jk}\right) = 0$, since $\mathrm{d}x^l\partial_j = \delta^l_j$. Therefore, $S^i{}_{jk} = C^i{}_{jk}$. That is, we have

$$\widetilde{\nabla}\partial_k = C^i{}_{jk}\partial_i \otimes \mathrm{d}x^j.$$

Then, for any vector field $v = v^i\partial_i$, by the Leibnitz rule,

$$\begin{aligned}
\widetilde{\nabla}v &= \widetilde{\nabla}\left(v^i\partial_i\right) = \widetilde{\nabla}v^i \otimes \partial_i + v^k\widetilde{\nabla}\partial_k\\
&= \partial_j v^i \mathrm{d}x^j \otimes \partial_i + v^k C^i{}_{jk}\partial_i \otimes \mathrm{d}x^j\\
&= \left(\partial_j v^i + v^k C^i{}_{jk}\right)\partial_i \otimes \mathrm{d}x^j.
\end{aligned}$$

That is, the components of $\widetilde{\nabla}v$ are

$$\widetilde{\nabla}_j v^i = \partial_j v^i + C^i{}_{jk}v^k.$$

Using the expression for $\widetilde{\nabla}\partial_i$ and $\widetilde{\nabla}\left(\mathrm{d}x^j\right)$ we can similarly compute $\widetilde{\nabla}T$ for any tensor

$$T = T^{i\ldots}{}_{j\ldots}\partial_i \otimes \ldots \otimes \mathrm{d}x^j \otimes \ldots,$$

and we can see that the resulted tensor has the components as in (8.2) with Γ replaced by C, that is,

$$\left(\widetilde{\nabla}T\right)^{i\ldots}_{kj\ldots} = \widetilde{\nabla}_k T^{i\ldots}_{j\ldots} = \partial_k T^{i\ldots}_{j\ldots} + C^i_{\ kh} T^{h\ldots}_{j\ldots} + \ldots - C^l_{\ kj} T^{i\ldots}_{l\ldots} - \ldots. \tag{8.3}$$

Therefore, the rest is to show that $C^i_{\ jk}$ must be equal to Γ^i_{jk}.

By the condition (5), metric compatibility, $\widetilde{\nabla} g = 0$. Now, by (8.3),

$$\widetilde{\nabla}_k g_{ij} = \partial_k g_{ij} - C^h_{\ ki} g_{hj} - C^h_{\ kj} g_{ih} = \partial_k g_{ij} - C_{jki} - C_{ikj},$$

where we use the notation of lowering indices. Therefore,

$$C_{ikj} + C_{jki} = \partial_k g_{ij}. \tag{8.4}$$

Rotating the indices, we also have

$$C_{jik} + C_{kij} = \partial_i g_{jk}, \tag{8.5}$$

$$C_{kji} + C_{ijk} = \partial_j g_{ki}. \tag{8.6}$$

Recall that $C^i_{\ jk}$ is symmetric at the lower indices. Add (8.4) and (8.6) and then subtract (8.5), we get

$$C_{ijk} = \frac{1}{2}\left(\partial_k g_{ij} + \partial_j g_{ki} - \partial_i g_{jk}\right).$$

Raising the index i again, we have

$$C^i_{\ jk} = g^{ih} C_{hjk} = \frac{1}{2} g^{ih}\left(\partial_k g_{jh} + \partial_j g_{kh} - \partial_h g_{jk}\right) = \Gamma^i_{jk}.$$

This completes the proof that ∇ is the unique operator satisfying the conditions (1)-(6) above.

8.5 Parallel Transportation, Geodesics and Curvature

Covariant derivative is used to define when a tensor field is parallelly transported along a curve. For a tensor field T, intuitively, ∇T is supposed to encode the rates of change of T in various directions. That is, for vector v^k, $v^k \nabla_k T^{i\ldots}_{j\ldots}$ gives the rate of change of T at the direction v^k. Recall that the tangent vector of a differentiable curve γ in the manifold M on an interval $[a,b]$ is the vector $\frac{\mathrm{d}\gamma^k}{\mathrm{d}t}$, where γ^k is the k-th component of the function $\mu \circ \gamma$ from $[a,b]$ to $\mathbb{R}^m$. In case

$$\frac{\mathrm{d}\gamma^k(t)}{\mathrm{d}t} \nabla_k T^{i\ldots}_{j\ldots}(\gamma(t)) = 0$$

for $t \in [a,b]$, we say that the tensor filed $T^{i\ldots}_{j\ldots}$ is parallelly transported along the curve γ, which means that the rates of change of $T^{i\ldots}_{j\ldots}$ at the direction of the curve are always 0. By (8.2), this is a system of first-order ordinary differential equations for the tensor components $T^{i\dots}_{j\dots}$:

$$\frac{\mathrm{d}}{\mathrm{d}t}T^{i\dots}_{j\dots} + \frac{\mathrm{d}\gamma^k(t)}{\mathrm{d}t}\left(\Gamma^i_{kh}T^{h\dots}_{j\dots} + \dots - \Gamma^l_{kj}T^{i\dots}_{l\dots} - \dots\right) = 0.$$

Parallel transportation is in turn used to define geodesics. A geodesic is a curve whose tangent vectors are parallelly transported along the curve. Since the tangent vector of a curve γ is $\frac{\mathrm{d}\gamma^k}{\mathrm{d}t}$, applying the above equation for parallel transportation we get

$$\frac{\mathrm{d}^2\gamma^i(t)}{\mathrm{d}t^2} + \Gamma^i_{kh}\frac{\mathrm{d}\gamma^k(t)}{\mathrm{d}t}\frac{\mathrm{d}\gamma^h(t)}{\mathrm{d}t} = 0. \tag{8.7}$$

This is a system of second-order differential equations. By the existential theorem on the solution of the initial value problem for a system of n-th-order differential equations, given any point $p \in M$ and any vector $v \in V_p$, there exists a geodesic γ passing through p and having the tangent vector v at p, that is, $\gamma(t_0) = p$, $\gamma'(t_0) = v$.

As in the classical theory, we can show that a geodesic is a curve with a stationary length. First, we have to introduce the notion of length. For a C^∞ curve γ in M on the interval $[a,b]$, recall that $\gamma'(t)$ denotes the tangent vector of γ at the point $\gamma(t)$, $t \in [a,b]$. When the metric g on M is a Riemann metric, we define the length of γ as

$$l_\gamma = \int_a^b g\left(\gamma'(t), \gamma'(t)\right)^{1/2}\mathrm{d}t.$$

When g is a Lorentz metric with the index 1 (that is, its component matrix in diagonal form has a single -1 on the diagonal), we say that γ is timelike if $g(\gamma'(t), \gamma'(t)) < 0$ for all $t \in [a,b]$, and it is spacelike if $g(\gamma'(t), \gamma'(t)) > 0$ for all $t \in [a,b]$, and it is null if $g(\gamma'(t), \gamma'(t)) = 0$ for all $t \in [a,b]$. For a spacelike curve, its length can be defined similarly as above. For a timelike curve, its length, also called proper time, is defined as

$$\tau_\gamma = \int_a^b \left(-g\left(\gamma'(t), \gamma'(t)\right)\right)^{1/2}\mathrm{d}t.$$

A curve $\gamma(t)$, $t \in [a,b]$, can be reparameterized. That is, let $t = t(s)$ be a C^∞ function that maps $[c,d]$ onto $[a,b]$ such that $t(c) = a$, $t(d) = b$. Then, the curve $\gamma_*(s) = \gamma(t(s))$ on $[c,d]$ is a reparameterization of γ. Note that $\frac{\mathrm{d}}{\mathrm{d}s}\gamma^k_*(s) = \frac{\mathrm{d}}{\mathrm{d}t}\gamma^k\frac{\mathrm{d}t}{\mathrm{d}s}$. Therefore, $\gamma'_*(s) = \frac{\mathrm{d}t}{\mathrm{d}s}\gamma'(t)$. Then, for spacelike curves or curves in Riemann space,

$$l_{\gamma_*} = \int_a^b g\left(\gamma'_*, \gamma'_*\right)^{1/2}\mathrm{d}s = \int_a^b g\left(\gamma', \gamma'\right)^{1/2}\frac{\mathrm{d}t}{\mathrm{d}s}\mathrm{d}s = l_\gamma.$$

That is, reparameterization does not change the length of a curve. The same holds for the length or proper time of a timelike curve. We can always reparameterize a curve using the length or proper time as the parameter. For the case of proper time, this means using

$$\tau = \tau(t) = \int_a^t \left(-g\left(\gamma'(t), \gamma'(t)\right)\right)^{1/2} \mathrm{d}t$$

as the new parameter. Note that $\frac{\mathrm{d}\tau}{\mathrm{d}t} = \left(-g\left(\gamma'(t), \gamma'(t)\right)\right)^{1/2} > 0$ on $[a,b]$. Therefore, its inverse function $t = t(\tau)$ exists and maps $[0, \tau_\gamma]$ onto $[a,b]$. After the reparameterization, we have $g\left(\gamma_*'(\tau), \gamma_*'(\tau)\right) = -1$ on $[0, \tau_\gamma]$.

Now, suppose that $\gamma(t), t \in [0,l]$, is a spacelike curve already parameterized with the length parameter. We want to find the condition that γ has a stationary length. For this we consider any C^∞ function δ from $[0,l] \times [-a,a]$ to M, such that $\delta(t,0) = \gamma(t)$ for $t \in [0,l]$, $\delta(0,s) = \gamma(0)$ and $\delta(l,s) = \gamma(l)$ for all $s \in [-a,a]$. This condition implies that

$$\frac{\partial \delta}{\partial s}(0,s) = \frac{\partial \delta}{\partial s}(l,s) = 0.$$

Let δ^k be the k-th coordinate component of δ, that is, the k-th component of the function $\mu \circ \delta$ from $[0,l] \times [-a,a]$ to $\mathbb{R}^m$. The curve $\gamma_s(t) = \delta(t,s)$ is a variation of $\gamma = \gamma_0$. The length $l_s = l_{\gamma_s}$ of the curve γ_s is

$$l_s = \int_0^l \left(g_{ij} \frac{\mathrm{d}}{\mathrm{d}t}\gamma_s^i \frac{\mathrm{d}}{\mathrm{d}t}\gamma_s^j\right)^{1/2} \mathrm{d}t = \int_0^l \left(g_{ij} \frac{\partial \delta^i(t,s)}{\partial t} \frac{\partial \delta^j(t,s)}{\partial t}\right)^{1/2} \mathrm{d}t.$$

The stationary condition for the length l_s at $s = 0$ is $\frac{\mathrm{d}}{\mathrm{d}s} l_s|_{s=0} = 0$. Denote

$$f(t,s) = g_{ij} \frac{\partial \delta^i(t,s)}{\partial t} \frac{\partial \delta^j(t,s)}{\partial t}.$$

Since $\gamma = \gamma_0$ is parameterized with the length parameter, $f(t,0) = 1$ for $t \in [0,l]$. Taking derivative under the integration, we see that at $s = 0$,

$$\begin{aligned}
\frac{\mathrm{d}}{\mathrm{d}s} l_s|_{s=0} &= \frac{1}{2} \int_0^l \frac{1}{f^{1/2}(t,0)} \frac{\mathrm{d}f}{\mathrm{d}s}|_{s=0} \mathrm{d}t \\
&= \int_0^l \left(\frac{1}{2} \frac{\partial g_{ij}}{\partial x^k} \frac{\partial \delta^k}{\partial s} \frac{\partial \delta^i}{\partial t} \frac{\partial \delta^j}{\partial t} + g_{ij} \frac{\partial \delta^i}{\partial t} \frac{\partial^2 \delta^j}{\partial t \partial s}\right)|_{s=0} \mathrm{d}t.
\end{aligned}$$

Note that $\frac{\partial \delta^i}{\partial t}|_{s=0} = \frac{\mathrm{d}\gamma^i}{\mathrm{d}t}$. We apply integration by parts to the second term,

$$\begin{aligned}
&\int_0^l g_{ij} \frac{\partial \delta^i}{\partial t} \frac{\partial^2 \delta^j}{\partial t \partial s}|_{s=0} \mathrm{d}t \\
&= g_{ij} \frac{\partial \delta^i}{\partial t} \frac{\partial \delta^j}{\partial s}\Big|_{s=0,t=0}^{s=0,t=l} - \int_0^l \frac{\mathrm{d}}{\mathrm{d}t}\left(g_{ij} \frac{\mathrm{d}\gamma^i}{\mathrm{d}t}\right) \frac{\partial \delta^j}{\partial s}|_{s=0} \mathrm{d}t \\
&= -\int_0^l \left(\frac{\partial g_{ij}}{\partial x^k} \frac{\mathrm{d}\gamma^k}{\mathrm{d}t} \frac{\mathrm{d}\gamma^i}{\mathrm{d}t} + g_{ij} \frac{\mathrm{d}^2 \gamma^i}{\mathrm{d}t^2}\right) \frac{\partial \delta^j}{\partial s}|_{s=0} \mathrm{d}t.
\end{aligned}$$

Therefore,

$$\frac{\mathrm{d}}{\mathrm{d}s} l_s|_{s=0} = \int_0^l \left(\frac{1}{2} \frac{\partial g_{ik}}{\partial x^j} \frac{\mathrm{d}\gamma^k}{\mathrm{d}t} \frac{\mathrm{d}\gamma^i}{\mathrm{d}t} - \frac{\partial g_{ij}}{\partial x^k} \frac{\mathrm{d}\gamma^k}{\mathrm{d}t} \frac{\mathrm{d}\gamma^i}{\mathrm{d}t} - g_{ij} \frac{\mathrm{d}^2\gamma^i}{\mathrm{d}t^2} \right) \frac{\partial \delta^j}{\partial s}|_{s=0} \mathrm{d}t$$
$$= \int_0^l \left(\frac{1}{2} \left(\frac{\partial g_{ik}}{\partial x^j} - \frac{\partial g_{ij}}{\partial x^k} - \frac{\partial g_{kj}}{\partial x^i} \right) \frac{\mathrm{d}\gamma^k}{\mathrm{d}t} \frac{\mathrm{d}\gamma^i}{\mathrm{d}t} - g_{ij} \frac{\mathrm{d}^2\gamma^i}{\mathrm{d}t^2} \right) \frac{\partial \delta^j}{\partial s}|_{s=0} \mathrm{d}t.$$

Obviously, a sufficient condition for $\frac{\mathrm{d}}{\mathrm{d}s} l_s|_{s=0} = 0$ is

$$\frac{1}{2} \left(\frac{\partial g_{ik}}{\partial x^j} - \frac{\partial g_{ij}}{\partial x^k} - \frac{\partial g_{kj}}{\partial x^i} \right) \frac{\mathrm{d}\gamma^k}{\mathrm{d}t} \frac{\mathrm{d}\gamma^i}{\mathrm{d}t} - g_{ij} \frac{\mathrm{d}^2\gamma^i}{\mathrm{d}t^2} = 0. \tag{8.8}$$

Multiply g^{hj} and contract the index j, we get

$$\frac{\mathrm{d}^2\gamma^h}{\mathrm{d}t^2} + \frac{1}{2} g^{hj} \left(\frac{\partial g_{ij}}{\partial x^k} + \frac{\partial g_{kj}}{\partial x^i} - \frac{\partial g_{ik}}{\partial x^j} \right) \frac{\mathrm{d}\gamma^k}{\mathrm{d}t} \frac{\mathrm{d}\gamma^i}{\mathrm{d}t} = 0.$$

This is exactly the geodesic equation (8.7).

Another application of covariant derivative is to define the Riemann curvature tensor. For any dual vector field $\mathrm{d}x^l$, $\nabla\nabla \mathrm{d}x^l$ is a tensor of the type $(0,3)$. Let $T_{ijk}{}^l$ be its components in the coordinate basis, that is,

$$\nabla\nabla \mathrm{d}x^l = T_{ijk}{}^l \mathrm{d}x^i \mathrm{d}x^j \mathrm{d}x^k.$$

We define

$$R_{ijk}{}^l = T_{ijk}{}^l - T_{jik}{}^l.$$

For any dual vector field $\omega = \omega_l \mathrm{d}x^l$,

$$\nabla\nabla\omega = \nabla \left(\nabla\omega_l \otimes \mathrm{d}x^l \right) + \nabla \left(\omega_l \otimes \nabla \mathrm{d}x^l \right).$$

Apply this to the vectors $\partial_i, \partial_j, \partial_k$, we have

$$\begin{aligned} &\nabla \left(\nabla\omega_l \otimes \mathrm{d}x^l \right) (\partial_i, \partial_j, \partial_k) \\ &= \left(\nabla\nabla\omega_l \otimes \mathrm{d}x^l + \nabla\omega_l \otimes \nabla \mathrm{d}x^l \right) (\partial_i, \partial_j, \partial_k) \\ &= \nabla\nabla\omega_l (\partial_i, \partial_j) \mathrm{d}x^l (\partial_k) + \nabla\omega_l (\partial_j, \partial_k) \nabla \mathrm{d}x^l (\partial_i, \partial_k). \end{aligned}$$

Note that in the second term above, ∇ is applied to $\nabla\omega_l$ first and to $\mathrm{d}x^l$ next. Therefore, the argument ∂_i, which is generated by the outer (i.e. the second application of) ∇ goes with $\nabla \mathrm{d}x^l$. Similarly,

$$\begin{aligned} &\nabla \left(\omega_l \otimes \nabla \mathrm{d}x^l \right) (\partial_i, \partial_j, \partial_k) \\ &= \nabla\omega_l (\partial_i, \partial_k) \nabla \mathrm{d}x^l (\partial_j, \partial_k) + \omega_l \nabla\nabla \mathrm{d}x^l (\partial_i, \partial_j, \partial_k). \end{aligned}$$

$\nabla\nabla\omega_l (\partial_i, \partial_j)$ is symmetric in i, j because ∇ is torsion-free. Then, we see that

$$\begin{aligned}&\nabla\nabla\omega(\partial_i,\partial_j,\partial_k)-\nabla\nabla\omega(\partial_j,\partial_i,\partial_k)\\&=\omega_l\left(\nabla\nabla \mathrm{d}x^l(\partial_i,\partial_j,\partial_k)-\nabla\nabla \mathrm{d}x^l(\partial_j,\partial_i,\partial_k)\right)\\&=R_{ijk}{}^l\omega_l.\end{aligned}$$

The components of the $(0,3)$ tensor $\nabla\nabla\omega$ are denoted as $\nabla_i\nabla_j\omega_k=\nabla\nabla\omega(\partial_i,\partial_j,\partial_k)$. Therefore, $R_{ijk}{}^l$ are the numbers such that for any dual vector ω_l,

$$(\nabla_i\nabla_j-\nabla_j\nabla_i)\,\omega_k=R_{ijk}{}^l\omega_l.$$

It is straightforward to verify that $R_{ijk}{}^l$ are the components of a type $(1,3)$ tensor field R. R is called the Riemann curvature tensor.

A direct calculation of $\nabla\nabla \mathrm{d}x^l$ gives

$$\begin{aligned}&\nabla\nabla \mathrm{d}x^l\\&=\nabla\left(-\Gamma^l_{jk}\mathrm{d}x^j\mathrm{d}x^k\right)\\&=-\nabla\left(\Gamma^l_{jk}\right)\mathrm{d}x^j\mathrm{d}x^k-\Gamma^l_{hk}\nabla\left(\mathrm{d}x^h\right)\mathrm{d}x^k-\Gamma^l_{jh}\mathrm{d}x^j\nabla\left(\mathrm{d}x^h\right)\\&=-\partial_i\Gamma^l_{jk}\mathrm{d}x^i\mathrm{d}x^j\mathrm{d}x^k+\Gamma^l_{hk}\Gamma^h_{ij}\mathrm{d}x^i\mathrm{d}x^j\mathrm{d}x^k+\Gamma^l_{jh}\Gamma^h_{ik}\mathrm{d}x^i\mathrm{d}x^j\mathrm{d}x^k.\end{aligned}$$

Therefore,

$$\begin{aligned}R_{ijk}{}^l&=\nabla\nabla \mathrm{d}x^l(\partial_i,\partial_j,\partial_k)-\nabla\nabla \mathrm{d}x^l(\partial_j,\partial_i,\partial_k)\\&=\partial_j\Gamma^l_{ik}-\partial_i\Gamma^l_{jk}+\Gamma^l_{jh}\Gamma^h_{ik}-\Gamma^l_{ih}\Gamma^h_{jk}.\end{aligned}\tag{8.9}$$

A similar direct calculation gives

$$\begin{aligned}&\nabla\nabla\partial_l\\&=\nabla\left(\Gamma^k_{lj}\mathrm{d}x^j\partial_k\right)\\&=\partial_i\Gamma^k_{lj}\mathrm{d}x^i\mathrm{d}x^j\partial_k-\Gamma^k_{lh}\Gamma^h_{ij}\mathrm{d}x^i\mathrm{d}x^j\partial_k+\Gamma^h_{lj}\Gamma^k_{ih}\mathrm{d}x^i\mathrm{d}x^j\partial_k.\end{aligned}$$

Therefore,

$$\begin{aligned}&\nabla\nabla\partial_l(\partial_i,\partial_j,\partial_k)-\nabla\nabla\partial_l(\partial_j,\partial_i,\partial_k)\\&=\partial_i\Gamma^k_{lj}-\partial_j\Gamma^k_{li}+\Gamma^h_{lj}\Gamma^k_{ih}-\Gamma^h_{li}\Gamma^k_{jh}\\&=-R_{ijl}{}^k.\end{aligned}$$

Then, for a vector field v^i, we can similarly conclude that

$$(\nabla_i\nabla_j-\nabla_j\nabla_i)\,v^k=-R_{ijl}{}^kv^l.\tag{8.10}$$

More generally, for any tensor $T^{i\cdots}{}_{j\ldots}$,

$$(\nabla_i\nabla_j - \nabla_j\nabla_i)\, T^{k\dots}_{h\dots} = -R_{ijl}{}^{k} T^{l\dots}_{h\dots} + R_{ijh}{}^{n} T^{k\dots}_{n\dots} - \dots. \tag{8.11}$$

We can show that the tensor $R_{ijk}{}^{l}$ captures information about curvature. Suppose that A^i, B^j, and v^k are vectors at a point p. Suppose the we first parallelly transport v along A for a short distance t (in the Euclidean distance of the coordinate system) to get v' and then parallelly transport v' along B for a short distance s to get v''. Suppose that we do the same parallel transportations with the directions A and B switched and get v^* and v^{**}. We can show that $R_{kih}{}^{j} A^k B^i v^h st$ is an estimate of the difference $v''^k - v^{**k}$, which signifies how parallel transportations depend upon the paths and shows that the Riemann curvature tensor is a measure of how the space is curved. We assume that v is actually a vector field and denote $v = v(p)$, $v' = v(p')$, $v'' = v(p'')$. We estimate the components v'^j and v''^j up to the second order. First, since v is parallelly transported along A, we have $A^j\nabla_j v^i = 0$. Therefore,

$$A^i\partial_i v^j = A^i\nabla_i v^j - A^i\Gamma^j_{ik} v^k = -A^i\Gamma^j_{ik} v^k,$$

and similarly,

$$B^i\partial_i v'^j = -B^i\Gamma'^j_{ik} v'^k,$$

where $\Gamma'^j_{ik} = \Gamma^j_{ik}(p')$. Expanding v'^j to the second order, we have

$$\begin{aligned} v'^j &\approx v^j + A^i\partial_i v^j t + \frac{1}{2}A^iA^k\partial_i\partial_k v^j t^2 \\ &= v^j - A^i\Gamma^j_{ih} v^h t + \frac{1}{2}A^iA^k\partial_i\partial_k v^j t^2. \end{aligned}$$

Then, we can expand v''^j to the second order:

$$\begin{aligned} v''^j &\approx v'^j + B^i\partial_i v'^j s + \frac{1}{2}B^iB^k\partial_i\partial_k v'^j s^2 \\ &= v'^j - B^i\Gamma'^j_{ih} v'^h s + \frac{1}{2}B^iB^k\partial_i\partial_k v'^j s^2 \\ &\approx v^j - A^i\Gamma^j_{ih} v^h t + \frac{1}{2}A^iA^k\partial_i\partial_k v^j t^2 - B^i\left(\Gamma^j_{ih} + A^k\partial_k\Gamma^j_{ih} t\right)\left(v^h - A^k\Gamma^h_{kl} v^l t\right) s \\ &\quad + \frac{1}{2}B^iB^k\partial_i\partial_k\left(v^j - A^i\Gamma^j_{ih} v^h t\right) s^2 \\ &\approx v^j - A^i\Gamma^j_{ih} v^h t + \frac{1}{2}A^iA^k\partial_i\partial_k v^j t^2 - B^i\Gamma^j_{ih} v^h s - A^kB^i\partial_k\Gamma^j_{ih} v^h ts + A^kB^i\Gamma^j_{ih}\Gamma^h_{kl} v^l ts \\ &\quad + \frac{1}{2}B^iB^k\partial_i\partial_k v^j s^2. \end{aligned}$$

We have a similar expansion for v^{**j}, by switching A and B and switching t and s. In $v''^j - v^{**j}$, many terms cancel each other, and we have

$$
\begin{aligned}
& v''^j - v^{**j} \\
\approx & -A^k B^i \partial_k \Gamma_{ih}^j v^h ts + A^k B^i \Gamma_{ih}^j \Gamma_{kl}^h v^l ts + B^k A^i \partial_k \Gamma_{ih}^j v^h ts - B^k A^i \Gamma_{ih}^j \Gamma_{kl}^h v^l ts \\
= & A^k B^i \left(-\partial_k \Gamma_{ih}^j + \Gamma_{il}^j \Gamma_{kh}^l + \partial_i \Gamma_{kh}^j - \Gamma_{kl}^j \Gamma_{ih}^l \right) v^h ts \\
= & R_{kih}{}^j A^k B^i v^h ts.
\end{aligned}
$$

Other symmetric properties of R can be easily proved. For instance, we have

$$
\begin{aligned}
R_{ijk}{}^l &= -R_{jik}{}^l, \\
R_{ijkl} &= -R_{ijlk}.
\end{aligned}
$$

The first follows from the expression (8.9) for $R_{ijk}{}^l$ directly. To see the second, note that since $\nabla g = 0$, by (8.11),

$$
0 = (\nabla_i \nabla_j - \nabla_j \nabla_i) g_{kl} = R_{ijk}{}^n g_{nl} + R_{ijl}{}^n g_{kn} = R_{ijkl} + R_{ijlk}.
$$

The Ricci tensor *Ric* is the trace of R at the second and the forth positions. Therefore, its components R_{ik} are

$$
R_{ik} = R_{ijk}{}^j.
$$

The scalar curvature, also denoted as R, is defined as

$$
R = R_i^i.
$$

Then, Einstein's equation in general relativity is

$$
R_{ij} - \frac{1}{2} R g_{ij} = 8\pi T_{ij},
$$

where T_{ij} is the stress-energy tensor. (See Wald [36], p. 72.)

8.6 Case Study: Spacetime and Singularity

This section will analyze one of Hawking's singularity theorems about spacetime. The common textbook proofs of the theorem are highly non-constructive. They typically use various non-constructive compactness arguments in topology. These proofs appear to rely on continuity of the spacetime manifold in an essential manner. However, on the one side, compactness in topology is actually a way to express the finitude of a topological space in some aspect. While arguments resorting to compactness are usually non-constructive, relying on compactness is not in itself a sign that the relevant properties and arguments are essentially beyond finitism. On the other side, we know that spacetime models in general relativity are merely approximations to real spacetime at the macroscopic scale. If a proof of the existence of singularities in a spacetime model logically indispensably relies on some infinity or continuity assumptions about the model, we will have reason to doubt the physical meaningfulness of the proof. This section will try to show, with the help of strict finitism, that we can transform one of the classical proofs of Hawking's singularity theorem into sound logical deductions (i.e., valid deductions with literally true premises) on statements about real spacetime, even if real spacetime is discrete at the microscopic scale. This explains, from the logical point of view, why the conclusion of Hawking's singularity theorem is reliable as an assertion about real spacetime.

We start with our definition of spacetime manifolds in strict finitism. A spacetime structure is a 4-dimensional semi-Riemann manifold with a Lorentz metric. We will assume that the index of the metric is 1. That is, in an orthonormal basis of the metric, the diagonal component matrix of the metric has a single -1 on the diagonal. We will fix a spacetime structure M with the Lorentz metric g in this section. A vector v is timelike if $g(v,v) < 0$, and it is null if $g(v,v) = 0$, and it is spacelike if $g(v,v) > 0$. A spacetime M is time orientable if there exists a smooth vector field F on M such that for every $p \in M$, F_p is timelike. In this section we will assume that M is time orientable and we will fix such a timelike vector field F and say that F_p points to the future direction. Then, at every point $p \in M$, it is meaningful to say that an arbitrary timelike or null vector v at p points to the future direction, by which we mean $g(v, F_p) < 0$.

A 3-dimensional submanifold S of M is called a hypersurface. Consider a hypersurface S. For $p \in S$, let V_p^S denote the space of tangent vectors in S at p. Recall that vectors in V_p^S can be seen as vectors in V_p and V_p^S then becomes a 3-dimensional subspace of V_p. We say that S is a spacelike slice of M, if $g_p(v,v) > 0$ for all $p \in S$, $v \in V_p^S$. Note that this is a point-wise condition. We also need a uniform condition on spacelikeness. Consider any coordinate system $\langle U, \mu \rangle$ of M such that $S \cap U$ corresponds to a 3-dimensional slice of $\mu(U)$ in $\mathbb{R}^4$. We assume that the coordinates are labeled from 0 to 3 and assume that the slice is

$$\{x = (x_0, ..., x_3) \in \mu(U) : x_0 = r\}$$

for some constant r. In this case, we say that the chart $\langle U, \mu \rangle$ is adapted to S. A vector $v \in V_p$ is a tangent vector of S just in case its components v^i in this coordinate systems are such that $v^0 = 0$. Consider the component matrix $(g_{p,ij})$ of the metric g_p at a point $p \in U \cap S$ in this coordinate system. For vectors $v_1, v_2 \in V_p^S$, $g_p(v_1, v_2) = \sum_{i,j=1}^3 g_{p,ij} v_1^i v_2^j$. That is, the component matrix of g_p restricted to V_p^S is the submatrix of (g_{ij}) obtained by deleting the first row and the first column. Denote this submatrix as G_p^{00}. We say that S is a uniformly spacelike slice of M, if it is spacelike and for any such chart and any basic regular compact subset $C \subset S \cap U$, there exists a constant $c > 0$, such that for any $p \in C$, we have $\left| G_p^{00} \right| \geq c$, where $|\cdot|$ means the determinant of a matrix. Note that in the classical theory, the existence of c follows from the compactness of C and the fact that G_p^{00} is positively definite and its entries are uniformly continuous on C.

This definition of uniform spacelikeness is independent of the chosen coordinate system. To see this, suppose that $x_i' = x_i'(x_0, ..., x_3)$ is another coordinate system adapted to S, and suppose that S corresponds to the slice $x_0' = r'$. Then, $\frac{\partial x_0}{\partial x_i'} = 0$ on

S for $i = 1,2,3$. Therefore, for $i, j = 1,2,3$,

$$g'_{ij} = \frac{\partial x_k}{\partial x'_i}\frac{\partial x_l}{\partial x'_j} g_{ij} = \sum_{k,l=1,2,3} \frac{\partial x_k}{\partial x'_i}\frac{\partial x_l}{\partial x'_j} g_{ij}.$$

That is, G'^{00} is resulted from G^{00} by the coordinate transformation $x'_i = x'_i(r, x_1, x_2, x_3)$, $i = 1,2,3$. Then, by a similar argument as the proof that uniform non-degeneracy of g is coordinate independent, we can see that uniform spacelikeness is coordinate independent.

In this section, we will say that a variable quantity q is bounded above if its absolute value $|q|$ is bounded above, and we say that it is bounded below, if there is a constant $c > 0$ such that $|q| > c$. Therefore, the determinant of a uniformly non-degenerated metric is bounded below on a regular compact subset, and the determinant of the space components of a metric on a uniformly spacelike hypersurface is positive and bounded below on a regular compact subset. Moreover, all metric components in any coordinate system are bounded above on a regular compact subset, since they are C^∞.

The definition of uniform spacelikeness has a consequence, which will be used later.

Corollary 8.12. *Suppose that S is a uniformly spacelike hypersurface of M and $\langle U, \mu \rangle$ is a chart adapted to S. For any basic regular compact subset $C \subset S \cap U$, there exists a constant $c > 0$, such that for any tangent vector v^i of S at $p \in C$,*

$$g_p(v, v) = \sum_{i,j=1}^{3} g_{p,ij} v^i v^j \geq c \sum_{i=1}^{3} \left(v^i\right)^2.$$

Moreover, c is a lower bound of $g_p(v, v)$ for all $p \in C$ and all $v \in V_p^S$ such that its Euclidean norm in the coordinate basis $|v|^2 = \sum_{i=1}^{3} \left(v^i\right)^2 = 1$.

Proof. Consider the diagonalization process for (g_{ij}) in Section 8.3 above and apply it to $G = G_p^{00}$ for $p \in C$. We obtain matrices $A_1, \ldots, A_k$ such that $A'_k \ldots A'_1 G A_1 \ldots A_k = G^*$ becomes a diagonal matrix. Carefully examining these matrices we can see that each A_l has a determinant ± 1. Therefore, $|G| = |G^*|$. Moreover, all matrix entries of G are bounded above by a constant for all $p \in C$, and uniform spacelikeness means that $|G| > c$ for a constant c. Then, in constructing the matrices $A_1, \ldots, A_k$, when choosing a non-zero matrix entry of G in the process, we can make sure that the entry is bounded below by some constant for all $p \in C$. This means that all matrix entries of $A_1, \ldots, A_k$ are also bounded above by a constant. Suppose that the diagonal entries of G^* are a_1, a_2, a_3. Then, a_1, a_2, a_3 are bounded above by a constant. Since $|G| = |G^*| = a_1 a_2 a_3 > c$, there is a constant $c' > 0$ such that each $a_i > c'$. Moreover, there is a constant $c'' > 0$ such that $|A_l v|^2 \leq c'' |v|^2$ for any column vector v in $\mathbb{R}^3$. That is, $\left|A_l^{-1} v\right| \geq |v|^2 / c''$. Then, it is easy to see that, for any column vector v in $\mathbb{R}^3$,

$$v' G v = \left(A_k^{-1} \ldots A_1^{-1} v\right)' G^* \left(A_k^{-1} \ldots A_1^{-1} v\right) \geq c' \left|A_k^{-1} \ldots A_1^{-1} v\right|^2 \geq c^* |v|^2$$

for some constant $c^* > 0$. This is exactly the conclusion of the corollary. □

In the rest of this section, we assume that S is a uniformly spacelike hypersurface of M. Let $\langle U,\mu\rangle$ be a coordinate system adapted to S as above again. Consider the coordinate vector fields $\partial_0,\partial_1,\partial_2,\partial_3$ and consider any point $p\in S\cap U$. The vectors $\partial_1,\partial_2,\partial_3$ at p constitute a basis for V_p^S, and ∂_0 at p has to be timelike. We may assume that ∂_0 points to the future. We want to construct a unit vector field n that is orthogonal to all $\partial_1,\partial_2,\partial_3$ and hence orthogonal to V_p^S at any point $p\in S\cap U$. Consider any basic regular compact subset $C\subset S\cap U$. Recall that the component matrix (g^{ij}) of g^{-1} is the inverse of (g_{ij}). Therefore, $g^{00}=\left|G^{00}\right|/\left|g_{ij}\right|$. By uniform non-degeneracy and uniform spacelikeness, $\left\|g_{ij}\right\|>c$ and $\left|G^{00}\right|>c$ on C for some positive constant c. From the diagonalization process for (g_{ij}) we can see that $\left|g_{ij}\right|$ is negative. Moreover, $\left\|g_{ij}\right\|$ is bounded above on C. Therefore, $-g^{00}>c$ on C for some positive constant c and it is also bounded above on C. Let

$$n_p^i=\frac{-g_p^{0i}}{\sqrt{-g_p^{00}}}$$

for $p\in S\cap U$. Then n is a smooth vector field on $S\cap U$. Moreover,

$$g(n,n)=g_{ij}n^in^j=-\frac{g_{ij}g^{0i}g^{0j}}{g^{00}}=-1.$$

That is, n is a unit vector. For $k=1,2,3$, we have

$$g(n,\partial_k)=-\frac{g_{ij}g^{0i}(\partial_k)^j}{\sqrt{-g_p^{00}}}=-\frac{\delta_j^0\delta_k^j}{\sqrt{-g_p^{00}}}=0.$$

Therefore, n is a unit vector orthogonal to $\partial_1,\partial_2,\partial_3$. Note that $g(n,\partial_0)<0$. Therefore, n is future directed. n is called the normal vector of S. It is easy to see that n is independent of the chosen coordinate system.

Consider a timelike geodesic through p with the tangent vector n_p at p. We can parallelly transport n_p along that geodesic (as its tangent vectors). In this way, we get a vector field n in a neighborhood of p in M. Consider the covariant derivative ∇n of this vector field in the neighborhood around p. Its trace at p divided by 3,

$$H_p^S=\frac{1}{3}tr(\nabla n(p))=\frac{1}{3}\nabla_i n^i(p),$$

is called the mean curvature of S in M at p, where $\nabla_j n^i(p)$ are the components of the $(1,1)$ tensor $\nabla n(p)$ in some basis at p. Suppose that $e_0=n_p,e_1,e_2,e_3$ constitute an orthonormal basis of V_p with e_1,e_2,e_3 constituting an orthonormal basis of V_p^S. We want to express H_p in this basis. First we have

$$tr(\nabla n)=\sum_{i=0}^{3}(\nabla n)\left(e^i,e_i\right)=\sum_{i=0}^{3}\left(e^i\right)_k(e_i)^j\nabla_j n^k.$$

Note that the vector field n is parallelly transported in the direction $e_0 = n_p$. That is, $(e_0)^j \nabla_j n^k = 0$. Therefore, we have

$$H_p^S = \frac{1}{3}\sum_{i=1}^{3} \left(e^i\right)_k (e_i)^j \nabla_j n^k. \tag{8.12}$$

A smooth curve γ is timelike, if the tangent vector $\gamma'(t)$ of the curve at each point $\gamma(t) \in M$ is timelike. Null curves and spacelike curves are defined similarly. A curve is causal if its tangent vectors $\gamma'(t)$ are such that $g(\gamma'(t), \gamma'(t)) \leq 0$. For $p \in M$, $I^+(p)$ denotes the set of points $q \in M$ such that there exists a future directed timelike curve from p to q. For a subset $A \subseteq M$, $I^+(A) = \cup_{p \in A} I^+(p)$. Similarly, $J^+(p)$ denotes the set of points $q \in M$ such that there exists a future directed causal curve from p to q. $I^-(p)$, $I^-(A)$, $J^+(A)$, $J^-(p)$ and $J^-(A)$ are defined similarly. We say that a curve γ_1 extends another curve γ_2, if after some necessary reparameterization, we have $\gamma_1 \in C^\infty((a,b), M)$ and $\gamma_2 \in C^\infty((a-c, b+d), M)$ for some $c, d \geq 0$, and $\gamma_1(t) = \gamma_2(t)$ for $t \in (a,b)$. If γ_1 is a future (or past) directed timelike curve and $c = 0$ (or $d = 0$), we say that this is a future (or past) extension.

A set A is achronal, if for any $p \in A$ and any timelike curve γ from p to q, if $p \neq q$, then $q \notin A$. For an achronal spacelike hypersurface A of M, $D^+(A)$ denotes the set of points $q \in M$ such that every past directed causal curve starting from q has a past extension into a causal curve that hits A. $D^-(A)$ is defined similarly, with 'past' replaced by 'future'. $D(A)$ is the set of points $q \in M$ such that every causal curve passing q has an extension into a causal curve that hits A. A Cauchy hypersurface is an achronal uniformly spacelike hypersurface S such that $D(S) = M$. A spacetime structure is globally hyperbolic if it has a Cauchy hypersurface. In the rest of this section we assume that M is globally hyperbolic with a Cauchy hypersurface S.

In the simplest version of Hawking's singularity theorem we make the following additional assumptions:

1. The mean curvature H^S of S is everywhere greater than a constant $k > 0$, which means that the universe is everywhere expanding toward the future.
2. M satisfies Einstein's equation together with a so-called strong energy condition, which implies that the Ricci curvature tensor R_{ij} is such that $R_{ij}v^i v^j \geq 0$ for all timelike vector v. (Wald [36], p. 219.)

We are then interested in a future directed timelike geodesic γ from a point p to a point $q \in S$. The simplest version of Hawking's singularity theorem claims that, under the above assumptions, the length of γ (from p to q) is bounded by $1/k$ independent of p and q. Therefore, no past extension of any past directed timelike geodesic from the points of S can be longer than $1/k$. This is called geodesic incompleteness and is by definition the existence of singularity. (Wald [36], p. 237, Theorem 9.5.1, O'Neill [27], p. 431, Theorem 55A, and Naber [25], p. 132, Theorem 3.8.1.) If q is the present event of a free-falling particle, this means that the local time of the particle can never be greater than $1/k$. That is, the particle has a finite lifetime not longer than $1/k$. Since no other restriction is put on the particle, this means that the universe must have a finite lifetime not longer than $1/k$.

We will examine the classical proof of this theorem in Naber [25]. The underlying idea of this classical proof is simple. Consider a 2-dimensional smooth surface Σ in the Euclidean space $\mathbb{R}^3$ and consider a straight line segment pointing upward from a point p to a point $q = q(x_1, x_2) \in \Sigma$. Let $L_q = L(x_1, x_2)$ be the length of this line segment. If L_q turns out to be the distance between p and Σ, that is, the minimum length of all line segments from p to points on Σ, then the function $L_q = L(x_1, x_2)$ must satisfy

$$\frac{\partial L}{\partial x_i}|_q = 0, \quad \frac{\partial^2 L}{\partial x_i^2}|_q \geq 0. \tag{8.13}$$

$\frac{\partial L}{\partial x_i}|_q = 0$ implies that the line segment from p to q has to be orthogonal to Σ. However, in case Σ is curved downward, toward p, $\frac{\partial^2 L}{\partial x_i^2}|_q \geq 0$ can be true only if the length L_q of the line segment does not exceed a bound depending on the curvature of Σ at q (but not on p). For instance, suppose that Σ is the upper half of the unit 2-sphere with its center is at the origin, i.e.,

$$\Sigma = \left\{ \left(x_1, x_2, \sqrt{1 - x_1^2 - x_2^2} \right) \in \mathbb{R}^3 : x_1^2 + x_2^2 < 1 \right\},$$

and suppose that $p = (0, 0, a)$ and $q = (0, 0, 1) \in \Sigma$ and therefore $L_q = 1 - a$. Then, $\frac{\partial^2 L}{\partial x_i^2}|_q \geq 0$ is true only if $a \geq 0$ and hence $L_q \leq 1$. For the case of spacetime, Σ becomes our Cauchy hypersurface S, and the upward line segments become future directed timelike geodesics from p to S, and the minimum length becomes the maximum length (i.e., proper time). Then, similarly, in case γ is a future directed timelike geodesic from p to $q \in S$ and its length turns out to be the maximum length of all timelike geodesics from p to S, its length function must satisfy a condition similar to (8.13) but with the inequality sign $\geq$ reversed (since now γ attains the maximum length, not the minimum length). One can show that the condition also implies that the length of γ cannot exceed a bound depending only on the mean curvature of S at q (but not on p). Since γ is the longest geodesic from p to S, it means that the length of any future directed timelike geodesic from any point to points in S is bounded by a constant depending only on the mean curvature of S. This is the conclusion of Hawking's singularity theorem.

This classical proof of Hawking's theorem in Naber [25] consists of two steps. The first step proves that there exists a continuous (not necessarily differentiable) timelike curve γ from p to S, such that the length of γ attains the maximum length of all continuous timelike curves from p to S. Moreover, such a curve γ must be a geodesic (and hence differentiable). This then implies a condition similar to (8.13) with the inequality sign $\geq$ reversed. The second step of the proof derives a bound for the length of γ from that condition. The first step of the proof is highly nonconstructive. It relies on compactness arguments to prove the existence of that continuous timelike curve γ with the maximum length. Recall that in strict finitism, for a continuous function f on $[a, b]$, we may not be able to find a point $t \in [a, b]$ such that f attains its maximum value at t. Similarly, even for a very regular compact smooth surface Γ in the Euclidean space $\mathbb{R}^3$ and a point p, we may not be able to find a point $q \in \Gamma$ such that the distance between p and q is the distance between p and Γ. That is, we may not be able to construct a straight line segment from p to Γ such that its length attains the minimum length of all straight line segments from p to Γ. For instance, suppose that Γ is the entire unit 2-sphere, $q_1 = (0,0,1)$, $q_2 = (0,0,-1)$, and $p = (0,0,\delta)$, where we cannot decide whether $\delta > 0$, or $= 0$, or < 0. Then, we cannot decide which of line segments $\overrightarrow{pq_1}$ and $\overrightarrow{pq_2}$ gives the shortest distance between p and Γ (although we can approximate that shortest distance value, which is $1 - |\delta|$). Therefore, for our Cauchy hypersurface S, we similarly do not expect that we can construct a continuous timelike curve from p to S to attain the maximum length of all continuous timelike curves from p to S.

However, we will see that, to derive a bound for the length of a timelike geodesic from p to S, we do not need to find a longest timelike geodesic from p to S, and we do not need an exact condition like (8.13). It suffices to find a timelike geodesic γ from p to a point $q \in S$ such that its length is sufficiently close to the maximum length of all timelike geodesics from p to S and such that some conditions in roughly the format

$$\left| \frac{\partial L}{\partial x_i}|_q \right| < \varepsilon, \ \frac{\partial^2 L}{\partial x_i^2}|_q < \varepsilon \tag{8.14}$$

are satisfied for some sufficiently small $\varepsilon > 0$, where L is the length function of some appropriate variations of γ. From these approximate conditions we can also derive a bound for the length of γ. This is obvious for a surface Σ in the Euclidean space $\mathbb{R}^3$. Let Σ be the upper half of the unit 2-sphere again and let p, q be as above again. If we have $\frac{\partial^2 L}{\partial x_i^2}|_q > -\varepsilon$ for some sufficiently small $\varepsilon > 0$, then a cannot be too much less than 0. That is, p cannot be too far below the origin. Similarly, we will show that an approximate condition like (8.14) is sufficient to derive a bound for the maximum length of all timelike geodesics from p to S. Moreover, in the example Γ above, while we cannot construct a line segment from p to Γ to attain the minimum distance exactly, we can construct line segments from p to Γ to approximate that minimum length arbitrarily and satisfy a condition like $\frac{\partial^2 L}{\partial x_i^2}|_q > -\varepsilon$ for arbitrarily small ε (just by approximating δ sufficiently). Therefore, we naturally expect that we can do the same for our Cauchy hypersurface S.

We will need an extra assumption about our spacetime model in order to do this, that is, to construct a timelike geodesic from p to S to approximate the maximum length (of all timelike geodesics from p to S) sufficiently and satisfy a condition like (8.14) for sufficiently small ε. The extra assumption is actually provable in the classical theory, but we have to make it an explicit assumption. We will argue that the new assumption is physically reasonable. Moreover, we will rely on our inductive belief on the consistency of classical mathematics to assure ourselves that a finitistic procedure for constructing a geodesic and verifying a condition like (8.14) will terminate with the result that the constructed geodesic does satisfy the condition. We will argue that this is sufficient to show that the classical proof can be transformed into sound logical deductions on statements about real spacetime, without assuming that real spacetime is literally continuous (or even that it is literally isomorphic with a classical, infinitely differentiable semi-Riemannian manifold).

We start with calculating the first and second order derivatives of the length of a future directed timelike geodesic with respect to some variations of the geodesic. We mostly follow Naber [25] here. In the following, we assume that the Cauchy hypersurface S has the mean curvature $H_p^S \geq k$ for some constant $k > 0$ and for all $p \in S$. We assume further that $R_{ij}v^i v^j \geq 0$ for any timelike vector v, where R_{ij} is the Ricci curvature tensor of M.

Let $p \in I^-(S)$. Since S is a Cauchy hypersurface, every future directed timelike curve starting from p hits S. Suppose that $\gamma : [-a, 0] \to M$ is a future directed timelike geodesic parameterized by its length such that $\gamma(-a) = p$ and $\gamma(0) \in S$. Therefore, the length of γ is a. Suppose that $\sigma : [-a,0] \times [-b,b] \to M$ is a C^∞ function such that $\sigma(t,0) = \gamma(t)$ for $t \in [-a,0]$, and $\sigma(-a,s) = p$ and $\sigma(0,s) \in S$ for all $s \in [-b,b]$. We call such a σ a variation of γ as a curve from p to S. The curves $\sigma_s(t) = \sigma(t,s)$ are the variation curves of γ as a curve from p to S, and the curves $\sigma^t(s) = \sigma(t,s)$ are the transverse curves of the variation. We assume that each σ_s is timelike for $s \in [-b,b]$. Let $T = T(t,s)$ denote the tangent vector of the curve σ_s at $\sigma_s(t)$ and $V = V(t,s)$ denote the tangent vector of the transverse curve σ^t at $\sigma^t(s)$. Therefore,

$$T^i = \frac{\partial \sigma^i(t,s)}{\partial t}, V^i = \frac{\partial \sigma^i(t,s)}{\partial s}.$$

Note that σ^0 is a curve on S, $V(0,s)$ is a tangent vector of S, and $V(-a,s)$ is the zero vector. Also note that

$$V^k \nabla_k T^i = V^k \partial_k T^i + \Gamma^i_{jk} T^j V^k = \frac{\partial}{\partial s} T^i + \Gamma^i_{jk} T^j V^k = \frac{\partial \sigma^i}{\partial t \partial s} + \Gamma^i_{jk} T^j V^k.$$

A similar expression holds for $T^k \nabla_k V^i$. Therefore,

$$V^k \nabla_k T^i = T^k \nabla_k V^i. \tag{8.15}$$

Let $L(s)$ be the length of the curve σ_s:

$$L(s) = \int_{-a}^{0} \left(-T_i T^i\right)^{1/2} \mathrm{d}t.$$

Note that σ_s may not be a geodesic for $s > 0$ and it may not be parameterized by its length. Denote $f(t,s) = \left(-T_i T^i\right)^{1/2}$. We have

$$\frac{\partial f}{\partial s} = -\frac{1}{2} f^{-1} \frac{\partial}{\partial s}\left(T_i T^i\right) = -\frac{1}{2} f^{-1} V^k \nabla_k \left(T_i T^i\right).$$

By the Leibnitz rule and the fact that $\nabla g = 0$,

$$\nabla_k \left(v_i v^i\right) = \nabla_k \left(v^i v^j g_{ij}\right) = 2 v^j g_{ij} \nabla_k v^i = 2 v_i \nabla_k v^i$$

for any vector v. Therefore, by (8.15),

$$V^k\nabla_k\left(T_iT^i\right) = 2T_iV^k\nabla_kT^i = 2T_iT^k\nabla_kV^i.$$

So we have

$$\frac{\partial f}{\partial s} = -f^{-1}T_iT^k\nabla_kV^i. \tag{8.16}$$

Now consider its value at $s=0$. Since $\gamma=\sigma_0$ is a geodesic parameterized by its length, we have $f=1$ and $T^k\nabla_kT^i=0$ at $s=0$. Then, by the Leibnitz rule, $T^k\nabla_k\left(T_iV^i\right) = T_iT^k\nabla_kV^i$. On the other side, $T^k\nabla_k\left(T_iV^i\right) = \frac{\partial}{\partial t}\left(T_iV^i\right)$. Therefore, $T_iT^k\nabla_kV^i = \frac{\partial}{\partial t}\left(T_iV^i\right)$. Finally, recall that $V=0$ at $t=-a$. Therefore,

$$L'(0) = \int_{-a}^{0}\frac{\partial f}{\partial s}|_{s=0}\mathrm{d}t = -\int_{-a}^{0}\frac{\partial}{\partial t}\left(T_iV^i\right)|_{s=0}\mathrm{d}t = -T_iV^i|_{s=0,t=0}. \tag{8.17}$$

This means that in case $|L'(0)|$ is very small, T is almost orthogonal to V.

Now we calculate the second order derivative. We will consider a special variation of γ. Take a unit tangent vector W of S at the point $q=\gamma(0)\in S$ and parallelly transport W along the geodesic γ down to $\gamma(t)$ for $t\in[0,-a]$ and get $W(t)$. Define

$$V(t,0) = \frac{a+t}{a}W(t).$$

Note that $V(0,0)=W$ and $V(-a,0)=0$. We can construct a variation σ of γ as a curve from p to S such that $V(t,0)$ are exactly the transverse vectors of σ at $s=0$. For this we have to use coordinate systems around the curve γ. Recall that by our definition of a differentiable curve, we can divide $[-a,0]$ into finitely many subintervals $[t_{i+1},t_i]$, $i=0,\dots,k$, $t_0=0$, and $t_{k+1}=-a$, such that the image of a subinterval under γ lies within a single local coordinate system. In the following, we will work as if the points in M were just their corresponding points in the coordinate systems. Suppose that the first local coordinate system containing $\gamma([t_1,t_0])$ is $\langle U,\mu\rangle$. Then, $q=\gamma(0)\in S\cap U$. We can assume that $\langle U,\mu\rangle$ is adapted to the hypersurface S and assume that $S\cap U$ corresponds to a slice $x^0=r$ of $\mu(U)$. Given a unit tangent vector W of S at q, let $\sigma^0(s)$ be the curve in S corresponding to the straight line

$$x^0=r,\ x^i=q^i+W^is,\ i=1,2,3,$$

where $\left(r,q^1,q^2,q^3\right)$ is the coordinate of q, and W^i are the components of W in the coordinate basis. This is the straight line with the tangent vector W at $\mu(q)$. Then, within this coordinate system, we can let

$$\sigma(t,s) = \gamma(t) + \frac{a+t}{a}W(t)s \tag{8.18}$$

for $t\in[t_1,t_0]$. (Here we treat the point $\gamma(t)$ as if it were $\mu(\gamma(t))\in\mu(U)$, which can be seen as a vector in $\mathbb{R}^4$, and the addition on the right hand side is the vector addition in $\mathbb{R}^4$.) For s in some interval $[-\delta,\delta]$, $\sigma(t,s)\in U$. Apparently, $\sigma(t,s)$ meets the requirements for a variation of γ for $t \in [t_1, t_0]$ within this coordinate system. When we move down to the next local coordinate system containing $\gamma([t_2, t_1])$, we already have the curve $\sigma^{t_1}(s)$ with $\sigma^{t_1}(0) = \gamma(t_1)$ and $\frac{d\sigma^{t_1}}{ds}|_{s=0} = \frac{a+t_1}{a} W(t_1)$ in that coordinate system. Then, let

$$\sigma(t,s) = \gamma(t) - \gamma(t_1) + \left(\frac{a+t}{a} W(t) - \frac{a+t_1}{a} W(t_1)\right) s + \sigma^{t_1}(s) \tag{8.19}$$

for $t \in [t_2, t_1]$. Similarly, for s in some interval $[-\delta, \delta]$, $\sigma(t,s)$ stays in the local coordinate system and it also meets the requirements for a variation of γ. Repeat this process we will get the required variation σ of γ. Note that we did not assume that the curve γ is orthogonal to S, and therefore $W(t)$ and $V(t,0)$ may not be orthogonal to $T(t,0)$. On the other side,

$$\frac{\partial \sigma}{\partial t} = \gamma'(t) + \left(\frac{1}{a} W(t) + \frac{a+t}{a} W'(t)\right) s. \tag{8.20}$$

Note that $g(\gamma', \gamma') = -1$. Therefore, for small s, the curves σ_s are still timelike.

Then, using (8.16) and the Leibnitz rule, we have

$$\begin{aligned}
\frac{\partial^2 f}{\partial s^2} &= f^{-2} \frac{\partial f}{\partial s} T_i T^k \nabla_k V^i - f^{-1} V^j \nabla_j \left(T_i T^k \nabla_k V^i\right) \\
&= -f^{-3} \left(T_i T^k \nabla_k V^i\right)^2 - f^{-1} \left(V^j \nabla_j T_i\right) \left(T^k \nabla_k V^i\right) - f^{-1} T_i V^j \nabla_j \left(T^k \nabla_k V^i\right).
\end{aligned}$$

By the Leibnitz rule and (8.15),

$$\begin{aligned}
V^j \nabla_j \left(T^k \nabla_k V^i\right) &= \left(V^j \nabla_j T^k\right) \left(\nabla_k V^i\right) + V^j T^k \nabla_j \nabla_k V^i, \\
T^k \nabla_k \left(V^j \nabla_j V^i\right) &= \left(T^k \nabla_k V^j\right) \left(\nabla_j V^i\right) + T^k V^j \nabla_k \nabla_j V^i \\
&= \left(V^j \nabla_j T^k\right) \left(\nabla_k V^i\right) + T^k V^j \nabla_k \nabla_j V^i.
\end{aligned}$$

Recall that

$$\left(\nabla_j \nabla_k - \nabla_k \nabla_j\right) V^i = -R_{jkl}{}^i V^l.$$

We have

$$V^j \nabla_j \left(T^k \nabla_k V^i\right) = T^k \nabla_k \left(V^j \nabla_j V^i\right) - R_{jkl}{}^i T^k V^j V^l.$$

Moreover,

$$T_i R_{jkl}{}^i T^k V^j V^l = T^i R_{jkli} T^k V^j V^l = T^i R_{kjil} T^k V^j V^l = R_{kji}{}^l T^k V^j T^i V_l.$$

Recall that $T_i T^k \nabla_k V^i = \frac{\partial}{\partial t}\left(T_i V^i\right)$. Similarly, we have

$$T_i T^k \nabla_k \left(V^j \nabla_j V^i\right) = \frac{\partial}{\partial t}\left(T_i V^j \nabla_j V^i\right).$$

Similar to (8.15) we have $V^j\nabla_j T_i = T^j\nabla_j V_i$. Therefore,

$$\frac{\partial^2 f}{\partial s^2} = -f^{-3}\left(\frac{\partial}{\partial t}\left(T_iV^i\right)\right)^2 - f^{-1}\left(T^j\nabla_j V_i\right)\left(T^k\nabla_k V^i\right) - f^{-1}\frac{\partial}{\partial t}\left(T_iV^j\nabla_j V^i\right) + f^{-1}R_{kji}{}^{l}T^kV^jT^iV_l.$$

Now we evaluate this at $s=0$. Recall that $f=1$. Since W is parallelly transported along γ, T_iW^i is a constant

$$c = T_i(t,0)W^i(t) = T_i(0,0)V^i(0,0) = -L'(0)$$

along $\gamma = \sigma_0$. Therefore, at $s=0$, we have $T_iV^i = \frac{a+t}{a}c$, $\frac{\partial}{\partial t}\left(T_iV^i\right) = \frac{c}{a}$. Moreover, since W^i is parallelly transported along γ, $T^k\nabla_k W^i = 0$. Then, since $V^i = \frac{a+t}{a}W^i$, by the Leibniz rule, $T^k\nabla_k V^i = \frac{1}{a}W^i$. Similarly, $T^k\nabla_k V_i = \frac{1}{a}W_i$. Note that W is a unit spacelike vector, that is, $W_iW^i = 1$. We finally have

$$\frac{\partial^2 f}{\partial s^2}|_{s=0} = -\frac{c^2}{a^2} - \frac{1}{a^2} - \frac{\partial}{\partial t}\left(T_iV^j\nabla_j V^i\right) + R_{kji}{}^{l}T^kV^jT^iV_l.$$

Moreover, note that $V(-a,s) = 0$. Therefore, $T_iV^j\nabla_j V^i|_{s=0,t=-a} = 0$. To estimate $T_iV^j\nabla_j V^i|_{s=0,t=0}$ we will use the unit normal vector n of the surface S. Let $A = T(0,s) - n$ on the transverse curve σ^0 in S and extend this into a vector field of M around q. Note that n is orthogonal to V on the transverse curve σ^0. Therefore, at $t=0$,

$$V^j\nabla_j\left(n_iV^i\right) = \frac{\partial}{\partial s}\left(n_iV^i\right) = 0.$$

Applying the Leibniz rule to the left hand side, we have

$$n_iV^j\nabla_j V^i = -V^iV^j\nabla_j n_i = -V_iV^j\nabla_j n^i.$$

Then, at $s=0$,

$$T_iV^j\nabla_j V^i = -V_iV^j\nabla_j n^i + A_iV^j\nabla_j V^i = -W_iW^j\nabla_j n^i + A_iV^j\nabla_j V^i.$$

Substitute these into the expression for $\frac{\partial^2 f}{\partial s^2}|_{s=0}$ and integrate, we finally have

$$\begin{aligned} L''(0) &= \int_{-a}^{0}\frac{\partial^2 f}{\partial s^2}|_{s=0}\mathrm{d}t \qquad (8.21)\\ &= -\frac{L'(0)^2+1}{a} + \int_{-a}^{0} R_{kji}{}^{l}T^kV^jT^iV_l|_{s=0}\mathrm{d}t + W_iW^j\nabla_j n^i|_q - A_iV^j\nabla_j V^i|_q. \end{aligned}$$

Now, we choose 3 unit tangent vectors W_1, W_2, W_3 of S at q so that they form an orthonormal basis of the tangent space of S at q. Let n be the unit normal vector field of S. n, W_1, W_2, W_3 constitute an orthonormal basis of the tangent space of M at q. We say that this orthonormal basis is adapted to S. We repeat the construction above for W_1, W_2, W_3 separately and get 3 parallel transportations $W_1(t)$, $W_2(t)$, $W_3(t)$ of W_1, W_2, W_3, 3 variations of γ, and 3 transverse vector fields V_1, V_2, V_3, and 3 first and second order derivatives $L_h'(0)$, $L_h''(0)$, $h = 1,2,3$, as in (8.17) and (8.21). We add up $L_h''(0)$, $h = 1,2,3$, and get

$$\sum_{h=0}^{3} L_h''(0) = -\frac{\sum_{h=1}^{3} L_h'(0)^2 + 3}{a} + \sum_{h=1}^{3} \int_{-a}^{0} R_{kji}{}^{l} T^k (V_h)^j T^i (V_h)_l \, \mathrm{d}t + \sum_{h=1}^{3} (W_h)_i (W_h)^j \nabla_j n^i|_q - \sum_{h=1}^{3} A_i (V_h)^j \nabla_j (V_h)^i |_q.$$

We estimate each term in this expression. First, by (8.12), at q,

$$\sum_{h=1}^{3} (W_h)_i (W_h)^j \nabla_j n^i = 3H_q.$$

We parallelly transport n along γ and get $n(t)$. $W_0 = n, W_1, W_2, W_3$ constitute an orthonormal basis for the tangent space of M at the points along γ. (But note that we did not assume that T is orthogonal to W_1, W_2, W_3). Let W^0 be the dual vector $-n_i$ and let W^h be the dual vector $(W_h)_i$. Then, $W^h(W_k) = \delta_k^h$. That is, W^h, $h = 0, ..., 3$, constitute the dual basis corresponding to W_h, $h = 0, ..., 3$. Therefore,

$$\begin{aligned} R_{ki} T^k T^i &= Ric(T,T) = \sum_{h=0}^{3} R\left(T, W_h, T, W^h\right) \\ &= -R_{kji}{}^{l} T^k n^j T^i n_l + \sum_{h=1}^{3} R_{kji}{}^{l} T^k (W_h)^j T^i (W_h)_l. \end{aligned}$$

Let $A = A(t) = T(t,0) - n(t)$. Note that since $R_{kji}{}^{l} = -R_{jki}{}^{l}$, we have

$$R_{kji}{}^{l} T^k T^j T^i T_l = R_{kji}{}^{l} T^k T^j T^i A_l = 0.$$

Moreover,

$$R_{kji}{}^{l} T^k T^j T^i A_l = R_{kjil} T^k T^j T^i A^l = R_{ilkj} T^k T^j T^i A^l = R_{kjil} T^i T^l T^k A^j = R_{kji}{}^{l} T^k A^j T^i T_l.$$

Therefore,

$$\begin{aligned} R_{kji}{}^{l} T^k n^j T^i n_l &= R_{kji}{}^{l} T^k (T-A)^j T^i (T-A)_l \\ &= -R_{kji}{}^{l} T^k T^j T^i A_l - R_{kji}{}^{l} T^k A^j T^i T_l + R_{kji}{}^{l} T^k A^j T^i A_l \\ &= -2R_{kji}{}^{l} T^k T^j T^i A_l + R_{kji}{}^{l} T^k A^j T^i A_l \\ &= R_{kji}{}^{l} T^k A^j T^i A_l. \end{aligned}$$

It means that

$$\int_{-a}^{0}\sum_{h=1}^{3}R_{kji}{}^{l}T^{k}\left(V_h\right)^{j}T^{i}\left(V^{h}\right)_{l}\mathrm{d}t$$
$$=\int_{-a}^{0}\left(\frac{a+t}{a}\right)^{2}\left(R_{ki}T^{k}T^{i}+R_{kji}{}^{l}T^{k}A^{j}T^{i}A_{l}\right)\mathrm{d}t$$
$$\geq\int_{-a}^{0}\left(\frac{a+t}{a}\right)^{2}R_{kji}{}^{l}T^{k}A^{j}T^{i}A_{l}\mathrm{d}t.$$

To estimate the last term $\left(V_h\right)^{j}\nabla_{j}\left(V_h\right)^{i}$ at q, we use the coordinate system around q above. First,

$$\left(V_h\right)^{j}\nabla_{j}\left(V_h\right)^{i}=\left(V_h\right)^{j}\left(\partial_{j}\left(V_h\right)^{i}+\Gamma_{jk}^{i}\left(V_h\right)^{k}\right)=\frac{\partial}{\partial s}\left(V_h\right)^{i}+\Gamma_{jk}^{i}\left(V_h\right)^{j}\left(V_h\right)^{k}.$$

By (8.18),

$$\left(V_h\right)^{i}(t,s)=\frac{\partial}{\partial s}\sigma_{h}(t,s)=\frac{a+t}{a}\left(W_h\right)^{i}(t).$$

Therefore, $\frac{\partial}{\partial s}\left(V_h\right)^{i}=0$, and

$$\left(V_h\right)^{j}\nabla_{j}\left(V_h\right)^{i}=\Gamma_{jk}^{i}\left(W_h\right)^{j}\left(W_h\right)^{k}.$$

Denote

$$B^{i}\left(W_h\right)=\Gamma_{jk}^{i}\left(W_h\right)^{j}\left(W_h\right)^{k}. \tag{8.22}$$

Then, in this basis, we have

$$\sum_{h=1}^{3}A_{i}\left(V_h\right)^{j}\nabla_{j}\left(V_h\right)^{i}=\sum_{h=1}^{3}A_{i}B^{i}\left(W_h\right).$$

Now we estimate A_i. Since $W_h(t)$ are parallel transportations, by (8.17),

$$g\left(T(t,0),W_h(t)\right)=g\left(T(0,0),W_h\right)=-L_h'(0).$$

Therefore, we have an expansion

$$T(t,0)-an-\sum_{h=1}^{3}L_h'(0)W_h$$

in the orthonormal basis n,W_1,W_2,W_3 at $\gamma(t)$. Since $T(t,0)$ is a timelike unit vector and it points to the future direction as the normal vector n does, we have

$$a=\left(1+\sum_{h=1}^{3}\left(L_h'(0)\right)^{2}\right)^{1/2}.$$

Therefore, in this orthonormal basis,

$$A_0 = -A^0 = 1 - a,\ A_h = A^h = -L'_h(0).$$

When $\left|L'_h(0)\right| < 1$, we have $|A_i| \le \sum_{h=1}^{3} \left|L'_h(0)\right|$, for $i = 0, ..., 3$.

Let

$$B_q = \sum_{h=1}^{3}\sum_{i=0}^{3} \left|B^i(W_h)\right|_q. \tag{8.23}$$

Then we have the estimation,

$$\left|\sum_{h=1}^{3} A_i (V_h)^j \nabla_j (V_h)^i\right| \le B_q \sum_{h=1}^{3} \left|L'_h(0)\right|.$$

Moreover, let

$$C_\gamma = \int_{-a}^{0} \sum_{j,l=0}^{3} \left|R_{kji}{}^{l} T^k T^i\right| \mathrm{d}t. \tag{8.24}$$

Then we have

$$\int_{-a}^{0} \left(\frac{a+t}{a}\right)^2 R_{kji}{}^{l} T^k A^j T^i A_l \mathrm{d}t \ge -\left(\sum_{h=1}^{3} \left|L'_h(0)\right|\right)^2 C_\gamma.$$

Finally, we have the estimate

$$\sum_{h=0}^{3} L''_h(0) \ge -\frac{\sum_{h=1}^{3} \left|L'_h(0)\right|^2 + 3}{a} + 3H_q - \sum_{h=1}^{3} \left|L'_h(0)\right| B_q - \left(\sum_{h=1}^{3} \left|L'_h(0)\right|\right)^2 C_\gamma.$$

Then, in case each $L''_h(0) < \varepsilon$ and $\left|L'_h(0)\right| < \varepsilon$ for some small $\varepsilon > 0$, we have

$$3\varepsilon > -\frac{3\varepsilon + 3}{a} + 3H_q - 3\varepsilon\left(B_q + C_\gamma\right).$$

Finally, since we assume that the mean curvature H_q on S is bounded below by a positive constant k, we have,

$$\frac{1}{a} > \frac{k - \varepsilon\left(B_q + C_\gamma + 1\right)}{\varepsilon + 1}. \tag{8.25}$$

Note that $B_q + C_\gamma$ depends on the chosen coordinate systems around γ. We have to show that it is uniformly bounded by a constant. Then, if for arbitrarily small ε we can find a timelike geodesic γ from p to S, such that the length of γ is an ε-approximation to the maximum length of all timelike geodesics from p to S, and such that γ satisfies the above inequality, we can conclude that $1/k$ is an upper bound for the length of all timelike geodesics from p to S. For this, we need some additional assumptions about the spacetime structure M.

First, choose a coordinate system around p such that the coordinate vector ∂_0 is timelike pointing to the future and $\partial_1, \partial_2, \partial_3$ are spacelike. By some coordinate transformation we may assume that $\partial_0,..,\partial_3$ constitute an orthonormal basis (in the metric g) of V_p. (Note that $\partial_0,..,\partial_3$ may fail to be orthonormal basis at other points near p.) Let

$$C^+ = \left\{ v \in \mathbb{R}^3 : \sum_{i=1}^{3} v_i^2 \leq \frac{1}{2} \right\}$$

be the closed ball of $\mathbb{R}^3$ centered at the origin with the radius $\frac{1}{\sqrt{2}}$. Each $v = (v_1, v_2, v_3) \in C^+$ corresponds to a vector $v = \sum_{i=0}^{3} v_i \partial_i \in V_p$ with $v_0 = \left(1 - \sum_{i=1}^{3} v_i^2\right)^{1/2}$. We denote the vector by v ambiguously. It is easy to verify that $g(v,v) \leq 0$. That is, v is causal and future directed. Moreover, for any non-zero future directed causal vector $v = \sum_{i=0}^{3} v_i \partial_i \in V_p$,

$$v' = \left(\sum_{i=0}^{3} v_i^2 \right)^{-\frac{1}{2}} v$$

corresponds to a point in C^+ in the above manner. Therefore, every future directed causal vector in V_p, after some rescaling, corresponds to a point in C^+. Note that C^+ is totally bounded in $\mathbb{R}^3$.

For each $v \in C^+$, construct a future directed causal geodesic γ_v starting from p with v as the tangent vector at p. Since S is a Cauchy hypersurface, γ_v can be extended into a geodesic hitting S at some point q. We consider γ_v a geodesic from p to q and assume that it is defined on the interval $[0,1]$. These cover all future directed causal geodesics from p, ignoring any reparameterizations. Recall that by our definition of smooth curves in M, $[0,1]$ can be divided into a finite number of subintervals $[a_i, a_{i+1}]$, $i = 0, ..., n$, such that $\gamma_v([a_i, a_{i+1}])$ is well-contained in the base set U_i of a local coordinate system $\langle U_i, \mu_i \rangle$ of M. Fix these local coordinate systems covering (the image of) γ from p to q. We can then treat γ_v as a smooth function from $[a_i, a_{i+1}]$ to $\mathbb{R}^4$. In the following, we will talk about points in $\mu_i(U_i)$ as if they were points in U_i. Then, there exists $d > 0$ such that

$$\Sigma_i = \left\{ x \in \mathbb{R}^4 : |x - \gamma_v(t)| \leq d \text{ for some } t \in [a_i, a_{i+1}] \right\}$$

is well-contained in U_i (actually, $\mu_i(U_i)$). It is easy to show that Σ_i is a compact subset in $\mathbb{R}^4$ (in the sense defined in Chap. 4). Moreover, $p \in \Sigma_0$ (actually, $\mu_0^{-1}(\Sigma_0)$) and $q \in \Sigma_n$. We say that $\Sigma(\gamma, d) = \langle \Sigma_0, ..., \Sigma_n \rangle$ is the compact tube of the radius d around γ and each Σ_i is a segment of the tube. Moreover, there exists $r > 0$ such that if $v' \in C^+$ and $|v' - v| \leq r$ then the image of $\gamma_{v'}$ is well-contained in $\Sigma\left(\gamma_v, \frac{d}{2}\right)$, since $\gamma_{v'}$ as the solution of some initial value problems for differential equations (in each coordinate system) is uniformly continuous in its initial value v'. In general, d and r depend on v. However, we make the assumption that d and r can be bounded below by positive constants for all $v \in C^+$. More accurately, we assume the following.

Geodesic Stability Assumption. *There exist $d, r > 0$ and an r approximation $v_1, ..., v_m$ to C^+, such that if $v' \in C^+$ and $|v' - v_i| \leq r$, then the image of $\gamma_{v'}$ is well-contained in $\Sigma\left(\gamma_{v_i}, \frac{d}{2}\right)$, and each segment of the tube $\Sigma(\gamma_{v_i}, d)$ is well-contained in one of the local coordinate systems around γ_{v_i}.*

In the classical theory, this follows from the compactness of C^+ and the fact that geodesics as solutions of initial value problems are uniformly continuous in the vectors in C^+ as their initial values. It is also reasonable from the physics point of view. Causal geodesics are the traces of free-falling particles in spacetime. This assumption means that if the initial 4-velocity of a free-falling particle at p is changed by a very small amount (represented by r), then its trace in spacetime shifts only by a small amount as well. Recall that general relativity is merely a macro-scale approximation to spacetime structure, above the Planck scale for instance. If r is of the Planck scale, our model of spacetime in general relativity should not be too sensitive to any change in the scale of r. In other words, the trace of a free-falling particle should actually be a small tube around a geodesic, not a single geodesic without width, and such a small tube should behave similarly as a single geodesic does.

In the rest of this section, we fix such an approximation $v_1, ..., v_m$, as well as the corresponding geodesics $\gamma_{v_1}, ..., \gamma_{v_m}$, their compact tubes, $\Sigma\left(\gamma_{v_i}, d\right)$, $i = 1, ..., m$, and their corresponding local coordinate systems implied in the assumption. Note that the estimations on various bounds below are performed in these chosen and fixed finitely many local coordinate systems, which now cover all causal geodesics from p to S. The final bound for the length of all timelike geodesics from p to S is of course independent of any coordinate system.

For $i = 1, ..., m$, suppose that $\Sigma\left(\gamma_{v_i}, d\right) = \langle \Sigma_0, ..., \Sigma_n \rangle$ and that γ_{v_i} hits S at q_i. The last compact tube segment Σ_n intersects S. We may assume that the coordinate system that contains Σ_n is adapted to S. Then, $D_i = \Sigma_n \cap S$ is a 3-dimensional closed ball centered at q_i with the radius d in S. Since S is uniformly spacelike, by Corollary 8.12, there exists a constant $c > 0$, such that

$$g_q\left(v, v\right) = \sum_{i,j=1}^{3} g_{q,ij} v^i v^j \geq c \sum_{i=1}^{3} \left(v^i\right)^2$$

for any $q \in D_i$ and any tangent vector v of S at q. This means that if W is a unit tangent vector of S at q (in the metric g), then $\sum_{i=1}^{3} \left(W^i\right)^2 \leq 1/c$ and therefore $\left|W^i\right| \leq \sqrt{1/c}$. Γ^i_{jk} is uniformly continuous and therefore bounded in D_i. Therefore, by the expression (8.22), (8.23) for B_q we see that B_q is bounded above for all $q \in D_i$. Since the compact tubes $\Sigma\left(\gamma_{v_i}, d\right)$, $i = 1, ..., m$, cover all future directed causal geodesics from p, B_q is bounded above by a constant K_B for all future directed timelike geodesics from p.

Similarly, consider any small $\delta > 0$ and

$$C_\delta^+ = \left\{ v \in \mathbb{R}^3 : \sum_{i=1}^{3} v_i^2 \leq \frac{1}{2} - \delta \right\}.$$

If $v \in C_\delta^+$, then $g\left(v, v\right) \leq -2\delta$. Note that γ_v is not parameterized by its length. To estimate L''_h as above, we must reparameterize γ_v by its length to get γ. We should have $T\left(-a, 0\right) = \gamma'\left(-a\right) = \left(-g\left(v, v\right)\right)^{-1/2} v$ in the coordinate system. Then $g\left(v, v\right) \leq -2\delta$ means that $\left|T^i\left(-a, 0\right)\right|$ are bounded above by a constant for all timelike geodesics γ_v, $v \in C_\delta^+$. The parallel transportation $T(t,0)$ of $T(-a,0)$ is again a solution to the initial value problem of a differential equation with the initial value $T(-a,0)$ and it depends on the initial value uniformly continuously. Therefore, $\left|T^i(t,0)\right|$ are bounded by a constant for all geodesics in a compact tube above. Since by the assumption, finitely many compact tubes cover all geodesics γ_v, $v \in C_\delta^+$, we see that $\left|T^i(t,0)\right|$ are bounded by a constant for all geodesics γ_v, $v \in C_\delta^+$. Then, by the expression (8.24) for C_γ, C_γ is bounded by a constant $K_{C,\delta}$ for all geodesics γ_v, $v \in C_\delta^+$.

Now, Let L_v denote the length of γ_v. For each i, L_v is uniformly continuous on $S(v_i,r) = \{v \in C^+ : |v - v_i| \le r\}$. Therefore, its maximum value in $S(v_i,r)$ exists. Since $S(v_i,r)$, $i = 1,\ldots,n$, cover C^+,

$$l_{\max} = \max\left\{L_v : v \in C^+\right\}$$

exists. Moreover, for any $\varepsilon > 0$, there exists $v \in C^+$ such that $L_v > l_{\max} - \varepsilon$. Note that $L_v = 0$ for $v \in \partial C^+$ the boundary of C^+ in $\mathbb{R}^3$. Therefore, there exists δ_1 such that whenever $L_v > \frac{1}{2} l_{\max}$, we have $v \in C_{\delta_1}^+$. Since we are interested in geodesics γ_v such that L_v sufficiently approximates $l_{\max}$, we can focus on $C_{\delta_1}^+$ in our constructions. Then, both B_q and C_γ are bounded by constants K_B and K_{C,δ_1}. Let

$$K_1 = K_B + K_{C,\delta_1} + 1.$$

With these bounds, the inequality (8.25) becomes

$$\frac{1}{a} > \frac{k - \varepsilon K_1}{\varepsilon + 1}. \tag{8.26}$$

We summarize the conclusion of the arguments so far in the following proposition.

Proposition 8.13. *Suppose that M is globally hyperbolic with a Cauchy hypersurface S such that its mean curvature $H_q^S \ge k$ for all $q \in S$, for some constant $k > 0$. Suppose that M satisfies Einstein's equation together with the strong energy condition, that is, $R_{ij}v^iv^j \ge 0$ for all timelike vectors v. Suppose further that M satisfies the Geodesic Stability Assumption above. Suppose that $\gamma : [-a,0] \to M$ is a future directed timelike geodesic parameterized by its length such that $\gamma(-a) = p$ and $\gamma(0) = q \in S$. Suppose that n, W_1, W_2, W_3 is an orthonormal basis of V_q with W_1, W_2, W_3 constituting an orthonormal basis of V_q^S. Suppose that σ_h, $h = 1,2,3$, is a variation of γ constructed as in (8.18), (8.19) above, whose transverse vectors along γ are $V_h(t,0) = \frac{a+t}{a} W_h(t)$, where $W_h(t)$ are parallel transportations of W_h. Suppose that $a = l_\gamma$, the length of γ, is such that $a > \frac{1}{2} l_{\max}$, where $l_{\max}$ is the maximum length of all future directed causal geodesics from p to S constructed above. Finally, suppose that for some $\varepsilon < k/K_1$, the length function $L_h(s)$ of the variation curve $\sigma_{h,s}$ from p to S satisfies the condition $\left|L_h'(0)\right| < \varepsilon$ and $L_h''(0) < \varepsilon$, where K_1 is the bound estimated above (in the chosen coordinate systems covering all timelike geodesics from p to S). Then, we have the estimates*

$$a < \frac{\varepsilon + 1}{k - \varepsilon K_1},$$
$$l_{\max} < \frac{\varepsilon + 1}{k - \varepsilon K_1} + (l_{\max} - a).$$

Moreover, if we can find such γ for arbitrarily small ε and such that $l_\gamma = a$ can approximate $l_{\max}$ arbitrarily, then we can conclude that $l_{\max} \le 1/k$.

Next, we examine the procedure for constructing the geodesic γ and its variations and for verifying the conditions $\left|L_h'(0)\right| < \varepsilon$ and $L_h''(0) < \varepsilon$. First, we have to find uniform upper bounds for $|L_v''(s)|$ and $|L_v'''(s)|$, $v \in C_{\delta_1}^+$. Suppose that $v \in C_{\delta_1}^+$. Now we assume that γ_v is already parameterized by its length. We have seen that $(\gamma_v')^i$ have a uniform bound. Consider the variation we use in (8.19). First, recall that we already have a uniform bound for W^i. Since $W(t)$ are parallel transportations along γ_v from W, they have a uniform bound as well. Note that being parallel transportations along γ_v means that

$$(W')^j(t) + \Gamma_{ik}^j (\gamma_v')^i W^k(t) = 0.$$

That is, the derivative W' can be expressed in terms of W. Therefore, $(W')^j$ have a uniform bound. Similarly, since γ_v is a geodesic, γ_v'' can be expressed in terms of γ_v'. Then, from the expressions (8.18), (8.19) for the variation $\sigma(t,s)$ we use, we see that all partial derivatives of $\sigma(t,s)$ have uniform bounds. Recall that

$$L(s) = \int_{-a}^{0} f^{\frac{1}{2}} \mathrm{d}t, \; f(t,s) = -\frac{\partial \sigma^i}{\partial t}\frac{\partial \sigma^j}{\partial t} g_{ij}(t,s),$$

where $g_{ij}(t,s)$ is the metric tensor at the point $\sigma(t,s) \in M$. Denote

$$X(t) = \frac{1}{a} W(t) + \frac{a+t}{a} W'(t).$$

From the expression (8.20) for $\frac{\partial \sigma}{\partial t}$ we see that

$$-f = (\gamma')^i (\gamma')^j g_{ij}(t,s) + 2(\gamma')^i X^j g_{ij}(t,s) s + X^i X^j g_{ij}(t,s) s^2.$$

γ' is an unit vector at $\sigma(t,0)$. That is, $(\gamma')^i (\gamma')^j g_{ij}(t,0) = -1$. g is uniformly continuous in the compact tubes and X^i have uniform a bound. Therefore, there are constants $\delta_2, \varepsilon_1 > 0$ such that $|f| > \varepsilon_1$ whenever $|s| < \delta_2$. This means that $\frac{\partial}{\partial s}\left(f^{\frac{1}{2}}\right)$, $\frac{\partial^2}{\partial s^2}\left(f^{\frac{1}{2}}\right)$, $\frac{\partial^3}{\partial s^3}\left(f^{\frac{1}{2}}\right)$ etc. have uniform bounds whenever $|s| < \delta_2$. Therefore, there are uniform upper bounds K_2, K_3 for $|L_v''(s)|$ and $|L_v'''(s)|$, $v \in C_{\delta_1}^+$, $|s| < \delta_2$. We will assume that $K_2, K_3 > 1$.

Then, let $\varepsilon > 0$ be a small number such that $\varepsilon < \min(k/K_1, \delta_2)$. We want to find $v \in C^+$ such that $L = L_v$ can approximate $l_{\max}$ arbitrarily and can satisfy $\left|L_h'(0)\right| < \varepsilon$ and $L_h''(0) < \varepsilon$ for some variation vectors W_h, $h = 1,2,3$. Let

$$\varepsilon_1 = \min\left(\frac{\varepsilon^2}{32K_2}, 2^{-11}\varepsilon^3/K_3^2\right).$$

We may assume that $\varepsilon_1 < \frac{1}{2}l_{\max}$. Choose any v such that $L_v > l_{\max} - \varepsilon_1$. Then we must have $v \in C^+_{\delta_1}$. Arbitrarily choose three orthonormal vectors W_h, $h = 1,2,3$, and approximate $\left|L'_h(0)\right|$ to decide whether $\left|L'_h(0)\right| < \varepsilon$ or $\left|L'_h(0)\right| > \varepsilon/2$. If we get $\left|L'_h(0)\right| < \varepsilon$ and $L''_h(0) < \varepsilon$ for $h = 1,2,3$, then the lemma above gives an estimate of $l_{\max}$.

Otherwise, suppose that we get $\left|L'_h(0)\right| > \varepsilon/2$ and suppose that $L'_h(0) > \varepsilon/2$. Since $\left|L''_h(s)\right| < K_2$ for $|s| < \delta_2$, we have $L'_h(s) > \frac{\varepsilon}{4}$ for $s \in \left[0, \frac{\varepsilon}{4K_2}\right]$. Therefore, at $s_1 = \frac{\varepsilon}{4K_2}$, we should have

$$L_h(s_1) > L_h(0) + \frac{\varepsilon}{4}s_1 = L_v + \frac{\varepsilon^2}{16K_2} > l_{\max} + \frac{\varepsilon^2}{32K_2}.$$

This means that the length of the variation curve σ_{h,s_1} exceeds $l_{\max}$ by ε_1. The case in which $L'_h(0) < -\varepsilon/2$ is similar. Similarly, suppose that $L''_h(0) > \varepsilon/2$. Let $\delta = 2^{-6}\varepsilon^2/K_3$. We can decide whether $L'_h(0) < \delta$ or $L'_h(0) > -\delta$. Suppose that $L'_h(0) < \delta$. Then, since $\left|L'''_h(s)\right| < K_3$ for $|s| < \delta_2$, we should have $L''_h(s) > \varepsilon/4$ for $s \in \left[-\frac{\varepsilon}{4K_3}, 0\right]$. Therefore,

$$L'_h\left(-2^{-4}\varepsilon/K_3\right) < L'_h(0) - \frac{\varepsilon}{4}\left(2^{-4}\varepsilon/K_3\right) < \delta - 2^{-6}\varepsilon^2/K_3 = 0,$$

and $L'_h(s) < \delta$ for $s \in \left[-2^{-4}\varepsilon/K_3, 0\right]$. This means that

$$L_h\left(-2^{-4}\varepsilon/K_3\right) > L_v - \delta\left(2^{-4}\varepsilon/K_3\right) = L_v - 2^{-10}\varepsilon^3/K_3^2.$$

Moreover, since $L''_h(s) > \varepsilon/4$ for $s \in \left[-\frac{\varepsilon}{4K_3}, 0\right]$,

$$L'_h\left(-2^{-3}\varepsilon/K_3\right) < L'_h(0) - \frac{\varepsilon}{4}\left(2^{-3}\varepsilon/K_3\right) < \delta - 2^{-5}\varepsilon^2/K_3 = -2^{-6}\varepsilon^2/K_3,$$

and $L'_h(s) < -2^{-6}\varepsilon^2/K_3$ for $s \in \left[-\frac{\varepsilon}{4K_3}, -2^{-3}\varepsilon/K_3\right]$. We have

$$\begin{aligned} L_h\left(-\frac{\varepsilon}{4K_3}\right) &> L_h\left(-2^{-3}\varepsilon/K_3\right) + \left(2^{-6}\varepsilon^2/K_3\right)\left(2^{-3}\varepsilon/K_3\right) \\ &> L_h\left(-2^{-4}\varepsilon/K_3\right) + 2^{-9}\varepsilon^3/K_3^2 \\ &> L_v + 2^{-10}\varepsilon^3/K_3^2 \\ &> l_{\max} + 2^{-11}\varepsilon^3/K_3^2. \end{aligned}$$

That is, for $s_1 = -\frac{\varepsilon}{4K_3}$, the length of the variation curve σ_{h,s_1} exceeds $l_{\max}$ by ε_1. The case in which $L'_h(0) > -\delta$ is similar. We summarize these in the following.

Proposition 8.14. *For any $\varepsilon > 0$, we can construct a geodesic γ_v such that $L_v > l_{\max} - \varepsilon_1$ with ε_1 defined above, and such that for arbitrarily chosen variation vectors W_h, $h = 1,2,3$, if the condition $\left|L'_h(0)\right| < \varepsilon$ and $L''_h(0) < \varepsilon$ is not satisfied in deciding whether $\left|L'_h(0)\right| < \varepsilon$ or $\left|L'_h(0)\right| > \varepsilon/2$ and whether $L''_h(0) < \varepsilon$ or $L''_h(0) > \varepsilon/2$, then we will have a timelike curve $\sigma_{h,s}$ from p to S whose length $L_h(s)$ exceeds $l_{\max}$ by at least ε_1.*

In the classical proof of Hawking's singularity theorem, one can show by some non-constructive compactness argument that there exists a timelike continuous curve γ from p to S such that its length attains the maximum length of all timelike continuous curves from p to S. This then implies that γ must be a geodesic. It means that $l_{\max}$ must also be the maximum length of all timelike smooth curves from p to S. Therefore, no $L_h(s)$ can exceed $l_{\max}$. This will ensure that the geodesic $\gamma = \gamma_v$ found above must satisfy the conditions $\left|L'_h(0)\right| < \varepsilon$ and $L''_h(0) < \varepsilon$ for any chosen W_h. However, this classical proof is highly non-constructive. We were not able to find a good finitistic substitute for this classical conclusion for arbitrary spacetime manifolds. The difficulty is perhaps due to the fact that the current definition of spacetime manifolds is still too abstract and it embodies too little computational content. In particular, while geodesics can be computed by solving the initial value problems for differential equations and can thus be represented by C^+ above, we do not have any simple representation of all smooth (or even continuous) timelike curves in the manifold. For instance, we do not have a general procedure for straightening an arbitrary timelike curve from a point p to another point q into a geodesic from p to q so that its length is not shrunk. If we were able to do this, we would immediately derive a contradiction from the assumption that $L_h(s)$ exceeds $l_{\max}$.

However, while we are not able to derive a contradiction within strict finitism from the assumption that $L_h(s)$ exceeds $l_{\max}$, we do have other reasons to believe that, after constructing the geodesic γ_v in the proposition above, when deciding whether $\left|L'_h(0)\right| < \varepsilon$ or $\left|L'_h(0)\right| > \varepsilon/2$ and whether $L''_h(0) < \varepsilon$ or $L''_h(0) > \varepsilon/2$ for each $h = 1,2,3$, the strictly finitistic and elementary recursive decision procedure will terminate with the result that $\left|L'_h(0)\right| < \varepsilon$ and $L''_h(0) < \varepsilon$ for all h. The reason comes from our belief in the consistency (not truth) of classical mathematics. Because, if we had found $\left|L'_h(0)\right| > \varepsilon/2$ or $L''_h(0) > \varepsilon/2$ for some h when the elementary recursive decision procedure terminates, we would have produced a contradiction in classical mathematics (by contradicting the classical proof of Hawking's theorem). In the introduction chapter we argue that the belief of the consistency of classical mathematics is essentially an inductive belief. Therefore, we do have a naturalistic justification for the belief that our finitistic procedure will terminate with the wanted result.

This then implies that we can transform our proof of Hawking's theorem into sound logical deductions on statements about real spacetime, even if real spacetime is discrete at the microscopic scale. We develop our spacetime manifold M within strict finitism. This means that our statements about our spacetime manifold M can be translated into true statements about real spacetime at the macroscopic scale, even if at the microscopic scale real spacetime is discrete. Continuity and differentiability conditions for our spacetime manifold M are translated into conditions about the smoothness of real spacetime at the macroscopic scale. They do not require that real spacetime is literally continuous, or even infinitely differentiable, in the classical sense. This is similar to using continuous functions to model discrete quantities such population growth. (See Sect. 3.7.) That is, our premises for deriving the bound for $l_{\max}$ can be interpreted into literally true assertions about real spacetime even if real spacetime is discrete. We show that we can actually construct a timelike geodesic from a point $p \in I^{-}(S)$ to S such that its length sufficiently approximates the maximum length of all timelike geodesics from p to S, and we can construct the variation curves $\sigma_{h,s}$. We can approximate $\left|L'_h(0)\right|$ and $L''_h(0)$ to see if they are less than ε or greater than $\varepsilon/2$. In case we get $\left|L'_h(0)\right| < \varepsilon$ and $L''_h(0) < \varepsilon$ for all variations h, we can derive a bound for $l_{\max}$ (independent of p). All these constructions and derivations are again within strict finitism, which means that they can be translated into literally sound logical deductions about real spacetime. The belief of the consistency of classical mathematics assures us that the strictly finitistic procedure for constructing the geodesic and the variation curves and for verifying $\left|L'_h(0)\right| < \varepsilon$ and $L''_h(0) < \varepsilon$ must terminate with a positive outcome. That is, the function of the belief of consistency is to predict the outcome of a strictly finitistic procedure and predict the result of a series of literally sound deductions about real spacetime, where real spacetime can very well be discrete. In this sense, we demonstrate that our proof of Hawking's theorem is sound for real spacetime and the conclusion of the theorem for real spacetime is reliable.

Note that being able to develop the spacetime manifold M within strict finitism is essential here. It allows us to present the derivation of the bound for $l_{\max}$ as valid logical deductions from literally true premises about real spacetime. In classical mathematics, we can also express our conclusion (not including the premises) of the singularity theorem as a finitistic claim. Then, the consistency of classical mathematics *and* the premises of the singularity theorem also implies the conclusion of the theorem within finitism (in Hilbert's sense). However, we have no explanation why the conclusion is true of real spacetime, which could be discrete, since the premises of the theorem are not literally true of real spacetime. The premises include assumptions about literal continuity and differentiability of spacetime. Developing the spacetime manifold M within strict finitism allows us to state our physics premises as literally true assertions about real spacetime (even if it is discrete). Then, when we use non-constructive arguments (to prove that our strictly finitistic procedure will terminate with the wanted result), we actually embed our strictly finitistic spacetime model into a richer (and fictional) classical model. The belief of the consistency of that classical (and fictional) model then implies the conservativeness of the embedding. That is, if φ is a strictly finitistic claim about our finitistic model and φ follows from the assumptions about the classical model, then φ should also follow from our strictly finitistic assumptions about the finitistic model alone. This is similar to the strategy for explaining the applicability of classical mathematics by nominalizing physics, which Field [13] first tried, but our method here is strictly finitistic and it respects the fact that our current physical theories are not committed to the existence of infinity in the physical universe.

References

1. Adams, F. 2003. Thoughts and their contents: Naturalized semantics. In *The Blackwell guide to the philosophy of mind*, eds. S. Stich and F. Warfield, 143–171. Oxford: Basil Blackwell.
2. Avigad, J., and S. Feferman. 1998. Gödel's functional ("dialetica") interpretation. In *Handbook of proof theory*, ed. S.R. Buss, 337–405. Amsterdam, The Netherlands: Elsevier.
3. Barendregt, H.P. 1981. *The Lambda Calculus, its syntax and semantics*. Amsterdam: North-Holland.
4. Benacerraf, P. 1973. Mathematical truth. *Journal of Philosophy* 70:661–679. [Reprinted in Benacerraf, P., and H. Putnam. 1983. *Philosophy of mathematics: Selected readings*, 2nd ed. Cambridge: Cambridge University Press].
5. Bishop, E. 1970. Mathematics as a numerical language. In *Intuitionism and proof theory*, eds. A. Kino, J. Myhill, and R.E. Vesley, 53–71. Amsterdam: North-Holland.
6. Bishop, E., and D.S. Bridges. 1985. *Constructive analysis*. New York: Springer.
7. Burgess, J.P. 2004. Mathematics and bleak house. *Philosophia Mathematica (3)* 12:18–36.
8. Burgess, J.P., and G. Rosen. 1997. *A subject with no object*. Oxford: Clarendon Press.
9. Chalmers, D. 1996. *The conscious mind*. Oxford: Oxford University Press.
10. Chihara, C. 1990. *Constructibility and mathematical existence*. Oxford: Oxford University Press.
11. Chihara, C. 2005. Nominalism. In *The Oxford handbook of philosophy of mathematics and logic*, ed. S. Shapiro, 483–514. Oxford: Oxford University Press.
12. Coddington, E.A., and N. Levinson. 1955. *Theory of ordinary differential equations*. New York: McGraw-Hill.
13. Field, H. 1980. *Science without numbers*. Princeton: Princeton University Press.
14. Field, H. 1998. Which undecidable mathematical sentences have determinate truth values? In *Truth in Mathematics*, eds. H.G. Dales and G. Oliveri, 291–310. Oxford: Oxford University Press.
15. Goodman, N., and W.V. Quine. 1947. Steps toward a constructive nominalism. *Journal of Symbolic Logic* 12:105–122.
16. Hellman, G. 1989. *Mathematics without numbers*. Oxford: Oxford University Press.
17. Hellman, G. 2005. Structuralism. In *The Oxford handbook of philosophy of mathematics and logic*, ed. S. Shapiro, 536–562. Oxford: Oxford University Press.
18. Hilbert, D. 1983. On the infinite, 183–201. [Reprinted in Benacerraf, P., and H. Putnam. 1983. *Philosophy of mathematics: Selected readings*, 2nd ed. Cambridge: Cambridge University Press].
19. Hoffman, S. 2004. Kitcher, ideal agents, and fictionalism. *Philosophia Mathematica (3)* 12: 3–17.

F. Ye, *Strict Finitism and the Logic of Mathematical Applications*, Synthese Library 355, DOI 10.1007/978-94-007-1347-5,

20. Lakoff, G., and R. Núñez. 2000. *Where mathematics comes from: How the embodied mind brings mathematics into being*. New York: Basic Books.
21. Leng, M. 2005. Revolutionary fictionalism: A call to arms. *Philosophia Mathematica (3)* 13:277–293.
22. Maddy, P. 2007. *Second philosophy: A naturalistic method*. Oxford: Oxford University Press.
23. Melia, J. 2000. Weaseling away the indispensability argument. *Mind* 109: 455–479.
24. Murawski, R. 1999. *Recursive functions and metamathematics*. Dordrecht: Kluwer.
25. Naber, G. 1988. *Spacetime and singularities: An introduction*. Cambridge: Cambridge University Press.
26. Neander, K. 2004. Teleological theories of content. In *Stanford encyclopedia of philosophy*, ed. E.N. Zalta. http://plato.stanford.edu/entries/content-teleological/.
27. O'Neill, B. 1983. *Semi-Riemannian geometry: With applications to relativity*. New York: Academic.
28. Papineau, D. 1993. *Philosophical naturalism*. Oxford: Basil Blackwell.
29. Quine, W.V. 1969. Epistemology naturalized. In *Ontological relativity and other essays*, ed. W.V. Quine, 69–90. Cambridge: Harvard University Press.
30. Quine, W.V. 1995. *From stimulus to science*. Cambridge: Harvard University Press.
31. Riesz, F., and B. Sz.-Nagy. 1955. *Functional analysis*. New York: Frederick Ungar Publishing Co.
32. Rosen, G., and J.P. Burgess. 2005. Nominalism reconsidered. In *The Oxford handbook of philosophy of mathematics and logic*, ed. S. Shapiro, 515–535. Oxford: Oxford University Press.
33. Tait, W. 1981. Finitism. *Journal of Philosophy* 78:524–546.
34. Troelstra, A.S. 1973. *Metamathematical investigation of intuitionistic arithmetic and analysis*. Lecture Notes in Mathematics, No. 344. Berlin: Springer.
35. Troelstra, A.S. 1990. Introductory note to 1958 and 1972. In *Kurt Gödel Collected works, volume II*, ed. S. Feferman, et al. 217–239. Oxford: Oxford University Press.
36. Wald, R. 1984. *General relativity*. Chicago: The University of Chicago Press.
37. Weidmann, J. 1980. *Linear operators in Hilbert space*. New York: Springer.
38. Yablo, S. 2001. Go figure: A path through fictionalism. *Midwest Studies in Philosophy* 25: 72–102.
39. Yablo, S. 2002. Abstract objects: A case study. *Noûs* 36:220–240.
40. Ye, F. 2000. Strict constructivism and the philosophy of mathematics. PhD diss., Princeton University.
41. Ye, F. 2000. Toward a constructive theory of unbounded linear operators on Hilbert spaces. *Journal of Symbolic Logic* 65:357–370.
42. Ye, F. 2009. A naturalistic interpretation of the Kripkean modality. *Frontiers of Philosophy in China* 4:454–470.
43. Ye, F. 2010. What anti-realism in philosophy of mathematics must offer. *Synthese* 175:13–31.
44. Ye, F. 2010. The applicability of mathematics as a scientific and a logical problem. *Philosophia Mathematica (3)* 18:144–165.
45. Ye, F. 2010. Naturalism and abstract entities. *International Studies in the Philosophy of Science* 24:129–146.
46. Ye, F. Introduction to a naturalistic philosophy of mathematics. Available online: http://sites.google.com/site/fengye63/. Accessed on 12 June 2011.
47. Ye, F. A structural theory of content naturalization. Available online: http://sites.google.com/site/fengye63/. Accessed on 12 June 2011.
48. Ye, F. On what really exist in mathematics. Available online: http://sites.google.com/site/fengye63/. Accessed on 12 June 2011.
49. Ye, F. Naturalism and objectivity in mathematics. Available online: http://sites.google.com/site/fengye63/. Accessed on 12 June 2011.
50. Ye, F. Naturalism and the apriority of logic and arithmetic. Available online: http://sites.google.com/site/fengye63/. Accessed on 12 June 2011.

Index

Symbols

A

B

C

M

N

O

P

R

S

T

MIX
Papier aus verantwortungsvollen Quellen
Paper from responsible sources
FSC® C105338

If you have any concerns about our products,
you can contact us on
ProductSafety@springernature.com

In case Publisher is established outside the EU,
the EU authorized representative is:
Springer Nature Customer Service Center GmbH
Europaplatz 3, 69115 Heidelberg, Germany

Printed by Libri Plureos GmbH
in Hamburg, Germany